EURO

COURSES

A series devoted to the publication of courses and educational seminars organized by the Joint Research Centre Ispra, as part of its education and training program.
Published for the Commission of the European Communities, Directorate-General Telecommunications, Information Industries and Innovation, Scientific and Technical Communications Service.

The EUROCOURSES consist of the following subseries:
- Advanced Scientific Techniques
- Chemical and Environmental Science
- Energy Systems and Technology
- Environmental Impact Assessment
- Health Physics and Radiation Protection
- Computer and Information Science
- Mechanical and Materials Science
- Nuclear Science and Technology
- Reliability and Risk Analysis
- Remote Sensing
- Technological Innovation

ENVIRONMENTAL IMPACT ASSESSMENT

Volume 1

Environmental Impact Assessment

Environmental Impact Assessment

Edited by

A. G. Colombo
Commission for the European Communities,
Joint Research Centre,
Institute for Systems Engineering and Informatics,
Ispra, Italy

Springer-Science+Business Media, B.V.

Based on the lectures given during the Eurocourse on
'Environmental Impact Assessment'
held at the Joint Research Centre Ispra, Italy, September 30–October 4, 1991

ISBN 978-94-010-5116-3   ISBN 978-94-011-2528-4 (eBook)
DOI 10.1007/978-94-011-2528-4

Publication arrangements by
Commission of the European Communities
Directorate-General Telecommunications, Information Industries and Innovation,
Scientific and Technical Communication Unit, Luxembourg

EUR 14206
© 1992 Springer Science+Business Media Dordrecht
Originally published by ECSC, EEC, EAEC, Brussels and Luxembourg in 1992

Published by Kluwer Academic Publishers,
P.O. Box 17, 3300 AA Dordrecht, The Netherlands.

Kluwer Academic Publishers incorporates the publishing programmes of
D. Reidel, Martinus Nijhoff, Dr W. Junk and MTP Press.

Sold and distributed in the U.S.A. and Canada
by Kluwer Academic Publishers,
101 Philip Drive, Norwell, MA 02061, U.S.A.

In all other countries, sold and distributed
by Kluwer Academic Publishers Group,
P.O. Box 322, 3300 AH Dordrecht, The Netherlands.

TABLE OF CONTENTS

FOREWORD

The Joint Research Centre of the European Communities, and in particular the Institute for Systems Engineering and Informatics (ISEI) at Ispra, have well established competences in risk analysis, uncertainty analysis and statistical data treatment. More recently, work on Environmental Impact Assessment (EIA), particularly on environmental indicators and indices and on a "system engineering approach" to EIA, has started. This approach attempts to move towards "unified" procedures to investigate normal operation and accidental risks; these are problems concerned within both the EIA Directive 85/337/EEC and the "Seveso" Directive 82/501/EEC.

In May 1990, a Workshop on "Indicators and Indices for Environmental Impact Assessment and Risk Analysis" was organized at the JRC, Ispra. The Proceedings of this Workshop (Report EUR 13060 EN, 1990) are a reference document in the field.

This book is based on the papers presented at the Eurocourse EIA/91 held at the JRC, Ispra in the Autumn of 1991. This was the first course on Environmental Impact Assessment given in the JRC's Eurocourse series at Ispra. It was a success because of both the high calibre of the lecturers and the well informed and numerous participants. The work focuses on the broader aspects of EIA, namely: legislation, indicators and indices, approaches and techniques, economic and sociological implications.

R.W. Witty, Director
Institute for Systems Engineering
and Informatics

vii

INTRODUCTION

Interest in Environmental Impact Assessment methods has greatly increased in the recent past throughout Europe, due in part to the newly promulgated European directives and national laws in the field of environmental protection. The aim of this book is to present methods and approaches which make the global cognitive process dealing with environmental impact assessment efficient and transparent. It includes seventeen papers grouped into four sections: legislation, EIA approaches and techniques, environmental indicators and indices, economic and sociological dimensions of EIA.

Section 1 includes three papers. The paper by Jörissen and Coenen gives an overview of the current state of implementation of EIA in the EC Member States. The different strategies of incorporation of the EEC directive into national law are analysed for central aspects of regulation and evaluated against the background of an ideal EIA concept. In the second paper, Amendola presents the directives aimed at controlling hazards linked with the processing, storage and use of dangerous substances. The "Seveso" directive is discussed extensively. The recent directives on biotechnology hazards are also considered. Muyselaar illustrates the Dutch integral environmental zoning project. In this project, the concept of risk management is used to establish quantitative acceptability criteria for different agents in the sectoral policy.

Section 2 includes six papers. The first paper, by Masera and Colombo, describes critically the phases that must be accomplished to carry out an EAI procedure. The objective of the paper is twofold. First, to emphasise the participation of the different actors of an EIA study, i.e.: developer, authority, technical advisers and the public. Second, to argue the comprehensive process that should support the relating procedure. The paper by Contini and Servida deals with risk analysis in the framework of environmental impact studies of industrial installations. Starting from the fundamental definitions of risk, hazard and accident, the main phases of the Probabilistic Risk Analysis procedure are described and the methods for implementing such a procedure are reviewed. Bro-Rasmussen discusses Danish experience in Environmental Impact Assessment and Risk Analysis. Two practical examples are illustrated. The first deals with the environmental stress from toxic chemicals that are discharged from a pesticide plant into the coastal marine ecosystem. The second is concerned with the planning and construction of permanent railway and motorway connections between two of the Danish islands. Kleinschmidt reports on the implementation of the EEC directive in Germany and illustrates a research project concerning environmental planning in the Ruhr area. Drogaris discusses the problems that arise in the evaluation of an environmental impact study for oil refineries and inorganic chemical processes. The last paper of this section is by Colorni and Laniado. They present a software package containing a methodological framework and a decision support system for EIA. The interactive procedure, adopted over all the steps, makes it possible to really involve all the people concerned and to obtain more rationality and transparency in decision making.

Section 3 also contains six papers. The paper by Volta and Servida reviews the nature of indicators and indices as knowledge carriers in the EIA process and looks again at the theory of measurement in the perspective of its application to knowledge relevant for EIA. Cernuschi and Giugliano discuss the main implications of air quality standards in the simulation of ground level pollutant concentrations by atmospheric transport and diffusion models. Particular emphasis is given to the Gaussian plume equation and to the criteria used for its application in deriving concentration values suitable for a direct comparison with the standards. Newman examines the various approaches to the assessment of surface water quality and reviews a range of such schemes currently in use in the Community. He also makes recommendations for possible future surface water classification schemes. The paper by van der Zee and de Haan deals with soil and ground water indicators. The present standards for soil and ground water quality are reviewed, considering the standards critically from the point of view of their aims. The main message of this contribution concerns the identification of appropriate quality indicators, according to the function of interest. Vighi and Calamari propose the scheme of an integrated ecotoxicological approach for hazard assessment and risk evaluation for the management of potentially harmful chemicals. They discuss the predictive capability of such an approach and the possibility of applying it to environmental impact studies. The paper by Romano et al. deals with indices to characterise the level of risk of an industrial plant. It illustrates the features of the most widely used risk indices regarding the scope of the analysis which can be performed using them.

Section 4 includes two papers. The paper by van Pelt et al. focuses on conflicts between short term economic development and long-term environmental problems. Definitions of the key criteria of sustainability, efficiency and equity are proposed by the authors. The last paper, by Pellizzoni, focuses on the sociological implications of different EIA models and on the associated problems which concern: the different functions of participation, the identification of the components of the public interested, their involvement in the procedure and the employment of objective and subjective indicators in the measurement of social impacts.

A. G. Colombo, Editor

INVITED LECTURER ADDRESSES

Dr. J. Jörissen
Department of Applied System Analysis
Nuclear Research Centre Karlsruhe
D - 7500 Karlsruhe
Federal Republic of Germany

Dr. A. Amendola
CEC, Joint Research Centre
Institute for Systems Engineering and Informatics
I - 21020 Ispra (Va)
Italy

Dr. A.J. Muyselaar
Ministry of Housing, Physical Planning and Environment
Directorate General for Environmental Protection
2260 MB Leidschendam
The Netherlands

Dr. M. Masera
CEC, Joint Research Centre
Institute for Systems Engineering and Informatics
I - 21020 Ispra (Va)
Italy

Dr. S. Contini
CEC, Joint Research Centre
Institute for Systems Engineering
and Informatics
I - 21020 Ispra (Va)
Italy

Prof. F. Bro-Rasmussen
Laboratory of Environmental Science and Ecology
Technical University of Denmark
DK - 2800 Lyngby
Denmark

Prof. V. Kleinschmidt
EIA Research Centre
University of Dortmund
D - 4600 Dortmund
Federal Republic of Germany

Dr. G. Drogaris
CEC, Joint Research Centre
Institute for Systems Engineering and Informatics
I - 21020 Ispra (Va)
Italy

Prof. A. Colorni
Systems Theory Centre of the National Research Council
and Department of Electronics, Milan Polytechnic
Via Ponzio 34
I - 20131 Milan
Italy

Dr. G. Volta
CEC, Joint Research Centre
Institute for Systems Engineering
and Informatics
I - 21020 Ispra (Va)
Italy

Dr. S. Cernuschi
D.I.I.A.R. - Environmental Section
Milan Polytechnic
Via Fratelli Gorlini 1
I - 20151 Milan
Italy

Dr. P. J. Newman
Water Research Centre Medmenham
Henley Road, Marlow, Bucks
UK - Buckinghamshire, SL7 2HD
United Kingdom.

Prof. S.E.A.T.M. van der Zee
Department of Soil Science and Plant Nutrition
Wageningen Agricultural University
6700 EC Wageningen
The Netherlands

Prof. Calamari
Institute of Agricultural Entomology
University of Milan
Via Celoria 2
I - 20133 Milan
Italy

Dr. A. Servida
TECSA, R & D Division
Via Caravaggi
I - 24040 Levate (Bg)
Italy

Prof. P. Nijkamp
Dept. of Economics
Free University of Amsterdam
PO Box 7161
1007 MC Amsterdam
The Netherlands

Dr. L. Pellizzoni
Institute of International Sociology
via Mazzini 13
I - 34170 Gorizia
Italy

THE EEC DIRECTIVE ON EIA
AND ITS IMPLEMENTATION IN THE EC MEMBER STATES

J. Jörissen and R. Coenen
Department of Applied System Analysis
Nuclear Research Centre Karlsruhe
D - 7500 Karlsruhe
Federal Republic of Germany

ABSTRACT. The following paper is predominantly based on a study carried out by the authors in 1988 on behalf of the German Federal Environmental Agency, but takes into account more recent developments also. It gives an overview of the current state of implementation of EIA in the EC Member States. The different strategies of incorporation of the EEC Directive into national law are analyzed with regard to central aspects of regulation and evaluated against the background of an ideal EIA concept. The comparison includes the following topics: area of application, responsibilities within the procedure, content of EIA, consultation of other authorities and the public, evaluation of the impact statement and EIA's role in decision making. Gaps and shortcomings in the EIA systems are outlined on the basis of first practical experience in different Member States. Finally, the issue of extending environmental assessment arrangements to cover higher tiers of action like policies, plans and programmes is discussed.

1. INTRODUCTION

Environmental Impact Assessment (EIA) is generally understood to be an instrument of preventive environmental management. It should provide an adequate information basis for decision making on activities affecting the environment. All environmentally relevant impacts of such activities should be identified, analyzed and evaluated before consent is given. The purpose of EIA is to ensure that appropriate attention is paid to environmental issues. It should not, however, reduce the competent authority's scope for discretion, nor should it necessarily give priority to environmental concerns over other concerns.

EIA as a procedure to support decision making was first introduced in the United States within the framework of the National Environmental Policy Act (NEPA), passed by Congress in 1969 and became law on January 1, 1970. NEPA requires Federal agencies to prepare a so-called Environmental Impact Statement (EIS), describing in detail the environmental impacts likely to arise from the proposed action. The preparation of the EIS is merely one part of the whole EIA procedure. The term Environmental Assessment (EA) is used for a brief document produced early in the EIA process, the purpose of which is to determine whether formal EIA is warranted. The British, however, have adopted Environmental Assessment (EA) to describe the whole process and use Environmental Statement (ES) as the title of the developer's EIA document. As other countries become involved confusion will certainly increase.

Therefore, the American terms will be used in this context, that is to say: EIA means the

1

A. G. Colombo (ed.), Environmental Impact Assessment, 1–14.
© 1992 *ECSC, EEC, EAEC, Brussels and Luxembourg.*

whole procedure including steps such as screening, scoping, consultation of other authorities, public participation etc., EIS means the final report produced by the developer or independent experts, and EA means a brief document on preliminary assessment.

After the enactment of NEPA, EIA had become an important issue on the political agenda of many European countries. Nevertheless, it took most EC Member States as long as twenty years to implement comprehensive EIA provisions.

In the seventies, France was the only country to follow the example of the USA introducing detailed EIA legislation within the framework of the Nature Protection Act of 1976. It must be said, however, that the French EIA-concept can hardly be regarded as an ambitious one and the efficiency of EIA in France is viewed rather critically.

As opposed to France, the Netherlands strived continuously to introduce an ambitious EIA concept from the mid-seventies. In 1986 it was implemented by law. Undoubtedly, the Dutch legislation comes quite close to the ideal concept of EIA. Both in scope and depth of regulation it clearly surpasses the EC requirements in many points. The introduction of EIA was preceded by several trial runs in order to test legal provisions in practice.

In other countries, attempts to introduce EIA failed. In 1974 a legislative proposal on EIA for public projects was presented in the FRG which did not even reach the stage of parliamentary debate. The only remaining result was an administrative order on preparing EIA for Federal activities, which produced hardly any effect at all.

In 1976, Ireland introduced EIA provisions within the context of planning legislation which, however, were quite restricted in their field of application. All public projects were exempted, and most private developments were also not subjected to EIA, because of the financial threshold which was high by Irish standards. Moreover, even for projects exceeding this threshold, the preparation of an EIA was not mandatory but within the discretion of the competent authority.

Early initiatives to introduce EIA in the United Kingdom failed to reach the stage of a legislative proposal. Although the UK did not have a formal EIA system, considerable experience had been gained by undertaking numerous non-mandatory EIA studies since the early seventies. It started with the development of on-shore oil exploration in Scotland. Because local planning authorities were not accustomed to handle such complex projects, the Scottish Department of Development decided to commission Aberdeen University to develop a systematic procedure for impact assessment. The manual produced (Clark et al. 1976) was distributed to every planning authority in the country, free of charge, and recommended for use. Furthermore, private developers tended to apply EIA on a voluntary basis as a tool to identify and clarify critical issues from the beginning in order to avoid delays in planning and opposition at the permission stage. The British government encouraged and supported such initiatives, e.g. by giving guidance, but refused to incorporate formal EIA provisions into the existing development control legislation (Wood and McDonic 1989).

Italy made numerous attempts to introduce EIA legislation, all of which failed. Up to now, an EIA-system is only partially implemented on the national level, whereas, because of the reluctance of the national legislator, some of the regions passed their own EIA laws. The first was the Region of Veneto providing comprehensive EIA provisions within the framework of the EIA Directive of June 1985 (referred to in the next chapter), followed by the autonomous Province of Trento in 1988 and Friuli in 1990.

For a general discussion of these problems see Coenen and Jörissen (1989).

2. THE EEC DIRECTIVE ON EIA

The breakthrough for the incorporation of EIA in Europe was finally brought about by the EEC Directive on the assessment of the effects of certain public and private projects on the environment (Council of the European Communities 1985). The Directive consists of 14 articles and three annexes. In order to harmonize already existing legal provisions, minimum requirements are established especially with regard to the type of projects to be assessed, the main obligations imposed on the developer and the content of the investigation. This, however, does not touch the right of national legislators to lay down stricter rules regarding the scope and procedure of the assessment process, which was done by the Member States more or less extensively.

A period of three years was scheduled to transpose the EEC Directive into national law. This period expired on July 3, 1988. Table 1 shows that most Member States failed to meet this deadline. Even now, six years after the approval of the Directive, the European provisions are not fully implemented in some countries.

In Belgium, the constitutional reform transferred EIA responsibility to the regions (i.e. Flanders, Wallonia and Brussels) except for nuclear installations which remain within the authority of the national government. As a result of the federalization process in Belgium, which disturbed and slowed down policy-making for many years, the implementation of the EEC-Directive is not yet completed. At the present, only the Walloon Region has enforceable EIA legislation. In Flanders, temporary provisions based on six administrative orders are in operation, until the final EIA decree is approved. In Brussels, where elections for a regional government and council were held for the first time in 1988, legal arrangements for EIA are still under discussion. Provisions for EIA of nuclear installations at the national level have not yet been implemented (Devuyst and Hens 1991).

Luxembourg has abandoned its original intention to implement the EEC Directive by a separate EIA law. A draft legislative proposal to this end which was submitted to the Chamber of Deputies in September 1988, has since been withdrawn. Present plans intend to apply the European provisions by means of a decree (project de règlement). This decree is expected to be adopted at the end of 1991.

In Italy, the government ought to have transposed the EEC Directive into a national law, fixing general guidelines for its application, whereas regional administrations ought to have carried out the actual procedures. The present situation, however, is quite different. The national law which should have pro
vided for formal implementation of the Directive is still at a preliminary stage. Currently, a new legislative proposal agreed upon by the Ministerial Council in October 1990 is under consideration by the Italian Parliament. Until the enactment of the national law, the interim arrangements for EIA based on two Prime Ministerial decrees remain in force (Tamburrino 1991).

The FRG as well as Portugal still lack administrative regulations for implementing the national EIA Acts. In the other EC Member States the process of incorporating the provisions of the Directive into the national legal system has been more or less completed.

Nevertheless it is already apparent, that EIA legislation in Europe has not reached its final state. The Commission of the European Community had stressed in the reasons given for its draft Directive of June 11, 1980 that although it first limited mandatory EIA to the approval of individual projects, the long-term aim was the extension of EIA to cover all tiers of decision making. A corresponding draft directive extending EIA arrangements to policies, plans and programmes is presently at the stage of internal consultation (Commission of the European Communities 1990). It will introduce a " Strategic Environmental Assessment" (SEA) which is to deal with alternatives and impacts not addressed adequately at the project level, including those of synergistic, trans-sectoral or global nature. As soon as this directive has been adopted, even though it may take some time, all the Member States will have to reform their legal provisions.

3. THE MAIN STEPS IN THE EIA PROCESS

The EIA is a procedural instrument, i.e. it is more than just a study on the potential environmental impacts of a project. An EIA is a formal, legal procedure, following a stipulated sequence of steps to be carried out within prescribed periods of time.

Many EIA practicians express the opinion that the procedural character of EIA is more important to its success than the scientific quality of the investigation. In other words, the findings of the scientific investigation achieve relevance for decision making only because they are incorporated in a procedure involving steps such as scoping, public participation, consultation with other authorities, external review by independent experts and monitoring. The experience gained in practice thus confirms the theoretical conviction which was the basis for introducing EIA in the USA: implementing policy through procedure. Caldwell, one of EIA's protagonists, clearly expressed this viewpoint: "The EIS was intended to set forth the environmental facts relative to proposed public action. It was conceived as a mandatory, action-forcing reorientation of planning and decision making. But it was never intended to preempt the decision making authority of responsible public officials. It was intended to influence the way in which this decision making authority was exercised." (Caldwell 1982). The American EIA process has been pertinently termed a "procedural strait-jacket", because it no longer lets authorities neglect environmental concerns.

The ideal course of a typical EIA procedure will consist of seven steps:
1. Deciding whether EIA is necessary (Screening)
2. Determining the scope of EIA (Scoping)
3. Preparing the EIA study or statement
4. Consultation of other authorities and public participation
5. Evaluating the EIA results and the findings from consultation
6. Reaching a decision
7. Monitoring the impacts after the project has been realized.

However, the European countries carry out these individual steps in very different ways, as will be pointed out in the following.

3.1. Deciding whether EIA is Necessary (Screening)

There are two basic ways of defining the area of application: either projects subject to EIA are listed in the act or an annex thereto, or legislation only stipulates a general clause, leaving the decision on whether or not an EIA is necessary to the competent authority on a case-by-case-basis. General criteria or thresholds can be established to facilitate this screening process, or as in the USA, a preceding Environmental Assessment can be provided for, in order to determine whether a formal EIA is warranted.

The EEC Directive takes a middle road in this context: Under Art. 2 (1) of the Directive all public and private projects which are likely to have significant effects on the environment by virtue of their nature, size or location are made subject to EIA. These projects are listed in two Annexes to the Directive. Whereas projects of Annex I mandatorily require the preparation of an EIA, Member States have a certain scope for discretion with regard to projects listed in Annex II. These projects shall be made subject to an assessment "where Member States consider that their characteristics so require". Most countries have drawn up criteria or thresholds to this end, referring either to the particular characteristics of the project (size, capacity, output) or the sensitivity of the location. But they differ greatly between Member States.

In France, for example, the thresholds are set so low that all Annex II projects should normally be subject to EIA; 4000-5000 EIA procedures are carried out each year. Furthermore, a similar number of simplified environmental assessments (notice d'impact) are carried out for projects with less harmful environmental impacts. In the Netherlands the list of projects requiring an EIA covers, in principle, all projects of Annex II. But, in contrast to France, the thresholds are set so high that, at least at the beginning, only large-scale projects are subject to EIA; 10-15 EIA procedures per year were expected. However, the number of EIAs prepared per year is increasing: from 40 in 1988 and 57 in 1989, to 81 in 1990 (Koning 1990). In view of the considerable problems caused by Dutch agriculture, it is remarkable that installations for large-scale animal rearing and food production are generally exempted from the EIA requirement. The need for a systematic procedure to counter the environmental hazards of increased factory farming has now been recognized by the EC Commission. Should the EC Commission carry out its intention to subject agricultural projects to mandatory EIA the Netherlands will have to reform their provisions for the application of EIA.

Diverging from the EEC Directive, the Federal Republic of Germany does not use physical but legislative features as selection criteria: EIA is mandatory for all projects whose licensing procedures require public participation under present law. This also applies to major alterations of existing facilities.

In Italy and Portugal EIA is mandatorily required only for projects of Annex I, whereas criteria or thresholds concerning Annex II projects are still lacking.

Only a few Member States have provided for the determination of whether or not EIA is necessary on a case-by-case basis. Great Britain and Ireland consequently belong to this group, giving wide discretionary powers to the competent planning authorities. The statutory regulations relating to different types of projects specify general criteria and thresholds, although these do not automatically trigger EIA but only serve as guidance. However, the early indications are that these criteria will be interpreted rigidly by the competent authorities in order to avoid litigation.

Belgium (Wallonia) and Luxemburg follow the American example and base decision making on whether EIA is warranted in the individual case on an Environmental Assessment (EA).

Most Member States have introduced EIA at the level of project approval procedures. An exception is Denmark, where EIA is implemented on the level of regional planning, thus having rather the character of a preliminary assessment, especially with regard to location and public acceptance of a planned project.

In the Netherlands, a two-tiered procedure has been adopted for certain projects. Waste disposal plans of the provinces, for instance, are subject to EIA as well as applications for individual large-scale waste treatment facilities. In order to avoid unnecessary duplication in such cases, the EIA on the project level should not repeat aspects which have already been dealt with at the preceding planning level. The German EIA act does not generally include plans and programmes. However, land-use plans or changes to regional plans are subject to EIA if they are a legal prerequisite to obtain the licence or permit for an individual project.

3.2. Determining the Scope of EIA (Scoping)

The content of the EIS is outlined in Art. 3 of the EEC Directive which defines the term "environment". According to this article the EIA must identify, describe and assess the direct and indirect effects of a project on human beings, fauna and flora, soil, water, air, climate and the landscape, the interactions between these factors, as well as the effects on material assets and the cultural heritage. Art. 5 (2) of the Directive specifies the information required from the developer: he must supply at least a description of the project, comprising information on the site, design and size of the project, a description of the main impacts which the project is likely to have on the environment, a description of the measures envisaged to avoid, reduce and remedy significant adverse effects, as well as a non-technical summary of the information mentioned above. These minimum requirements are further specified in Annex III to the Directive.

Nevertheless the actual requirements of the Directive still remain rather abstract and must, therefore, be translated into concrete action for each type of project. Here, in turn, various different approaches can be used. The scope of investigation can be set by means of mandatory requirements concerning methods and content, by using project-specific check-lists or by single-case coordination within a scoping process or, of course, by any given combination.

All Member States essentially adopt the content requirements of the Directive in accordance with Art. 5 and Annex III, or use those as a basis for project-specific check-lists. In some countries, including the Federal Republic of Germany, the aim is to set mandatory requirements concerning content and methods for the different project types; in the long term these will be compiled to become an "EIA Technical Instruction Manual".

While defined requirement catalogues of this type indeed have many advantages from the point of view of legal security and uniform application, they also entail disadvantages. They can lead to excessive efforts being placed on minor aspects. On the other hand, the specific local conditions may not be given sufficient attention, as a result making the findings of EIA of but little value for the decision making process. In the light of this situation, the USA introduced scoping as a formal step in the procedure. Scoping aims to create early consensus between all parties involved (project proponent, agencies, public) as to which issues are particularly relevant in the given case and consequently require extensive investigation, and which are less significant and can hence be neglected.

Even though the EEC Directive does not include scoping as a formal procedural step, the positive experience with this instrument in the USA and Canada has encouraged some Member States, for example Belgium, the Netherlands, Portugal and Spain to introduce a scoping process. The other countries did not bring into operation mandatory scoping. However, early negotiation between the developer and the competent authority, including consultation of other authorities and the public, is also considered to be useful, but is left to the discretion of the competent authority. In Ireland and the United Kingdom an early scoping on a voluntary basis is explicitly recommended, in order to reduce public resistance and to avoid delays at the permission stage.

Particularly in the USA, a detailed examination of alternatives to the proposed action is considered the heart of EIA, thus providing a clear basis for choice among options by the decision maker and the public. The discussion of alternatives not only widens the information potential of the impact assessment, but also has an important methodological function. As there is a lack of standards, criteria and norms, an absolute evaluation of a project generates considerable problems. If several alternatives are examined, however, a relative statement will be sufficient, such as: Alternative A is preferable to alternative B from the environmental point of view. Thus discussion of alternatives at least supplies a yardstick for comparison.

Nevertheless, contrary to earlier drafts, the EC directive attaches little importance to the consideration of alternatives. According to Annex III, only an outline of the main alternatives studied by the developer and an indication of the main reasons for his choice, taking into account the environmental impacts, should be given "where appropriate". Submitting an outline of this type would probably serve to enhance the assessment of the alternative selected by the developer, but not, as actually intended, to extend the basis of choice for the decision maker. With a few exceptions the provisions of the Member States do not exceed the minimum requirements of the Directive. In Belgium the study of alternatives does have a certain significance with regard to public projects and in Denmark with regard to location.

The Netherlands, whose statutory regulations are closely in line with the American example, is the only EC Country that places greater importance on the consideration of alternatives. In any case, the EIS must deal with two types of alternatives: first, the alternative providing the best technical means to protect the environment; second, the "no action" alternative, i.e. the total abandonment of the proposed project. In addition, an examination must be made of whether the total abandonment of the project would not shift the burden to other areas. Thus, for example, the decision to abandon the construction of a by-pass road could lead to increased inner-urban traffic and so aggravate the environmental problems there. Other "reasonable" alternatives must also be described, i.e. those the developer or the competent authority could choose to carry out. Thus, for example, the costs of an alternative must be of the same order of magnitude as the costs of the proposed action (Ministry of Housing, Physical Planning and Environment 1984)

The discussion of alternatives will probably play a far more significant role in the context of Strategic Environmental Assessment mentioned above than in the context of project-related EIA.

3.3. Preparing the EIA Study

Art. 5 of the EEC Directive places the burden for providing the necessary information and preparing the impact study on the developer. In this way, major tasks, such as identifying and forecasting environmental impacts, become a primary responsibility of the developer, who must also bear the costs incurred, whereas responsibilities of the competent authority remain rather limited. This created a new splitting of tasks not in conformity with the existing administrative law of many European countries and was a matter of controversial discussion.

Originally, Lee and Wood, who had carried out preliminary work on implementing EIA in Europe on behalf of the EC Commission, proposed another division of tasks following the NEPA model. According to this, the competent authority was to prepare the EIS or, at least, had to take full responsibility for the scope, objectivity and content of the EIS (Lee and Wood 1976). This proposal was based on the consideration that the proponent, having a primary interest in implementing the project, would hardly be capable of unbiased assessment. He might be inclined to present the data in a manner most favourable to his case ("sweetheart statements").

On the other hand, the developer has better access to the project-specific data. The second, more significant advantage is, however, that the proponent is forced to take into account any possible environmental consequences of his project as early as first planning preparations start. The target is for project planning and environmental impact assessment to be carried out interactively and run parallel to each other.

Considering these advantages, all Member States, with the exception of Belgium, have adopted the distribution of roles between the developer and the competent authority as outlined in the EEC Directive. In order to compensate for any deficiencies or shortcomings which may result from biased and incomplete information, most countries have introduced an external review of the study by an independent body, even though this is not explicitly required by the Directive.

In view of the disadvantages mentioned above, Belgium has arranged a task-splitting which differs from those in the EEC Directive. In the Walloon region the developer must only supply the information necessary to carry out the analysis, whereas the environmental impact study is prepared by state-recognized experts or organisations. In the Flemish region, the EIA report must be prepared by the developer but with the active collaboration of certified independent experts. This explicit legal requirement is designed to balance the need to integrate EIA within the actual design process and the need to promote the completeness, objectivity and quality of the EIA reports (Schreurs 1990).

3.4. Consultation and Participation

3.4.1. Consultation of other authorities.
The defective coordination of environmental policy and the poor exchange of information between different authorities was an important impulse for introducing EIA both in the USA and in the EC. The requirement that other authorities be involved which have jurisdiction by law or special expertise is therefore a logical result from the trans-sectoral approach of EIA. Consultation is the only way to ensure that the competent

authority also takes into account environmental impacts outside its scope of responsibility.

Consequently, the EEC Directive gives those authorities an opportunity to comment. However, these consultations take place after the developer has submitted the application, i.e. when the project has already been designed in detail and the impact study has been completed. There is no guarantee, therefore, that other authorities will be able to incorporate their specific expertise at the early stage of determining content and scope of the EIS.

Recognizing this deficit, certain Member States have adopted regulations which go beyond the minimum requirements of the Directive. In Greece, the Netherlands and Portugal consultation of other authorities is mandatory during the scoping process. In other countries, for example the Federal Republic of Germany and the United Kingdom, early involvement of other authorities is left to the discretion of the competent authority.

3.4.2. Public participation. The public has several roles to play throughout the EIA process: they may figure as a source of information, as a barometer of public opinion and social acceptance and finally as a "watch-dog" during the implementation of a project, ensuring that the conditions linked to the licence are met (Davies 1989). To satisfy these different functions, public participation should start at the earliest stage possible and continue throughout the entire planning and decision making process up to monitoring. As experiences indicate, participation beginning only at the permission level can create considerable delays, because new environmental issues and new alternatives first enter the scene when the planning process has already been completed.

The public should have access to all important information and the definition of the term "public" should be kept as wide as possible, including not only persons directly affected but also associations, environmental organisations, citizens' groups, and the general public. The results of participation should be adequately taken into account by the competent authority. Furthermore, the introduction of a record of decision should guarantee feedback between the decision maker and the public.

Measured against these criteria, the provisions of the Directive are quite far from assuring effective public participation. Similar to the issue of consulting other authorities, the Directive does not provide for public participation before the application has been submitted and the impact study has been completed. The data obtained must be made available to the public together with the application, but only the "public concerned" is to be given the opportunity to express an opinion. The "public concerned" shall be determined by the Member States, depending on the particular characteristics of the project or site.

The importance of public consultation to improve the information basis and to increase the acceptance of the decision is taken into account by the EIA provisions of most Member States. In some cases they clearly exceed the minimum requirements of the EEC Directive with regard to the timing of public participation as well as to the definition of those entitled to comment.

The EIA act in the Netherlands undoubtedly contains the most far-reaching provisions concerning public participation that fulfil in practice all the criteria mentioned above. In Belgium (Wallonia Region), the public is involved as early as the scoping process, and again after the Environmental Impact Statement is prepared. Integration of EIA into the regional planning procedure in Denmark ensures comprehensive and early participation.

In Italy, Portugal and Spain possibilities for public consultation are also provided in principle before the application is submitted. However, account should be taken of the fact that these countries have little tradition of public participation, which was first introduced within the framework of EIA legislation. In Italy, for example, thirty projects were subjected to EIA in the first year of national operation. But the Ministry only received objections from the public in two cases.

Great Britain and Ireland recommend early public involvement, although this is not a binding requirement. The practice to date has indicated, however, that many private developers seek contact with the public at an early stage in order to avoid later controversies. In France and the Federal Republic of Germany, public participation depends on the type of project but, in line with the EEC Directive, it generally does not take place until the application is submitted.

Most EC countries do not restrict the right of participation to persons concerned , but rather every individual and organization interested is given the opportunity to comment. The Federal Republic of Germany and Spain are exceptions, where only persons whose rights are affected are allowed to lodge objections. By this limitation the function of advanced protection of legal rights is attributed primarily to participation neglecting the different roles of the public within the EIA procedure.

3.4.3. External review. An external control of the environmental impact study put forward by the project developer is not required by the EEC Directive. This seems to be particularly problematic, because the interests of the project developer and the competent (development-oriented) authority often converge, or may even be identical in the case of public projects. Many member states have thus provided for external review by an independent body, in line with the example of the USA where the Environmental Protection Agency evaluates all EISs for scientific quality, completeness and adequacy.

In Belgium and the Netherlands this task has been assigned to a special commission established for this purpose. In Italy, Portugal and Spain the control function is carried out by the authorities responsible for environmental affairs. Ireland intends to form an Environmental Protection Agency which will also play a central role in the EIA process. France, the Federal Republic of Germany and the United Kingdom do not provide for control by an independent institution. The lack of external review has proven to be a major weak-point of the EIA system both in France (Ministère de l'Environnement 1984) and in the United Kingdom; for the Federal Republic of Germany sufficient information is not yet available. Lack of external control hinders the development of uniform yardsticks and leads to great variations in the quality of the submitted studies.

3.5. Evaluation of the EIA Study by the Competent Authority

The EEC Directive does not provide for a final evaluation of the environmental compatibility of the project as a separate procedural step, despite the fact that the trans- sectoral, holistic evaluation of the results is indeed the final and most important step in the decision-making, or, better, decision-preparation process. Most Member States have given special attention to this fact, but individual regulations differ greatly.

On the basis of the developer's data, the comments received and its own investigations, the competent authority in the Federal Republic of Germany draws up a summary of the environmental impacts of the project, which serves as a basis for evaluation. In Italy and Portugal the review of the report put forward by the developer and the evaluation of the environmental compatibility of the projects is in the competence of the Ministry of Environment. In Greece, these tasks are carried out jointly by the Ministry for Environmental Protection, Regional Planning and Public Works and the ministry responsible for the approval of the project. In Spain, the competent permitting authority is only responsible for organizing the EIA procedure. However, the evaluation of the findings and the preparation of the Environmental Impact Statement, which also includes binding conditions attached to permission, is the task of the authority competent for environmental affairs at the given governmental level.

In the Netherlands, the holistic evaluation of all environmental impacts is carried out by the Lead Agency which must be appointed in cases where several permits are required by different authorities. The Lead Agency has, above all, the task of examining whether the total impact on all the different media would lead to an unreasonable burden on the environment and whether adverse impacts would be felt in areas which are not covered by environmental legislation. To this end, the Dutch EIA-Act explicitly provides for an extension of its quite limited evaluation possibilities. However, this does not change the decision making competences. Each authority involved must decide separately, on the basis of the results of the EIA.

3.6. Reaching a Decision

The provisions of the EEC Directive regarding the linkage of EIA and decision making are relatively vague. Art. 8 requires that the information supplied by the developer and the comments received are to "be taken into consideration" within the consent procedure. As the only indication of the extent to which the competent authority has complied with this obligation, Art. 9 demands that the content of the decision and the conditions attached must be made available to the public concerned. The reasons and considerations on which the decision is based must be published only where the Member States' legislation so provides.

In order to ensure that the results of EIA flow into decision making, most Member States have adopted provisions which considerably exceed the minimum requirements of the EEC Directive. In Belgium and the Netherlands, the competent authority must prepare a "Record of Decision" describing the considerations on which the decision is based and explaining what influence the findings of EIA had on decision making.

In some countries, EIA even has predetermining effects on the final decision. This applies particularly to Portugal where the permission procedure is not initiated when the minister responsible for environmental affairs gives a negative judgment on the project with regard to environmental concerns. In Italy and Spain, the final decision must be taken by the Council of Ministers if there is disagreement on the environmental compatibility of a project between the Minister of the Environment and the minister in charge of the permission procedure. In Greece, the ministry responsible for environmental affairs and the ministry which has to grant permission for the project must take a joint decision on the measures envisaged in order to avoid, reduce and remedy significant adverse effects. In these countries, EIA de facto departs from the function of being a procedure to support decision making and takes on the character of an advanced permission procedure. However, it remains to be seen whether EIA can fulfil this important role in practice.

The term, "the results of EIA must be taken into consideration", means that appropriate attention is paid to environmental issues. It does not mean, however, that environmental concerns should necessarily be given priority over other concerns in the decision making process. However, it is an essential prerequisite that the competent authority is legally entitled to respond to the results of EIA in an adequate way. This precondition is not always given in countries with a distinct sectoral environmental law.

In the Federal Republic of Germany in particular, integrating the trans-sectoral approach of EIA into the existing administrative procedures is rather problematic (Soell 1990). This applies primarily to procedures in which the authorities have no scope for discretion. The most important case is approval according to the German Clean Air Act. In those cases the applicant has a legal right to be granted permission as long as he has satisfied the statutory requirements. It remains to be seen whether the holistic overview can be guarantied by modifying the current practice of execution, or whether the existing law will have to be supplemented and extended in order to satisfy the goals of the Directive.

3.7. Monitoring

Under the EEC Directive, the EIA procedure is completed when the decision on the project has been taken. Contrary to early drafts, monitoring is not required.

However, introducing follow-up monitoring of projects that have been subjected to an EIA seems to be useful for various reasons. On the one hand, it allows a review of whether the developer has correctly implemented the measures envisaged to avoid, mitigate, and remedy environmental damages. On the other hand, it allows the decision to be revised if adverse impacts come up which had not been anticipated. Finally, a comparison of actual environmental impacts and those predicted can serve to improve EIA methods.

In most EC countries, the existing environmental legislation allows for follow-up monitoring as to whether the project developer continues to satisfy the conditions linked to the permit. However, this monitoring usually only covers the facilities and equipment and not the environment affected. To date specific monitoring related to the findings of EIA is only required under the Greek and Dutch legislation. In both countries, the consent given can be revoked if monitoring proves that the project produces significant adverse impacts which not have been forecast in the EIS.

4. CONCLUSIONS

In the last three years an increasing number of EISs have been prepared in all Member States, even in those where the legal basis has not yet been completed. Although first practical experience has been gained, it is too early to give final indications of the efficiency of different implementation strategies pursued by the Member States and of the changes in decision making on environmentally relevant projects caused by the introduction of EIA. Nevertheless, some important conclusions can already be drawn at the present time.

It has become obvious in all countries that the legal provisions can hardly be enforced unless detailed administrative regulations exist. There is need for further guidance on operating the new system, both for more general guidance directed to the various steps of procedure such as scoping, public participation, etc. and for more specific guidance directed to particular types of projects. Clear guidance on screening and scoping procedures, on identification of all the relevant impacts, on determination of the significance of these impacts, on the choice of methods used for prediction and survey, on existing data sources, and on the communication of the findings, would improve practice (Wood and McDonic 1989).

Evaluation studies undertaken by countries with lengthier EIA practical experience, like France, the Netherlands and the UK, have indicated the significance of external review in order to obtain comprehensive and sound assessments. In the absence of a clear definition of adequacy, individual developers will take either a comprehensive or a minimalist approach, depending upon their own attitudes, and the degree of guidance or encouragement offered by the competent authority (Nelson 1990). The result is considerable variation in the quality of EISs, both in terms of structure and content. It contradicts the spirit of the Directive that it is not possible for a developer to obtain permission without submitting an environmental statement, but that it may be possible for him to do so by submitting an inappropriate and insufficient statement. In order to ensure a uniform quality level, it therefore appears essential to introduce an external review by an independent body which carries out a standardized evaluation of all Environment Impact Statements submitted.

Furthermore, the findings of evaluation studies demonstrate the need for monitoring after implementation of a project with regard to both enhancement of environmental quality as well as improvement of EIA methods. These findings agree with the results of US studies, where the lack of monitoring proved to be a major deficit for the efficiency of the EIA procedure. For this reason, the forthcoming NEPA amendments contain wide-ranging regulations on monitoring which cover both the review of the proposed mitigation measures and also the accuracy of the forecasts made in the EIS.

Practical experience has shown that judicial review of EIA is crucial to its enforcement for two reasons. Firstly, if the agency's compliance with the legal requirements cannot be controlled by the courts, EIA runs the danger of having rather poor effect in practice, creating new deficits in administrative execution. Secondly, EIA is a new instrument, i.e. the application of the procedural and material requirements has not been tested before or only to a small extent. Appeal to the courts offers the potential for further judicial interpretation of the statutory regulations, thereby raising their efficiency. In the USA and France (Prieur 1991) and also in the United Kingdom after just two years of practice (Buxton 1990), court review has considerably contributed to the discovery of gaps and shortcomings in the legal provisions and to developing criteria for evaluation of the adequacy of EISs.

A precondition for an effective judicial control of EIA is, however, - as the USA example shows - that standing is not restricted to persons affected in their own rights but also conceded to recognized environmental organisations. This is a crucial demand, because decision making on environmentally relevant projects enlarges the category of injuries which the public and not the individual may sustain, e.g. harm to the public's right to know, to participate and to have the interests of future generations protected. Thanks to an increasing liberalization regarding access to the courts, this precondition has been satisfied in most EC Member States. The Federal Republic of Germany and Spain are exceptions, where the right to challenge administrative decisions is granted only to persons affected in their own rights. Therefore, the contribution of the courts to further developing the EIA system will remain small in these two countries.

An EIA which starts only at the permission stage, i.e. at the lowest level of a multi-tiered planning process, often comes too late, because binding, or at least predetermining decisions have already been made at previous levels. The extension of EIA arrangements to cover policies, plans and programmes is, therefore, a logical and essential step within the framework of a seriously meant preventive environmental policy. As already mentioned above, the EC Commission intends to introduce a "Strategic Environmental Assessment" (SEA). This would deal with alternatives and impacts which cannot be adequately addressed at the project level.

Although there is wide acceptance of the concept of SEA, its incorporation into national law will face EC Member States with considerable problems. There are significant differences between the environmental assessment of policies, plans, programmes, on the one hand, and of projects, on the other hand. The extension of EIA to higher tiers of action would probably cover a wide range of quite different actions ranging from research programmes through measures to improve agriculture, regional economics and coastal protection, land-use planning, and other multi-sectoral actions, including economic and fiscal policy. Whereas more or less similarly-structured procedures exist for licensing individual projects in all EC Member States, the adoption of policies plans and programmes are subject to very different decision making processes. The consent of one or several ministries or the Council of Ministers may be necessary, or even the formal adoption by local, regional or national parliaments. Furthermore, the issue of public participation for such actions must still be clarified. Experience from the lengthy deliberations on Project-EIA give rise to the fear that passing the Directive on the EIA for policies, plans and programmes will take decades.

REFERENCES

Buxton, R. (1990), "Interpretation of EIA Regulations in the UK Courts", EIA Newsletter 5, 12-13.

Caldwell, L.K. (1982), Science and the National Environmental Policy Act: Redirecting Policy through Procedural Reform, The University of Alabama Press, Second Printing (1985), Alabama.

Clark, B.D., Chapman, K., Bisset, R. and Wathern, P. (1976), Assessment of Major Industrial Applications: A Manual, Research Report 13, Department of the Environment, London.

Coenen, R. and Jörissen, J. (1989), Umweltverträglichkeitsprüfung in der Europäischen

Gemeinschaft, Erich Schmidt Verlag, Berlin.

Commission of the European Communities (1990), Proposal of a Directive on Environmental Impact Assessment of Policies, Plans and Programmes (Draft), Doc No. XII/194/90-DE, Brussels.

Council of the European Cummunities, Directive of 27 June 1985 on the Assessment of the Effects of Certain Public and Private Projects on the Environment (85/337/EEC), Official Journal of the European Communities, No. L 175/40.

Davies, G. (1989), "Environmental Impact Assessment, Public Participation and the Project Cycle", in H. Paschen (ed.), The Role of Environmental Impact Assessment in the Decision Making Process, Erich Schmidt Verlag, Berlin, pp. 239-252.

Devuyst, D. and Hens, L. (1991), "Environmental Impact Assessment in Belgium", Environmental Impact Assess Rev 11, 157-169.

Koning, H. (1990), "EIA in the Netherlands", EIA-Newsletter 5, 10-11.

Lee, N. and Wood, Ch. (1976), The Introduction of Environmental Impact Assessment in the European Communities, EEC Document ENV/197/76-D, Brussels.

Ministère de l'Environnement (1984), Enquête sur la pratique des études d'impact d'unités industrielles, Paris.

Ministry of Housing, Physical Planning and Environment (1984), Environmental Assessment in the Netherlands, The Hague.

Nelson, P. (1990), "Current Practice of EIA: the Consultant's View", paper presented at the Conference on Environmental Assessment: How Well Is It Working?, Manchester , 11 January 1990.

Prieur, M. (1991), "Impact Studies and Judicial Review in France", paper presented at the International Symposium on Environmental Impact Assessment, Genova 16-18 May 1991.

Schreurs, P.J. (1990), "EIA in Belgium", EIA Newsletter 5, 5-6.

Soell, H. (1990), "Wieviel Umweltverträglichkeit garantiert die UVP?", Neue Zeitschrift für Verwaltungsrecht 8, 705-723.

Tamburrino, A. (1991), "EIA Training in Italy from an Academic Viewpoint", in Ch. Wood and N. Lee (eds), Environmental Impact Assessment Training and Research in the European Communities, Occasional Paper No. 27, EIA Center, Department of Planning and Landscape, University of Manchester, pp. 131-141.

Wood, Ch. and McDonic, G. (1989), "Environmental Assessment: Challenge and Opportunity", The Planner 1, 12-18.

Table 1. Implementation of the Directive on EIA in the EC Member States. (State: November 1990).

Member State	Present State of Legislation	Date
Belgium (national level)	EIA provisions for nuclear installations (first draft)	under discussion
- Wallonia	EIA Decree Executive Order	11 September 1985 10 December 1987
- Flanders	Decree on antipollution measures 6 Administrative Orders	28 June 1985 23 March 1989
- Brussels	EIA Decree (draft)	under discussion
Denmark	Amendments to the National and Regional Planning Act, the Regional Planning Act for the Metropolitan Area and the Environmental Protection Act (Act No. 216) Executive Order No. 446	5 April 1989 23 June 1989
Federal Republic of Germany	National EIA Act Regulations for implementing the national EIA Act (draft)	12 February 1990 under discussion
France	Natural Conservation Act Decree No. 77-1141	10 July 1976 12 October 1977
Greece	Environmental Protection Act (1650/1986) - Art. 3,4,5 and 21 effective through Ministerial Decision (69269/5387)	25 October 1990
Ireland	European Communities (Environmental Impact Assessment) Regulations, 1989 Local Government (Planning and Development) Regulations, 1990 Regulations relate exclusively to motorways, 1988	1 February 1990 1 February 1990 1 September 1988
Italy	Interim arrangements based on Decree 377 (list of projects) and on Decree (EIA procedures) National EIA Act (draft)	10 August 1988 28 December 1988 under discussion
Luxembourg	Decree to implement the EC directive (draft)	under discussion
The Netherlands	Environmental Protection (General Provisions) Act - Chapter 4A Environmental Impact Assessment Decree	13 June 1986 20 May 1987
Portugal	Decree Law (No. 186) introducing EIA on the national level Regulations for implementing the EIA provisions (first draft)	6 June 1990 under discussion
Spain	Royal Legislative Decree 1302/86 introducing EIA on the national level Royal Decree 113/88	28 June 1986 30 September 1988
United Kingdom	19 different sets of statutory regulations on EIA concerning different types of projects	by November 1990 all but two effective

THE EEC DIRECTIVES ON ENVIRONMENTAL HAZARDS

A. Amendola
CEC, Joint Research Centre
Institute for System Engineering and Informatics
I - 21020 Ispra (Va)
Italy

ABSTRACT. Hazards to the environment are presented by a multiplicity of natural phenomena and human activities. The paper is restricted to the description of directives aimed at controlling hazards linked with the processing, storage and use of dangerous substances. In particular the so-called "Seveso" Directive, for which application experience already exists, is discussed extensively: attention is focused on some particular critical problems such as "information to the public " and land use policy. Furthermore, more recent directives on biotechnology hazards are described in brief and new trends in regulations for environment protection are commented upon.

1. INTRODUCTION

Council Directives controlling environmental hazards have two major goals: to protect[(+)] man and the environment by implementing an adequate risk management policy with respects to industrial activities; and to harmonise the obligations of manufacturers in Member States in order to avoid the distortion of competition within the Community. No industrialist should profit from investing in a Member State where the overall costs are lower because of laxer safety constraints.

Directives are formally adopted by the EEC Council of Ministers after a Commission proposal, and taking into account the opinion of the European Parliament and the Economic and Social Committee (ECOSOC). This Committee represents all categories of economic and social life, industrialists, professionals, workers, farmers, consumers, and representatives of other public interests. Directives establish the objectives and basic principles to be complied with by all Member States: each State must transpose them into its own national legislation. This allows various cultural traditions, institutional structures, and regulatory styles to be accommodated [1]; on the other hand, this also may result in a variety of criteria and procedures, which can contradict the ultimate goal of the harmonization of the national approaches. Therefore, the Commission and competent national bodies must constantly monitor the implementation process aiming at a substantial convergence.

Under these aspects Directives differ substantially from EEC "Regulations" since the latter are directly applicable in all Member States, without undergoing national legislative procedures.

(+) Since experience has shown that in general implementation tends to be anthropocentric by paying less attention to hazards affecting only the environment, the rewording "to protect the environment including man" may be preferred in future texts.

A. G. Colombo (ed.), Environmental Impact Assessment, 15–30.
© 1992 *ECSC, EEC, EAEC, Brussels and Luxembourg.*

Many Directives have aimed at controlling the environmental hazards of nuclear materials and dangerous chemical substances. They cover control radiological protection, and the requirement that no new chemical substance should be introduced on the market before their possible dangerous potential has been identified. Information is required on the harmfulness of the materials, the way of identifying them by appropriate labels etc... and cannot be described in the limits of this paper.

Other Directives are more linked with industrial processes or commercial activities involving the use of dangerous substances. Such activities may present

- environmental hazards in their routine operation (emissions in air, water, soil, wastes, etc...) which should be the subject of the environmental impact assessment procedures and are regulated by the Directive described elsewhere in this book;
- accident hazards which are controlled by ad-hoc Directives, which are the principal subject of this paper.

Of course the two types of hazards are not independent of each other: this should call for better coordination in national practice as far as both the rules and the competent authorities are concerned.

An effective way of regulating industrial hazards consists not only in "controlling" procedures by more or less stringent licensing or operational permits, but also in creating incentives for industry and society for a self-organizing process towards safe and environment-friendly technologies. This has been recognised by the EEC in its very recent regulatory action. Relevant examples are the "right to know" Directive and the proposed ECO-Auditing Regulation.

2. THE "SEVESO" DIRECTIVE

The Directive on the Major Accident Hazards of Certain Industrial Activities, more commonly known as the Seveso Directive, specifies the minimal requirements for regulation of the hazards connected with process and storage facilities involving dangerous chemical substances.

The Directive, issued in 1982 [2] has been already amended twice [3,4] in order to cover certain storage installations that after the occurrence of some new severe accidents (e.g. Basel) appeared not to be sufficiently covered by the original Directive, and to stress the need for information for the public exposed to the risk (all amendments have been included in a comprehensive text [5] to facilitate understanding of the present state of the Directive).

Recommendations issued by the Council of Ministry [6] to include provisions for Land Use Planning are being taken into account in the fundamental revision of the Directive, which is being improved according to the experience gained with its practical implementation in these years as well.

The control of the implementation of the Directives is ensured by the CEC Directorate-General "Environment, Nuclear Safety and Civil Protection" together with the Committee of the National Competent Authorities, whereas technical support for a harmonised implementation is being given by the CEC-JRC/ISEI.

2.1. The Contents of the Directive

The major accidents that stressed the need for a Directive regulating hazardous industry (e.g. Seveso, Flixborough) had in common the features that local authorities did not know what chemicals were involved and in what quantities; they did not know enough about the processes to understand what chemicals/energy could be produced or released under accident conditions; and there was a lack of planning for emergencies. With this background, the Directive is largely concerned with the generation and the control of a correct information flow [7] among the different actors in the risk management procedure. It applies to industrial activities involving processing or storage of substances which, in the case of an accident, could provoke major toxic releases or fire and explosion events. The nature and inventories of such substances are specified according to classes or nominal lists.

The principal requirements of the Directive can be summarized as follows:
- each Member State must appoint a Competent Authority (CA);
- at any time the manufacturer shall prove to the CA that major hazards connected with the installation have been identified and adequate safety measures have been taken to prevent accidents;
- when dangerous substance inventories exceed specified thresholds, the manufacturer shall provide the competent authority with a written notification on the installation hazards (Safety Report), shall prepare an internal emergency plan, and give the information needed by the CA for the preparation of off-site emergency plans;
- the CA should be notified of major modifications;
- CA shall provide for external emergency planning;
- Member States shall ensure that people liable to be affected by an accident are "actively" informed of the safety measures and how to behave in the event of an accident;
- the manufacturer shall report to the CA any major accident which occurs, and the national authorities should notify major accidents to the Commission;
- the Commission shall keep a register of accidents so that Member States can benefit from this experience for prevention purposes.

The Commission organizes quarterly meetings of the Committee of Competent Authorities (CCA), during which questions concerning the Directive and its implementation are discussed so that common approaches can be adopted, and the experience gained with the implementation can be used to ameliorate the Directive itself.

As a result of this experience, and by considering the need to reinforce control of installations which appeared to present particularly high hazards after the occurrence of some new major accidents (Bhopal, Mexico City, Basel) the original Directive was amended a first time to adjust threshold quantities for safety report notification; and a second time, for better coverage of storage activities and active information of the public.

Furthermore, the Commission has organized the Community Documentation Centre on Industrial Risk (CDCIR), located at Ispra, which collects, classifies and diffuses information on published accident investigations, regulations, safety codes of practices, risk studies etc. The Centre is generally accessible, and the information is diffused by bulletins which are widely available. It also promotes publications of studies performed or sponsored by the Commission on the technical aspects of the Directive's application in the Member States. This action, which also contributes to an increased availability of information on all questions concerning risk, provides policy makers and safety analysts with a wide basis of knowledge on national practices and therefore accelerates and improves the harmonization process.

2.2. The Safety Reports

Before the Directive was adopted, many EC countries already required the submission of safety reports for certain classes of installations.The Directive extended this obligation to all countries: for existing installations the final date for submission of safety reports was July 1989. In most Member States the requirement to produce these reports is considered an improvement on earlier practice. In some countries the reports are accessible to the public, except for sections containing commercially sensitive information. The philosophy implied by the safety reports varies from country to country, especially with respect to the role of risk analysis and whether or not a licensing procedure exists.

2.2.1. United Kingdom. In the U.K. [8] the essence of the "safety case", and the reason behind the choice of this term, was that the onus lay on manufacturers to assess their own hazards, take measures to control them adequately, and then to present their arguments to the HSE, which is the competent authority for the U.K. There is no licensing procedure as such, whereas the "safety report" required under the notification obligation of art. 5 of the Directive has assumed to some extent a meaning fairly similar to that in use in the other countries. In any case it is a tool to enable the HSE to judge whether all significant risks have been identified and are being properly managed. It is used by HSE inspectors as a "living" inspection tool.

Probabilistic safety assessment is not mandatory in the safety case. However, the HSE "... may well find it easier to accept conclusions which are supported by quantified arguments. A quantitative assessment is also a convenient way of limiting the scope of the safety case by demonstrating either that an adverse event has a very remote probability of occurring or that a particular consequence is relatively minor." [8]. Quantitative criteria are being introduced as guidelines for land use planning [9].

2.2.2. The Netherlands. In the Netherlands, different requirements are imposed by legislation according to whether worker safety or that of the general public is at issue. Indeed, the occupational safety report must include a descriptive identification of hazards, organizational information and on-site emergency information. It may include reliability assessments, but more emphasis is given to the human factors and management issue. The report is established in close collaboration between representatives of the Ministry of Labour and Industry [10].

External safety is on the other hand the responsibility of the Ministry of Housing, Physical Planning and the Environment (VROM). For external safety acceptable risk criteria have been established [11]. These criteria are part of the environmental policy approved by the Dutch Parliament in 1985, and are mainly used to assess the compatibility of siting proposals with existing land use, as well as the need for backfitting or relocation of existing installations.

2.2.3. France. In France the licensing application is subject to a public enquiry. It includes an "étude des dangers" (literally, study of dangers) which does not imply a probabilistic evaluation. Rather, it compiles an inventory of possible failure sequences, their consequences, and a description of the preventive measures adopted. Even reference and envelop scenarios are established for land use planning and emergency planning. Only in certain particular risky situations are an independent assessment or the analysis of certain particular hazards required. This is called an "étude de sûreté" (safety study) [12]. Since no account is given of probability even certain rather improbable scenarios are fully evaluated and depicted in the report. This certainly contributes to increasing awareness of risk in the industry and is an incentive towards safer technologies.

2.2.4. Italy. Italy has issued detailed guidelines not only for the compilation of the safety reports, but also for the "safety declaration" which is required for installations involving inventories of hazardous substances below the notification obligations [13]. Hazards are identified by the evaluation of a very comprehensive fire, explosion and toxicity index. Afterwards, scenarios are described by using probabilistic techniques as well.

2.2.5. Federal Republic of Germany. In the Federal Republic of Germany there is a mandatory licensing procedure, based on a deterministic philosophy, which implies that a safe facility will have essentially zero risk, which is achieved by the use of the best safety technology and practices; and by adequate design of redundancies in the safety barriers. Therefore, the safety report is essentially limited to the identification of possible hazards and a description of technical measures taken to prevent failures or to contain their consequences within the establishment.

2.2.6. Some remarks. Even if not complete, this comparison demonstrates the effects of national values, institutional structures, and regulatory styles in the practical implementation of a common requirement. A priority activity for the Commission in 1989 was the performance of a comparison in depth of the national approaches to the safety reports which resulted in a comprehensive report included in the CDCIR [14]. This comparison also showed a rather strong link which exists between the safety report requirement and the environmental impact assessment. Environmental effects of accidents must be identified in the safety report: however this is only required for "major events". Minor malfunctions of the same systems will also affect the environment even if with smaller consequences (but possibly with greater frequency). In certain cases different authorities are responsible for control of the same systems according to the magnitude of their possible accidents. This may contribute to increasing the burden to industry without improving the control, because a grey area between accidents and malfunctions may not be covered by any control. At the same time duplications of efforts may occur. There is a need to link the two decisional process as better. Without making a merit judgment of the effectiveness of the control, it is useful to mention that when as in France for a new installation a unique dossier including the four volumes:

- installation description,
- environmental impact assessment,
- safety report and
- work health and safety

is presented to just one authority (the prefecture which involves all competent technical and administrative bodies in the licensing procedures), the basis for a rational cost effective control of all linked aspects appear to be possible.

2.3. Land Use Planning with Respect to Major Accident Hazards

The control of risk is not a simple technological problem. Whichever technological measures are taken, risk cannot be reduced to a zero level. Each system can fail. Excessive reliance on the quality of design, on the safety systems, even a long operation time without major failures can decrease the operator's attention towards the fact that a major accident may occur and therefore contribute to the organizational factors, which are among the most frequent underlying causes of accidents.

However, in the case of an accident risk decreases with the distance from the source. Greater distances between installations and inhabited zones may also result in more time being available for implementing emergency plans. Therefore, effective mitigation of the consequences of an accident, effective risk regulation, can only be achieved if, as well as safety assessments and emergency preparedness, the use of the land around an hazardous installation is also properly controlled.

For the time being the "Seveso" directive does not include any obligation with respect to proper land use planning. Before this directive, in countries without complete legislation in this field, authorities did not have an adequate description of the risk connected with the operation of a given plant. This may have created the existence of intolerable situations of proximity of houses, schools, hospitals etc. to risky installations. Now the authorities can find out about the risk, therefore they should control the actual situation and plan future land uses adequately.

In the autumn of '89, the problem was considered by the EEC Council of Ministers, which recommended that land use planning be included in the work of the next revision of the directive [6]. Since this recommendation, the Commission is working to include the land use planning among the new obligations of the "Fundamental Revision of the Directive", even asking that the public should be informed and possibly should participate in the decision.

This then created problems which are difficult to solve not only for existing plants, but even for new ones, which often for economical reasons tend to be located on sites already occupied by old ones (because of replacement of obsolete production, expansion of production, functional links, land ownership, etc.).

Tendencies in countries like UK, NL and F show that the problem is already being approached in different ways whereas a pilot study in Italy is resulting in useful indications for planners in a large industrial area like that of Ravenna.

2.3.1. United Kingdom. Risk analysis for land use planning is separated from the submission of the safety report according to the notification in art. 5 of the Directive. The pragmatic approach to regulation in the UK generally results in "advice or recommendations" of the safety authorities (HSE) rather than in legislative acts.

New Installations. The planning of a new activity must be approved by the local administrative authorities. These, if the activity involves major risks, ask for the HSE's advice. The HSE then performs a risk analysis on the conceptual design of the installation and gives its advice case by case, on whether the proposed installation can be located on the site with respect to the risk. In theory the local authority might decide against this advice according to other parameters, for instance benefits from increased employment. The safety report is then produced by the industrialist on the final design and submitted to HSE at least three months before the operation of the plant.

Existing installations. After the submission and the revision of the safety reports, one can assume that if a plant continues to be operated this means that it is sufficiently safe. The duty of the authorities is (in addition to inspections) to control the developments around the site to avoid the creation of new risky situations. For this purpose the HSE has produced a relevant guidance document [9] which states criteria mostly of a probabilistic nature for the land use around a hazardous installation.

This document recognizes that when risk is not due to toxic releases, safety distances can be evaluated according to anticipated consequences of typical fire and explosions events (i.e. BLEVE in the case of LPG storage). Where toxic releases are concerned, a deterministic criterion would be much more difficult to apply, since effects may be considerable even over rather large distances (kilometres). Therefore one should move towards risk criteria (i.e. the probability of given consequences at a given distance). Quantitative targets have been fixed in such a way that the following criteria should be matched:

- "HSE will advise strongly against any developments which introduce a substantial number of people into an area where their individual risks are "significant" when compared with other risks to which they are exposed in everyday life." This should be considered as "involuntary exposure" for people who gain little direct benefit from the risky activity and sets an upper level of risk criterion;
- "HSE will not advise against developments where the individual risks seem to be very small in comparison with everyday life, unless the development makes a significant contribution to societal risk." This sets a lower level criterion for individual risk;
- between the extremes above, advice is given case by case according to the dimension of the proposed development.

To understand how the principles stated above are translated in quantitative terms it is necessary to refer to the arguments and also to the overall HSE control system [9]. However just to give an idea about the conclusions these are briefly summarised in the following. The HSE will usually advise against developments near major hazards in the circumstances shown below:

1. Housing developments providing for more than about 25 people where the calculated individual risk of receiving a defined "dangerous dose" of toxic substance, heat or blast overpressure exceeds 10 in a million per year.
2. Housing developments providing for more than about 75 people where the calculated individual risk of receiving a "dangerous dose" exceeds 1 in million per year.

3. According to their dimensions stores, restaurants etc.. will be judged case by case between the above limits.
4. Hospitals, schools because of their dimensions and particular vulnerability may see the 1 in a million per year criterion extended downwards.

2.3.2. The Netherlands. Again when speaking of quantitative criteria one must refer not only to the method employed, but also to the overall control philosophy in each country. Criteria should be seen as relative to their utilization context.

In the Netherlands, risk analysis is a part of the safety report: it is mandatory for the so-called external safety report. Furthermore risk acceptance criteria have been adopted after a discussion in Parliament [11]. They therefore have a compulsory value, rather than an advisory one, even if the experience with their application is still small because not all reports for existing plants have been analysed to draw final conclusions.

The basic criterion has been the definition of an unacceptable risk value for a person as a consequence of his/her "involuntary" exposure to any kind of man-induced hazards (transport, storage, process of dangerous substances including radioactive ones). It is not possible to accept risks comparable to those voluntarily accepted in every day life (10^{-4}/y).

Therefore, individual risk for a member of the general public, connected with any kind of industrial activities, should at least be an order of magnitude less (therefore should be less than 10^{-5}/y calculated by assuming a continuous presence "outdoors" in a certain location). Since this represents a global exposure to any kind of source of hazards, another value is defined for each source which should not be accepted (one order of magnitude less). Concerning major accident risk from a single fixed installation for people outside the fence (workers are subjected to voluntary risk, are well informed and well prepared for emergencies):

- unacceptable individual risk > 10^{-6} events/year,
- negligible risk < 10^{-8} events/year.

Between these limits the authority can negotiate measures for risk reduction. Both individual risk and group risk criteria have been fixed.

The philosophy which is intended to guide decisions concerning the existing situations can be summarised as follows: if the external risk calculated in the safety report exceeds the values described, the owner should implement measures to bring this figure below the limits. When these measures correspond to bringing the plant to the updated state of technology, then the costs are to be covered by the owner. If, because of the particular inhabitant density and proximity, exceptional measures must be implemented than society must not forget the benefits from the industrialization and contribute to the cost. If these are the main orientations of the physical planning policy, the relatively short time which has elapsed from the submission of the safety reports does not allow us to discuss how the criteria have worked in practice. Certainly no permit will be given for new houses at distances less than those corresponding to acceptable risk values.

2.3.3. France. In France probabilistic criteria have not been adopted. The authorities preferred to require the evaluation of accident consequences with respect to two different scenarios: a "reference scenario" for the evaluation of safety distances and an "envelope scenario" which is useful for emergency planning.

Reference scenarios take account of serious accidents which can reasonably be hypothesized. However there is always a certain degree of subjectivity in defining a reference scenario. One can refer to past accidents as well as to possible events like the break of a piping connected with a toxic storage facility.

The French approach to physical planning [15] shows a strong cooperative effort between authorities, local administrators, elected representatives and industries involved in discussing together the results of the safety studies and agreeing on exclusion distances or on areas with limitations for new developments.

2.3.4. The ARIPAR Project. In Italy, although at the moment there is no legislative act regulating the issue, a pilot experiment has been attempted for the Ravenna industrial area, for which risk to the population induced both by fixed installations and all type of transportation of hazardous material has been evaluated as a basis for the administrators to plan both urban developments and infrastructure improvements. The ARIPAR project [16] has demonstrated that planning with respect to accident hazards cannot ignore risks connected with the transport of dangerous goods.

2.3.5. Some remarks. From the previous description it appears that criteria being introduced in different EC countries are not uniform.

The probabilistic criteria are compared with the problems of uncertainties. Different analysts, different procedures, assumptions, data and models may result in a rather wide spread of calculated risk.

The deterministic approach is also compared with the difficulty of choosing the reference scenarios, and therefore of the exclusion distances in the case of toxic releases.

In the end both approaches should result in practical "safety distances" between industrial activities and urban developments. It will be interesting to assess if, even starting from different philosophical bases, safety distances with respect to similar activities might in the end be similar.

Finally, one should not forget that even emergency planning will benefit from correct land use planning especially as far as access ways and the allocation of physical resources are concerned, and from the avoidance of particularly vulnerable development being allowed too near to hazard sources.

Certainly, the introduction of these principles will lead to results in the long term, even if it does not immediately solve existing risky situations.

Again, the need to link planning with respect to accident hazards with more effective environmental planning is obvious. Once again it appears that EIA must be integrated with accident impact analysis in a joint procedure. Especially, it should be noted that the criteria described essentially refer to the direct risk to man, by thus confirming the anthropocentric interpretation until now given to the Directive.

2.4. The MARS System

To store and retrieve the accident reports notified by the Member States under Articles 10-12 of the Directive, the MARS (Major Accident Reporting System) data bank has been established at the JRC/ISEI. The content of the information to be supplied has been defined in consultation with the CCA, so that:
- all countries can use a uniform reporting procedure;
- the information supplied is consistent with the requirements of the Directive;
- the information is adequate for understanding both the primary and underlying causes and the accident circumstances;
- Information can be easily processed and stored in the MARS informatics structure.

Special accident reporting forms [17], designed to meet the above requirements, have been adopted and are now in use. To date, 130 accidents have been reported and entered in MARS. Since the concept of "major accident" is rather fuzzy, the Commission and the CCA have worked to establish provisional guidelines to define it better. Under this perspective an accident gravity scale has been adopted for a two year trial period. In practice, the tendency is to report not only accidents which have caused major damage, but also incidents that might have led to serious consequences had emergency systems not functioned properly.

Accidents notified and stored in MARS are analysed and characterized by various indices, e.g. activity concerned, substances involved, type of accident (fire, explosion, toxic release, etc.) ,gravity indices, causative factors, etc.

A structured exchange of experience among the CCA members is ensured through regularly issued reports:

- at each CCA meeting. a summary of major accidents, indicating causes, consequences and measures that could prevent recurrence, and/or mitigating consequences;
- a yearly report on "Lessons Learned" .

Complete information is available only to the CCA members, however a first report on lessons learned purged of confidential information has been published [18]. This report also includes the provisional scale for measuring the severity of consequences of accidents for man and the environment.

The report is valuable in improving information on accidents and therefore improving prevention. Many results of the accident analysis might merit being discussed more extensively. But at least two principal outcomes need to be mentioned here, namely:

- in more than 80% of the cases the underlying causes of the accidents notified were found to be in inadequate management either in the design phase or in the operational one. Therefore in the fundamental revision of the Directive more emphasis should be put in the control of safety management;
- again in the reporting of the accidents an anthropocentric application of the Directive has been identified. Damage to the environment is poorly described, and accidents have been reported mostly when they presented hazards to man. An investigation of environmental accidents from open sources sponsored by the JRC and performed by the Danish Water Quality Institute confirmed that only rather scarce and poor information can be found on the environmental consequences of accidents. In only one case were measurements of ground water quality given. (The report is now being printed within the CDCIR publication series). This result also confirms the need that in the revision of the Directive more emphasis should be put on environmental protection.

2.5. Information to the Public

The problem of risk communication is becoming very crucial in any planning concerning activities involving environmental risks. The second amendment defines the content of the information to be provided to the public. For example: a simple explanation of the activity, information concerning the substances involved, potential effects of major accidents, information on warning systems and media used during the course of an accident, how to behave should an accident occur etc..

However, the impact of risk information on the public has not yet been assessed in all countries. The Commission has organized a conference on risk communication [19] which resulted in a broad picture of the problems involved and is now active in promoting studies on selected sites to issue some guidelines on how to give information to the public. Application examples are available for most countries. In some cases the safety reports are public documents. However, the problem of communicating with the public about risk has a wider dimension than just that linked with the Directive and will therefore discussed to a certain extent in the following.

2.6. Main Effects of the Implementation of the Directive

Even if the state of implementation is not completely satisfactory in all member countries, certain significant irreversible improvements have been realised:

- in all countries safety notifications have been submitted to the competent authorities. Even when the reports have not been analysed or approved, their simple performance have compelled industrialists to take awareness of the risks in a structured way, in some cases improvements have been suggested and implemented immediately by the industrialist himself;
- CA now have comprehensive information on major accident hazards sites. It is easier for them to control and to plan;
- in most countries emergency plans have been implemented and people are becoming aware of how to behave in emergencies;
- the exchange of information on accidents has started a common learning process and reminded industry of the need to reinforce safety management in general;
- a social process (conferences, workshops, public debates) has been promoted in which industry is confronted with public opinion. This kind of control may be "per se" more productive that any bureaucratic imposition.

3. TRANSPORTATION HAZARDS

The Seveso Directive concerns only fixed installations. Transport facilities of dangerous substances are covered only when functionally linked with the installations. Now there are facilities such as temporary stores in docks, marshalling yards and similar which become in practice a store of multiple chemicals in transit, whose hazards are comparable with those of fixed installations and may present even larger risks because of their proximity to the centre of large towns [16]. It is not necessary to retrieve accident statistics [20] even if these have the merit of focusing the attention of policy makers on "facts" to be aware of the accident hazards due to the transport of dangerous chemicals. It is sufficient to recall the oil spills in the sea in Alaska and in Liguria, the deaths in the camp site in Spain and in the trains involved in the explosion caused by the break of the Russian pipeline, just to mention rather recent and impressive events in the USA, the EEC and the USSR. The nature and the extent of the damages are similar to accidents in fixed installations. The emergency operations are of the same nature, even if in most cases the site of the possible event is random and in some cases the content of the transported substance difficult to identify because identification labels can be obscured in tanks engulfed in fires.

Despite this awareness and recommendations formulated after accidents [6], the complexity of the regulations involving transboundary factors and a multiplicity of scattered administrative authorities so far has impeded the establishment of provisions similar to those established in the Seveso Directive, apart, possibly, from increased international cooperation in response to major events (EEC, OECD, ECE).

However, at least as far as emergency planning and response is concerned the Seveso Directive has contributed in some cases to the preparation of emergency planning on an all hazard concept, also including transportation accidents and oil spills near to coasts [21]. The approach towards a general preparedness for chemical accidents, independently of whether they arise from fixed or transport facilities, must also be recommended as far as public information is concerned: people are indeed exposed to risks from mobile sources and people move from one site to another more and more frequently. Preparing people to behave correctly in an emergency cannot be limited to residents near fixed installations.

4. BIOTECHNOLOGY HAZARDS

According to the principle that side effects of new technologies should to be assessed before their introduction on the market (a principle which has not been followed in the past as the history of industrial development shows), two Directives aimed at controlling the hazards of biotechnology industry have been adopted recently [22,23]. To which extent biotechnology may present hazards for the environment and for health has not yet been clarified. The nature of the hazards and of the safety assessments is different from that of chemical installations. Voluntary or involuntary release of genetically mutated organisms has the theoretical potential to provoke effects over a geographical area whose extent is not easily foreseeable. In the following the contents of the Directives will be briefly summarised, whereas the reader is referred to [24] for a wide-ranging discussion of the scientific background of the regulation. The Member States should implement the Directives by October 1991, so that it is too early to discuss any experience with their implementation.

4.1. The Directive on the Contained Use of Genetically Mutated Micro-Organisms (GMMOs)

The Directive defines "contained use" as
"any operation in which micro-organisms (any microbiological entity, cellular or non-cellular, capable of replication or of transferring genetic materials) are genetically modified or in which such GMMOs are "cultured", "stored", "used", "transported", "destroyed" or "disposed of" and for which physical barriers, or a combination of physical barriers together with chemical and or biological barriers, are used to limit their contact with the general population and the environment".
For the first time the complete life cycle of an activity has been taken into account, including any form of transportation. Of course provisions are different according to the nature of the activity phase.
Article 6 requires that "the user shall carry out a prior assessment of the contained uses as regards the risks to human health and the environment". The assessment must take into account parameters specified in Annex III, a summary of which is given as follows:
1. Characteristics of the donor, recipient or (where appropriate) parental organism(s).
2. Characteristics of the modified micro-organism.
3. Health considerations.
4. Environmental considerations, which concern:
- factors affecting survival, multiplication of the modified micro-organism in the environment;
- techniques available for detection, identification and monitoring of the GMMOs;
- techniques available for detecting transfer of the new genetic material to other organisms;
- known and predicted habitats of the GMMOs;
- description of ecosystems to which the GMMO could be accidentally disseminated;
- anticipated mechanism and result of interaction between the GMMO and the micro-organism, or the organisms, which might be exposed in the case of a release into the environment;
- known or predicted effects on plants and animals such as pathogenity, infectivity, toxicity, virulence, vector of pathogen, allergenicity, colonization;
- known or predicted involvement in biogeochemical processes;
- availability of methods for decontamination of the area in the case of a release to the environment.
The (incomplete) list of parameters previously described gives a sufficient idea of the

complexity of the assessment required. This assessment must be performed by the user and a record of the assessment must be kept at the disposal of the competent authorities or submitted to them in the notification required in articles 8,9 or 10 which are not applicable to transportation.

Notifications are requested when a particular installation is to be used for the first time and for particular groups of GMMOs and/or operations.

Emergency planning, information of the public, accident notification, communications between member countries are articles similar to those included in the Seveso Directive.

4.2. The Directive on the Deliberate Release Into the Environment of Genetically Mutated Organisms(GMOs).

"Deliberate release" means any intentional introduction into the environment of a GMO or a combination of GMOs without provisions for containment to limit their contact with the general population and the environment. Therefore, the Directive covers both the deliberate release for research and development (part B) and for the placing on the market of products containing GMOs (part C).

Before any experimental release an environmental risk assessment must be performed, a notification including an emergency response plan and impact/risk statement must be submitted to the Competent Authority of the country where the experimental release is to be performed and the CA must give written consent. Simplified procedures can be agreed upon with the Commission if an authority has obtained sufficient experience with the release of certain GMOs. After the release the notifier shall send the result concerning any risk to man and the environment.

The novelty of the Directive with respect to the others previously described consists in the fact that the CA shall send the Commission a summary of each notification, the Commission shall forward the summary to the other Member States which may ask for information or present observations through the Commission, or directly and the CA shall give information on the final decision taken (art.9).

This procedure is reinforced when a product is put on the market for the first time. When a product is put on the market in one country it is no longer submitted to similar obligations in another country provided that the Directives' provisions have been respected. In this case, if within 60 days after being informed by the Commission the competent authority of another country raises an objection, a procedure is established which through a technical committee, or through the Council acting by qualified majority, allows the Commission to take the final decision (art.21).

5. PLANNING, UNCERTAINTIES AND RISK COMMUNICATION

All the Directives described prescribe that the public liable to be affected by risk should be informed; the public might be involved in the land use planning provisions being established within the fundamental revision of the Seveso Directive. Information between member states is a dominant characteristic of the biotechnology regulation. Furthermore, the right to know about the state of the environment and about the activities having an impact on the environment was established in a new Directive in June 1990 [25]. EIA procedures and decisional models foresee public participation to an increasing extent. At what level planning is performed, or new activities are to be authorised, an effective risk communication process needs to be implemented in the relevant social context.

At the conference already quoted, Otway [27] summarised a view common to many papers and discussions: "Risk communication is not just about the narrow issue of providing technical information; it is about relationships. If society is to enjoy the full benefits of technology, wisely selected and applied, new relationships must be established amongst government, industry and public".

Communication is not a one way process. It implies that a subject giving information accepts that feedback information from other subjects can modify his perception of the world. This is particularly relevant when communicating about risk where by definition one communicates about what is uncertain as far as both likelihood of an event and its possible outcomes are concerned.

However, for a number of years even risk assessors who assumed probability as a measure of uncertainty tended to forget the subjective character of probability: i.e. to forget that a probability value is not "calculated" but is "assigned" according to the status of knowledge and by strict "coherence" rules [28]. As demonstrated by recent "benchmark exercises" [29,30] the largest uncertainty in quantitative assessments is not caught by the so-called confidence bounds calculated by the assessors. Indeed a risk assessment consists of two parts:
- a codified part based on defined algorithms and data, and
- a part, more qualitative, which includes the overall delineation of the problem, the choice of the starting hypotheses, the choice of what is relevant or not relevant and the choice of the most suitable models of physical phenomena for which competing models exist.

This second part is the main source of uncertainty and, therefore, of conflicts [31]. If expert knowledge is uncertain, how can this uncertainty go through the communication process without undermining confidence in experts, planners and emergency coordinators? [26].

Furthermore, in the communication process knowledge is exchanged through flows of information. These may result in an increase of the overall uncertainty because of the introduction of noise, the lack of reliable quality indicators for information sources [32], the bias created by interest groups and connected media, or perceived as such by the multiple actors in the process, etc.

Since for effective planning, environmental friendly decisions can only be taken if there is a social consensus (how many wrong choices are made because of emotional reactions to unclear debates!), the incentives derived by the active/passive public information obligations of the EEC Directives should be seen not as a barrier to technological developments but as beneficial in the long term even to industry: experience has shown that when the communication process has been implemented clearly, people learn to live with the risk and at the same time technology and management become "safer".

6. INCENTIVES FOR ENVIRONMENT FRIENDLY ACTIVITIES

Safety was first regulated by compliance with safety standards: this was the case for pressure vessels, fire protections etc. These norms still exist and are useful. But moving from components to complex systems it is more difficult to apply standard and norms legislatively with respect to environmental risk: accidents and involuntary releases are caused mainly by human and organizational factors. Rating systems are being introduced for management performance. Probabilistic acceptability criteria have been discussed with their limits, but are not generally accepted as replacing the simple previous requirement that components should comply with legislative standards. Implementation of the Directives calls for a direct assumption of responsibility by the "industrialist" or the "user".

On the other hand, many authors (see e.g. [33]) have identified the self-organizing structure of complex industrial systems. It is then possible to put "stimuli" into the system which can attract processes to move towards planned objectives, and, afterwards to supervise the development in a kind of dynamic control loop.

The public information provisions are already stimuli for safer technologies. The fact that most of the Seveso Directive authorities require consequence scenarios to be fully evaluated and inserted into the safety reports when these are public documents is also a stimulus towards safer technology and management. The structured dissemination of information on accidents is another example. The assumption of inventory thresholds for the notification obligation has also in some cases achieved the result that inventories of dangerous substances have been decreased to the level really needed by the process, or processes have been developed in which dangerous material is consumed as soon as it is produced to avoid the store of large quantities.

This regulatory way seems to have been followed explicitly by the on-going proposal for a Community environmental auditing scheme for certain industrial activities. The scheme, which should be implemented by a Regulation and not by a Directive, has the objective of promoting the use of environmental auditing as a voluntary tool for systematic, periodic and objective evaluation of environmental performances of certain industrial activities.

ECO-AUDIT is defined as a management tool comprising a systematic, documented, periodic and objective evaluation of how well organizations, management and equipments are performing with the aim of contributing to safeguarding the environment. The companies adopting the system will gain in image and bureaucracy should be as small as possible. If these are the intentions, the difficulties will begin as soon as "objective" evaluation is better defined: however independently of an "objective" rating system to compare different audited organizations (this problem will certainly deserve R&D), progress towards environmental friendly processes will be realised by the establishment within an organization of an environmental programme whose success can be "measured" periodically.

7. CONCLUSIONS

The paper gives an overview of the problems connected with the implementation of EEC Directives linked with the control of industrial installations with respect to environmental protection. The control of complex activities is giving increasing importance to public participation and planning. The complexity of the systems is shifting the tendency from prescriptive regulations towards goal setting approaches: this can be seen both in the tendency in certain countries to prepare concepts of tolerable risk, and in the tendency of Directives to stress the importance of management and environmental auditing. Correspondingly both the authorities and R&D organizations are being faced with the rather new challenge of rating environmental and managerial improvement programmes as well as risk reduction policy.

REFERENCES

[1] H. Otway and M. Peltu (eds), Regulating Industrial Risks: Science, Hazards and Public Protection, Butterworths, London, 1985. (See in particular T.O'Riordan contribution pp 20-39).

[2] Council Directive of June 24, 1982 on the major-accident hazards of certain industrial activities (82/501/EEC). Official Journal of the European Communities, L230, Vol.25, August 5, 1982.

[3] Official Journal of the European Communities, L 85, March 3, 1987.

[4] Official Journal of the European Communities, L 236, December 7, 1988.

[5] EUR 12705 EN / ISBN 92-826-1456-5, CEC, 1990.

[6] Official Journal of the European Communities, C 273, October 26, 1989.

[7] H.Otway and A.Amendola, "Major Hazard Information Policy in the European Community: Implications for Risk Analysis", Risk Analysis, Vol.9, No 4, 1989, 505-512.

[8] HSE, "A Guide to the Control of Industrial Major Hazard Regulations", Health and Safety Executive HS(R) 21, 1984.

[9] HSE, "Risk Criteria for Land-Use Planning in the Vicinity of Major Industrial Hazards", (1989).

[10] J.I.H. Oh and C.A.W.A. Husman,"Major Hazard Regulation in the Netherlands: The Organisational Safety Aspects", The Institution of Chemical Engineers, Symposium

[11] C.J. van Kuijen, "Risk Management in The Netherlands: A Quantitative Approach" in B. Segerstahl and G. Kroemer (eds.), Issues and Trends in Risk Analysis. IIASA, Laxenburg (A), WP-88- 34, pp.41-57.

[12] Ministry of the Environment (France), Guide d'Application de la Directive Seveso, (1985).

[13] The Official Journal of the Italian Republic, No 93, April 29 ,1989.

[14] A. Amendola and S.Contini, "National Approaches to the Safety Report. A Comparison", CDCIR No 677-EAb19-I.2, SP-I.91.07, CEC-JRC, ISEI, 1991.

[15] J.Mansot, The case of Lyon, OECD workshop on the role of the authorities in research into major hazards and land use planning, London, 19-22 February, 1990.

[16] D. Egidi et al., The Aripar project. Preliminary conclusions. (In Italian). Relation presented in Ravenna on 28 February 1991.

[17] A. Amendola, S. Contini and P. Nichele, "MARS: The Major Accident Reporting System", Proceedings of the Conference on Preventing Major Chemical and Related Process Accidents, Institute of Chemical Engineers Series 110, 445-458, 1988.

[18] G. Drogaris, Major Accident Reporting System. Lessons Learned from Accidents Notified. CDCIR No 658-EAb5-IV.3. EUR 13385 EN ,1991.

[19] H.B.F. Gow and H. Otway (eds), Communicating With The Public About Major Accident Hazards, Elsevier applied science publisher, London. 1990.

[20] P.Haastrup and P. Brockhoff, "Severity of accidents with hazardous materials. A comparison between transportation and fixed installations", Loss Prevention, vol 3, October, pp 395-405, 1990.

[21] E.J. Smith and G. Purdy, Lessons Learnt from Emergencies After Accidents in the United Kingdom Involving Dangerous Substances. CDCIR No 625-UKb3-I.3, EUR 13322 EN , (1990).

[22] Council Directive (90/219/EEC) of 23 April 1990 on the contained use of genetically mutated micro-organisms. Official Journal of the European Communities. L117, 8 May 1990.

[23] Council Directive (90/219/EEC) of 23 April 1990 on the deliberate release into the environment of genetically mutated organisms, Official Journal of the European Communities, L117, 8 May 1990.

[24] F. Campagnari and V. Sgaramella (eds.), Proceedings of the Eurocourse "Scientific and technical backgrounds for biotechnology regulation", 1992, (to be published by Kluwers Academic Publisher).

[25] Council Directive (90/313/EEC) of 7 June 1990, on the freedom of access to information on environmental matter, Official Journal of the European Communities, L 158/56, 23 June 1990.

[26] M. Peltu, Executive Summary : Conference Overview and Highlight (in the book at ref.19).

[27] H. Otway, Communicating with the Public about Major Hazards: Challenges for Research (in the book at ref. 19).

[28] B. De Finetti, Theory of Probability, John Wiley and Sons,1979.

[29] A. Poucet and A. Amendola, "State of the art in PSA reliability modelling as resulting from the international benchmark exercises project", NUCSAFE 88 Conference, Avignon(F), October 2-7, 1988.

[30] S. Contini, A. Amendola and I. Ziomas, Benchmark Exercise on Major Hazards Analysis. Vol 1: Description of the project, discussion of the results and conclusions, EUR 13386 EN, 1991.

[31] G. Volta, "Structural safety assessment as collective cognitive process", Introductory Lecture to the JRC-ISEI / HPA course on "Expert systems in structural safety assessment", Stuttgart (FRG), October 2-4, 1989.

[32] S. Funtowicz and J.R. Ravetz, "The communication of quality of hazard information",(in the book at ref.19).

[33] G. Volta, "Safety control and new paradigms in systems science", World Bank Workshop on Safety Control and Risk Management, Washington, D.C. (USA), October 18-21,1988.

THE DUTCH INTEGRAL ENVIRONMENTAL ZONING PROJECT

A.J. Muyselaar
Ministry of Housing, Physical Planning and Environment
Directorate General for Environmental Protection
2260 MB Leidschendam
The Netherlands

ABSTRACT. In the Netherlands, environmental zoning of sources of pollution is becoming more accepted. The concept of risk management is used to establish quantitative acceptability criteria for different agents in the sectoral policies. An example is the sectoral external safety policy for hazardous activities, with established limit values for (un)acceptable risk levels. In a recent project called "Integral Environmental Zoning" a zoning system is developed for areas under multiple loads from different agents. Experiences in this project shows that the formation and application of a quantitative integral approach meets many constraints. The application of a provisional system in a number of regional pilot projects provides the tool to make the necessary decisions to obtain an acceptable environmental quality in the living areas. Whether the proposed system or a variant will be applied in future is partly dependent on the development of the sectoral policies and the political forces.

1. INTRODUCTION

The policy of the Dutch ministry for the Environment is based on two approaches (or tracks) to handle environmental problems. In the source-oriented track, general principles are applied to known sources of pollution to prevent unnecessary environmental pollution. This includes sources of potential hazards. In addition, the effect-oriented approach introduces initiatives to deal with the effects of contaminants. The effect-related policies are directed towards avoiding deleterious effects on man, animals, vegetation, environmental functions and property, from sources of emission, both now and in the future. It is directed towards minimising the chance of harm to a negligible level. Modern assessment techniques make it possible to quantify the effects (immission levels) of agents by means of measurement or calculation. For a number of agents, such as hazards, noise, odour and (environmentally hazardous) substances from industrial installations, the effects are confined within a relatively small area and the dangerous effect levels (risks) diminish with the distance from the source(s) of emission. With the instrument of environmental zoning we can now solve sectoral environmental problems at local or regional level.

The concept of risk management is fundamental in implementing related policies and serves to establish satisfactory environmental protection levels for the objects to be protected. Risk management was first introduced in the "1986-90 Multi-year Programme for Environmental Management " [1] and has been further explained in the document "Premises for Risk Management" [2] (published as an annex to the Dutch National Environmental Policy Plan "To Choose or to Lose"). It has led to the development of methods to quantify the consequences and estimate the risks and threshold values of the impacts of the following agents: hazardous

31

A. G. Colombo (ed.), Environmental Impact Assessment, 31–51.

activities including transportation, (new) substances, genetically modified organisms, radiation, noise and unpleasant odours.

The development of sectoral policies regarding the agents given do not proceed in parallel and the Dutch government has only established limit values for maximum permissible risk and negligible hazard levels for some of the agents. The policy regarding noise from stationary installations (industry) and transportation has the longest history and has been developed furthest to date, with a separate Noise Nuisance Act (1979) stipulating measures to be taken for new or existing activities (traffic noise and industrial noise). Measures have been formulated for the noise source itself or to its surroundings in terms of spatial planning restrictions and zoning, based on permissible (safe) noise levels.

The risk management philosophy and its adoption of the quantitative approach with approved criteria for acceptable risk levels for exposed vulnerable objects, for example the nearby population of a hazardous activity, has also gradually become more firmly established in the sectoral policy field of external safety. Criteria for risk limits are set to establish acceptable safe distances to hazardous sources (zoning). In 1988, the Major Hazards Decree for stationary hazardous installations [3] was implemented in Dutch law. This decree fulfils most of the requirements of the EEC Directive on Major Accident Hazards, also known as the Seveso Directive of 1982 [4]. Industrial risks must now be quantified in a Quantitative Risk Analysis (QRA). The QRA must be part of the so-called External Safety Report which is required, under the Major Hazards Decree mentioned, for each activity falling under the working influence of the Seveso directive.

In the project "Cumulation of Sources and Integral Environmental Zoning (IEZ)" an attempt is made to develop a multi-sectoral policy. Effort is put into the development of methods to cumulate the effects from different agents and different sources for vulnerable objects and living areas. In addition, effort is put into the development of guidelines for the classification of areas under multiple loads and into possible variants for sanitation in order to obtain the required environmental quality in living areas. The project should end with a proposal for a bill by the end of 1991.

2. ENVIRONMENTAL POLICY

Measures to obtain a required environmental quality can concern either the source of pollution (source oriented) or its surroundings (effect oriented). A sometimes confusing aspect in the Dutch two-track policy is what kind of measures are to be taken, because in both policies measures can be formulated for the source of pollution.

In the source-oriented approach, the starting point is the limitation and prevention of environmental pollution. General premises are the application of the stand-still principle, the ALARA principle (As Low As Reasonably Achievable) and the promotion of clean technologies. This is mainly achieved by applying the best practical or technical means (BPM and BTM), which are translated into guidelines for risk reduction at the source. These guidelines are applied in the relevant permits.

Generally required quality aims for the environment, relevant in the effect-oriented track, sometimes cannot be obtained with measures at the source. This can be due to a number of causes:
- the partial dependence on the present state of technology to reduce risks,
- the combination of effects of agents,
- the measured levels originate from more than one source, and
- trans-boundary pollution.

Effect-oriented measures could be exercised by either taking additional measures at the source (even total prohibition), known as backcoupling by risk-control, or by minimising the effects to sensitive objects and living areas. The latter can be achieved by additional measures at the receiving end, e.g. sound barriers to minimize noise nuisance, or safe zoning distances. It must be stated, however, that measures are preferably oriented towards the source of pollution, so as to keep the amount of "living" space as large as possible. Figure 1 gives a graphical presentation of the relations mentioned.

3. RISK MANAGEMENT

The elements which play an important role in risk management are objects (man, environment), effects (death, disease, loss of environmental function or welfare) and agents (hazardous activities including transportation, (new) substances, genetically modified organisms, radiation, noise and unpleasant odours). For each combination of an object, effect and agent, guidelines and acceptance criteria can be developed. Currently, the policies concerning the health risks to humans have been developed furthest with established risk limits. Continued effort is being put into research and development of policy for the other objects/effects/agents combinations mentioned. There is an increasing awareness that ecological and economic implications of large-scale environmental pollution should receive more attention. Typical examples are agricultural land and drinking water reservoirs, which rely on water intake from major rivers. The Sandoz accident near Basel in Switzerland in 1986, with a release of toxic substances into the river Rhine, causing severe ecological damage downstream of the river, focused public attention on these risks. It induced additional European legislation for the protection of the environment. Restoring damage to environmental functions is usually very expensive and frequently must be paid from public funds. This is indeed the situation if the effects of emissions only become apparent after a long period of time.

Maximum permissible and negligible risk levels are defined for each type of agent. On the basis of these limits, criteria may be set for the individual agents which can be translated into more meaningful units such as concentrations in mg/l for substances in water or dB(A) sound levels for noise.

Risks above the maximum permissible level are unacceptable. In this situation, the disadvantages of the activity are considered to be greater than the (economic) advantages. Stringent (source-oriented) measures are in place to achieve acceptable levels or even prohibition of the activity.

Risks below the negligible level are acceptable. No further measures are required, unless perception or sustainable development (e.g. raw materials) necessitates different views. A grey zone is created between the two levels, in which the advantages of the activity must be weighed up against the risks and disadvantages involved. These activities are in principle allowed. However, the aim is still to reach the negligible level over the years with source-oriented measures.

An advantage of risk quantification is the provision of information about the cost effectiveness of different sets of risk reducing measures. It also provides a tool for land use planning. Application of risk analysis techniques has allowed limits which society can tolerate to be defined.

3.1. The Risk Concept

In the process of risk management, four stages can be distinguished:
   1. Identification of the hazards to man and the environment;
   2. Quantification of the extent of potentially detrimental effects and their likelihood of occurrence, by estimation or calculation;

    3. Decision on the acceptability of the risks of the activity and about the risk reducing
       (effect-oriented or source-oriented) measures that can or must be taken;
    4. Control, to maintain a situation of acceptable risk and, where possible, aiming to
       achieve a state of negligible risks by introducing further measures.

For the agents mentioned, the translation of these stages into policy have been separately developed within the sectoral framework of the Ministry for the Environment. It will be apparent that, depending on the subject, different problems arise in the different stages. With regard to the other sectoral policy fields, the quantification of the effects of hazardous installations brings a difficult element into focus. Whereas we can assume that many sources of nuisance, e.g. noise and odours and chemicals (substances such as NO_x, benzene, ozone), are continuously present and can be sensed, hazardous events are probabilistic in nature and the resulting risk cannot be felt directly.

The four stages for the sectoral policy of external safety from hazardous activities, used in the decision making process, are explained in the following paragraphs.

3.1.1. Risk identification. As a first step in the risk management process, hazards which are caused by the installation or the activity under consideration must be identified. A representative set of failures for the facility under consideration must be produced. There are various techniques to do this, including hazard and operability studies (HAZOP), check list methods and process safety analysis.

3.1.2. Risk quantification. A major contribution to the development of the risk quantification method for chemical plants was given in the COVO study, which was undertaken in 1979, in the industrialised area near Rotterdam. The risks from six hazardous activities (storage installations holding liquid nitrogen gas, chlorine, propylene, ammonia, acrylonitrile and a hydrodesulphurizer) were examined [5]. Although the conclusions of the study were that such a risk analysis was useful to decision-making, it also became clear that the method was much too laborious, time consuming and expensive to be used as a standard tool. Therefore, it was decided in 1981 to sponsor a research project aimed at the development of methods to quantify risks in a more cost-effective way, while at the same time improving the accuracy of the calculations. This work has taken some 30 man-years of effort. The result is the SAFETI package, a series of computer programs to perform complete quantitative risk analysis (QRA) of chemical process plants [6]. It enables the user to calculate the effects of chemical releases such as explosions, pool-fires, fireballs and the dispersion of gases with toxic loads. It combines these with data on local meteorology and population density and calculates effect distances and the impact on the population (lethality probability and numbers killed). The final result is the required individual risk plot and a societal risk curve (F-N curve). The accuracy of this model has been systematically analyzed. The inherent uncertainty in the physical modelling was reported to be a factor 3 and the uncertainty in the calculated frequencies about 10. Although this uncertainty is significant, the results of the model were judged to be close enough to observed effects and frequencies to be usable in decision making.

3.1.3. Risk acceptability. The document "Premises for Risk Management" [2], which has already been mentioned in the introduction, gives quantitative levels of acceptability for risks from new industrial plants. By accepting this document, the Dutch parliament also approved the proposed criteria. The starting point for determining these criteria was the statement by the Advisory Committee on Major Hazards that the risk from an activity which is hazardous to a member of the public should not be significant when compared with risk in everyday life.

Following this approach, for new major hazardous installations the maximum permissible level for individual risk has been taken as the risk level which increases the risk of death by all other causes with a maximum of one percent. The individual "natural death" risk run by the population group of 10 to 14 years old, which is 10^{-4} per year, has hereby been taken as the basic risk. The maximum acceptable individual risk has thus been established as 10^{-6} per year. In other words, the risk of a fatal accident to which an individual is exposed because of his continuous presence (365 days per year) in the neighbourhood of a hazardous activity shall be less than one in a million years. The individual risk of 1% of the maximum acceptable level (10^{-8}/yr) has been taken as the negligible level.

As a maximum permissible level of societal risk, a chance of 10^{-5} per year of an incident with a maximum of 10 deaths has been set. A chance of 10^{-7} per year with a maximum of 10 deaths has been taken as the negligible level. Furthermore, a heavier weight must be assigned to the more severe consequences of accidents. With this in mind, it has been decided that a consequence "n" times greater must correspond to a chance "n^2" smaller.

If the individual or societal risk is found to be unacceptable, risk reducing measures need to be taken. The starting condition is that the normal "best practical means" should have already been applied (source-oriented policy). This may not be the situation for older existing facilities. In addition "best technical" measures (effect-oriented policy) can be taken to obtain the required environmental quality beyond plant boundaries. They can be related to the layout of the plant, the application of additional safety devices, such as detection and alarm systems, and the use of smaller storage containers or less hazardous activities. Also, if necessary, safe distances from the source to the receivers (zoning) can be maintained. Often, a combination of both types of reduction is needed. Limiting the size of the zones by measures at the installation and maintaining sufficient distance to sensitive areas promote both prudent space use and the good "fitting in" of installations where activities with dangerous substances take place.

3.1.4. Risk control. The aim is to achieve a state with negligible levels at the receiving end. In order to obtain such a situation, the policy is to follow the line of progressive standard setting with more stringent limits in the future. This can only be achieved with continuous effort in developing risk reducing measures.

3.2. External Safety Policy

In the last decade, the risk management philosophy has gradually been introduced in the sectoral policy for "external safety". Together with the obligation to implement the Seveso Directive [4], this resulted in 1988 in the Major Hazards Decree for stationary hazardous installations [3]. The adoption of the quantitative approach, with approved criteria for acceptable risk limits in the external safety policy, has become more firmly established over the years. The obligation under the Seveso Directive to submit a notification report is translated in Dutch law by requesting an external safety report from hazardous plant owners. Apart from a general chapter, with information on location, hazardous materials and a qualitative description on risks to the environment, the external safety report must also contain a Quantitative Risk Analysis (QRA). In a QRA for a representative set of possible release scenarios, at specific failure frequencies, the risks outside the plant boundaries are calculated. The results are presented in individual risk contours and societal risk curves. The individual risk gives the location specific risk by depicting plots of iso-risk contours of the individual risk over the area. It thus gives the geographical spread of risk to individuals. The societal risk curve gives insight into the total group of the nearby population at risk. The plots with the individual risk contours combined with the societal risk curves will be used to determine whether risk reducing measures need to be taken to establish acceptable safe distances to hazardous sources (zoning) and to establish the effects to (future) spatial developments in the areas of concern. An evaluation of the already submitted external safety reports of existing hazardous plants is now under way. Preliminary results show that, on average, risk levels to the population are within the preliminary acceptable limits.

3.3. Risk Limits

For most agents, the negligible level is set at 1% of the maximum permissible level. This substantial margin has been deliberately set in view of the uncertainties associated with risk assessment methods, the natural background values and possible consequences of cumulative exposure from multiple sources. Risk assessment methods and risk limits have been developed or are under development for the following agents: major hazards, new and existing chemicals, genetically modified organisms, radiation, noise and unpleasant odours. These limit values are subdivided into values for existing and new activities and for different (sensitive) objects. It is felt that already existing nuisance due to an activity should be judged less severely than a similar new activity within an area. Next, to limit values for one activity, there are also limit values for the cumulative "risk" due to more activities with the same agent. Of the sectoral policy areas which take part in the Integral Environmental Zoning project, only the agents with the furthest developed sectoral policies have been (provisionally) selected in order to keep the duration of the project within practical limits.

3.3.1. External Safety. Only the concept of the individual risk is used in Integral Environmental Zoning because, in contrast to the societal risk, it displays the spatial impact of risk. It is expressed in the year average lethality probability due to the (to be) licensed activity. The values of the individual risk limits to be applied for external safety around major hazardous activities are depicted in table 1.

3.3.2. Chemicals. Regarding the policy field "continuous exposure to concentrations of chemicals (substances)", a distinction is made between existing and new substances and between two different possible effect modes on humans. The established limit values are depicted in table 2. We distinguish between substances with and without a threshold, i.e. the existence of a specific dose (=concentration time relationship) below which an effect does or does not occur. Carcinogenic and mutagenic substances are for instance chemicals without a threshold and toxic substances are substances with a threshold. In assessing the risks of substances without thresholds, it has been assumed that all induced cancers are lethal. The continuous exposure to a certain concentration level can be translated into a lethality probability (stochastic effect). With standard analysis techniques, maximum permissible concentration levels, known as the No Effect Levels (NEL), can be obtained for each substance. In view of the variety of effects, it has not yet proved possible to establish a risk limit for the cumulative effect of substances with threshold levels.

3.3.3. Unpleasant odours. The limits set for the nuisance by odours (stench) from companies (industrial, agricultural) are expressed in the year average fraction (percentile value) which may not be exceeded for those concentrations causing a nuisance (expressed in odour units ou/m^3). They are depicted in table 3. These limits have a provisional status awaiting the outcome of an extensive evaluation of their application in practice.

3.3.4. Noise. The total noise load from all the possible sources of an industrial site is expressed in an equivalent noise level "L_{Aeq}" in decibel units (dB(A)) per twenty-four hour period, for possible impulse noise, tonal noise, low frequency noise and other industrial noise. The acceptability limits are depicted in table 4. Noise from other sources like road and rail traffic can be added in proportion by means of their nuisance coefficients.

In contrast to the other agents mentioned, the noise criteria are derived for a complete industrial site, which may comprise several different companies and noise emission sources. The local authorities are primarily responsible for the application of the noise policy. Because of this, there is no need for cumulative criteria. This is in contrast to the other policy sectors mentioned which are directed towards individual sources or the permit-applicant.

4. INTEGRAL ENVIRONMENTAL ZONING

In March 1988, the project "Cumulation of Sources and Integral Environmental Zoning (IEZ)" officially started. At that time, some regional projects had already started and were busy developing local integral environmental policies. One example is the integral zoning project of the Dutch State Mines (IZ-DSM), a large chemical complex with several neighbouring cities. Another project is the Integral Environmental Policy Maastricht (PIM Maastricht) involving several scattered sources within the city limits.

Intuitively, one feels that the environmental quality will be worse if under multiple loads, although the cumulation of effects from different sources seems to be inconsistent (as in the Dutch saying "adding apples and pears"). It was felt that there was a need for a national uniform system for integral zoning, with optimal solutions for the three parties involved: housing, physical planning and environmental quality. This system should be applicable in situations with foreseeable loads of different components (agents) in living areas and where zoning can be used as an instrument for the agents involved. Matters of zoning distances, costs for source and effect oriented measures (sanitation) and the desirability of future changes regarding developments in exposed living areas should be dealt with. To support the development of the IEZ system a comprehensive research program for policy support was initiated. Parallel to this central project, experience was to be gained in 15 different regional IEZ pilot projects, such as IZ-DSM and PIM-Maastricht, managed by the local authorities and subsidised by the ministry. The IEZ project should enable the drafting of an IEZ-Bill in 1991.

From the start of the project, three parties belonging to the same ministry, but with different policy tasks related to zoning, were involved: the directorates-general for Housing, Physical Planning and Environment. From the directorate-general for the Environment, several departments were involved, including the sectoral policy departments. The project management was and still is performed by the IEZ unit, which is part of the "noise abatement department".

Acceptance of the draft programme took more than a year, which can mainly be attributed to the different interests of the parties involved. This is not as strange as it may seem. In a small, densely populated and highly industrialised country like the Netherlands, land is a rare commodity which needs to be distributed with care. It is generally accepted that there is a need for a system to establish "safe" distances between sources of pollution and living areas, in order to ensure an acceptable quality of the environment. It is however tempting to be able to diverge from a decision if considered necessary and let other (economical) arguments prevail in the decision making process. Establishing a safe zoning distance around a source means that depending on the environmental load, for the area involved, removal of sensitive objects may be considered or restrictions may be formulated for existing sensitive objects, like houses, and for possible (new) plans for spatial development.

A number of choices were made at that time to make the project and the project duration feasible. The IEZ project would be provisionally restricted to fixed installations, living areas and the compartments "odour", "noise" and "external safety", on account of their advanced developed zoning policies and experience with zoning with (preliminary) criteria for fixed installations. Furthermore, the consequences would be restricted to the cumulated "grey" areas.

For the different possible grey areas under multiple loads, land use planning restrictions would be formulated for (future) housing developments.

This approach gave the possibility of also considering other zoneable agents, such as the immission levels from continuous emission sources of hazardous substances, while developing an IEZ method. Furthermore, the system could even be extended with other activities like transportation. In fact, toxic and carcinogenic substances have been part of the developments from the start of the project. Awaiting the results of the research programme and the pilot projects for the policy support in drawing up a definite IEZ method, a provisional method would be worked out with a set of reference rules for the grey areas of the agents mentioned in industrial areas. This would be used in regional pilot projects.

In the next sections, the present state of the research programme, the IEZ Provisional System for handling multiple load situations and the present status of the project will be explained.

4.1. Provisional System for IEZ

The manual for a Provisional System for Integral Environmental Zoning [7] of 1990 was intended as an additional guideline for the IEZ pilot projects. It deals with technical and procedural aspects. The technical aspects cover the provisional method of determining the integral environmental quality of the area neighbouring an industrial site. The procedural aspects cover the necessary steps, including establishing the integral and sectoral zones, in an IEZ pilot project, to reach the required plans for (future) developments and an acceptable environmental quality in living areas.

4.1.1. Technical aspects. To obtain an overview of the integral environmental quality of an area, the system build on the established (preliminary) standards of the sectoral policies for stationary industrial sources and transportation. The integral noise level from different industrial sources such as impulse and tonal noise are obtained by using the prescribed method of the equivalent noise level (L_{Aeq}). The odour load is determined by using the Long Term Frequency Distribution (LTFD) model, which calculates the distribution of average percentile values per year from the given continuous emissions of (dangerous) substances of multiple sources. Methods for the cumulation of the relevant environmentally dangerous substances have been specifically developed for the IEZ pilot projects. The concentration levels calculated for the carcinogenic substances are related to the known unacceptable level, which is the human lethality probability of 10^{-6}/yr. The resulting probabilities can than easily be added to give the cumulative risk. For toxic substances, the calculated levels are related to their respective No Observed Adverse Effect Levels (NOAEL), which are considered equal to the unacceptable level. The different toxic substances cannot be cumulated this way, so the numbers are indexed in the same quality classes as the provisional system. The indexed numbers can then be summed to give the cumulated index. The cumulation of the (lethality) risks from different hazardous activities gives no difficulties since the separate calculated probabilities can be added to give the cumulated risk. In addition to industrial sources, the noise and risks from transportation (road, rail) within the enveloping zone are incorporated into the system. The associated risk and noise levels are cumulated with the sectoral component according to the methods mentioned. This way the relevant sectoral zones to the relevant boundary values are established for the sources on an industrial site.

The cumulated sectoral levels can then be compared to criteria (or standards) formulated in sectoral environmental quality classes. The standards used in the provisional system are in principle based on the official published (draft) standards. In addition, supplementary values, not (yet) published by the sectoral policies, are introduced to obtain a uniform system of standards for all the components involved. The provisional system follows the classification of the Noise Nuisance Act with standards for four categories. Furthermore, a distinction was made for new and existing situations, which resulted in two classification schemes. A new situation is defined as a new activity or the expansion of an already existing activity (large stationary installations) resulting in an extension of the integral zone, or the construction of (not yet built) new sensitive objects within the zone. In table 5 the sectoral environmental quality classes are given as an example of the classification schemes. The severity of the levels for each agent in the same classes are equal, although expressed in different units. For example, the 50 dB(A) noise level falls in the same class as the 10^{-8} risk level. The classes rank from A to E in ascending degree of severity. The classification of existing situations is made comparable to that for new situations by using the same classes, but with different levels. Only class C implies the same load in both situations.

Next to the application of the provisional system, the sectoral policies must also be considered in the pilot projects, with appropriate actions according to the sectoral policies. For instance, the societal risk is not considered in the IEZ project because it can not be made visible as a load on the map and therefore can not be fitted into the IEZ system. The consequences of exceeding the limits for the societal risk will have to be considered by the local authorities. These considerations also need to be made if sectoral policies are more lenient or stricter than the provisional system.

In the absence of a method for the accumulation or integration of dissimilar environmental loads, a system was selected based on the classes of standards previously mentioned. After the cumulation of the sectoral loads around an industrial site, a map is created with combinations of sectoral classes assigned to locations, or similarly the assignment of a location to an integral class. Six integral environmental classes (ranking from I to VI) with possible combinations are provided for both new and existing situations. The integral class III for existing situations covers for instance the combination of "at most 3 x C, no D or E". Tied to these classes are the consequences, for the area covered, in terms of land use restrictions. The land use restrictions formulated in the sectoral policies remain in force for situations with only one sectoral load.

4.1.2. Procedural aspects. The provisional system also gives guidelines along which an IEZ project should develop. The integral environmental zone must be established in two phases. Phase 1 results in a provisional environmental zone based on the inventory projects of the cumulated sectoral loads. On the basis of this picture, the local authorities pursue a restrained policy with respect to residential development plans and the granting of environmental permits. In this phase, research into appropriate abatement measures and into possible sanitation variants of sources for the problem (grey and black) areas is also commissioned. Phase 2 is completed with the formal establishment of the integral environmental zone in the local zoning plan and the regional land use plan. In addition, a programme for revising the environmental permits must be initiated. Based upon this, the sanitation plan is executed in order to reduce the environmental load, with choices between (above-normal) source-oriented measures and effect-oriented measures such as land use restrictions.

4.2. Research programme

The research programme is divided up into four blocks covering different aspects of the IEZ project and must provide the basis for the support of possible policy development:
- inventory and analysis of zoning,
- cumulation of environmental loads,
- use and management of information, and

- costs for obtaining high environmental quality.

The investigations formulated for each block are performed by external institutes under the supervision of members of the IEZ project team.

4.2.1. Inventory and analysis of regulations on zoning. In this block, three different study subjects were formulated. An inventory and analysis of regulations about zoning in the environmental and physical planning legislation was initiated as the first study object. The work for this has already been done and the results provide the point of departure for possible IEZ legislation [8]. The second study was an inventory of potential IEZ areas in the Netherlands [9]. The third one a survey with a comparative description on the experiences gained in the IEZ pilot projects, which differ considerably in character and size, with respect to managerial and juridical aspects [10].

The second inventory concerns the extent of all the areas of influence due to zoneable environmental loads from the sectoral compartments noise, odour, external safety and toxic and carcinogenic substances in industrial areas. The influence areas are derived from the measured or calculated distances from the source of pollution to established limit values for the negligible level, set by the sectoral (draft) criteria. In the first phase of the study, an overview of the areas involved was generated. Later on, more information would be gathered about actual environmental loads, about the relations in size of the different sectoral environmental loads and about the sensitive objects within the areas of concern.

For the first overview, the information required for hazardous activities (Seveso installations) was not yet available and also some of the information needed for odour levels was not always available. All the relevant areas with hazardous activities were therefore given a standard distance from the source to the target immission level of 1000 m. The following data in the interim report are worth noting. In 85% of the identified 1300 locations there is mention of an influence area of one or more environmental components. For about 260 of these industrial locations, two or more influence areas exist. These locations can in principle be considered for integral environmental zoning. Of these locations, 199 must be zoned according to the Noise Nuisance Act, but the noise influence area is only significant in 84 cases. Although some information on several odour sources was lacking, it appears that odour (8 % of cases) can produce quite considerable influence areas (> 2000 m). It must be said that the established areas of influence may change considerably as a result of sanitation plans. As mentioned, the survey will be updated and complemented with the results of the external safety reports of hazardous installations and new relevant information about odour.

4.2.2. Cumulation of environmental loads. Two parallel research studies have been initiated in order to develop methods of aggregating different types of environmental loads. The first one was the continuation of scientific research into the area of the cumulation of different noise types and sources and the cumulation of noise and odour. This fundamental research can in future be expanded to cover cumulation of environmental loads from other agents. The second is a feasibility study to evaluate three methods of arriving at one indicator, called the environmental load index (ELI) for the integral quality of the cumulated loads of noise, odour, substances and hazardous activities [11]. This study is finished and a follow up has been initiated to test the method and to compare results with the provisional system IEZ in three IEZ pilot projects.

The feasibility study started with a workshop, where scientific experts together with officials of the ministry determined the scope of the study. The three ways in which an index can be generated reflects the compromises in trying to establish a sound scientific basis although there is little scientific information available, and the social/political desire for concise information. It is evident that a manageable system for such complex information can only be developed at the expense of loss of information. To be able to verify and validate information it is important to lose as little information as possible in the derivation of the environmental indicator. From this point of view it is best that the aggregation of information, of the different effects from different agents, into an index or indices takes place as late as possible in the method. Obviously, the opposite applies to the system where the aggregation is done early in the procedure. This implies that scientifically valid models need to be available for the cumulations of the effects from different agents.

For the derivation of the ELI, the researchers choose to reflect the approaches in matrices (see tables 6,7 and 8), in which the different steps of the methods become clear. The feasibility of the cumulation steps is qualitatively assessed. The different agents are presented along the rows as the parts of the sectoral environmental compartments they belong to. The possible adverse effects are presented down the columns.

For each agent, one effect is taken as representative, for instance "nuisance" for the compartments noise and "odour" and "lethality" for carcinogenic substances and hazardous activities. The differences of the approaches appear when the (first) aggregation step is done. In the scientifically most valid method A, first the effects in each column for all of the agents are cumulated and than the different effect values are combined into an ELI. In method B, the effects for each environmental compartment are first cumulated, for example road, rail, air and industrial noise, and the separate values for the different compartments are then aggregated. Method C has the sectoral environmental criteria for each agent as the point of departure. The effects for each of the agents are expressed in their standard units and then aggregated to give an ELI. The advantage of this approach is that the calculated or measured values can be compared to the sectoral criteria before the aggregation step. The cumulation of effects from different agents is not based on scientifically valid relationships as in method A, but hidden somewhere in the composition of the ELI index. The ELIs of the three methods are found by using combination rules with weights attached to each parameter. For instance, weights are assigned to the cumulated effects (method A) which are then used in the combination rule to produce the environmental load index.

The three cumulation methods are still being investigated and scientific research for better solutions will continue.

4.2.3. Use and management of information. Integral environmental zoning leads to the collection of considerable amounts of information. For each industrial area, data about the environmental loads (emission and immissions) from each agent must be collected as well as local spatial data with information on relevant sensitive objects. Account must be taken of different data types, for instance data of environmental loads and sensitive objects before and after possible sanitation. To keep this stream of information up to date and within manageable limits computer automation seemed the solution. The systems developed could be tried and tested in practice with the regional IEZ pilot projects.

Two research surveys were initiated resulting in IEZ computer codes. The first one covers the central data collection of information on project management and status of IEZ (pilot) projects. An automated system for the management of IEZ projects called the "control system for integral environmental zoning (MIEZ)" [12] was the result of the second project. MIEZ is tested in three of the IEZ pilot projects and will be implemented later in the other pilot projects.

MIEZ comprises a database management tool, which is needed in order to keep track of the huge amount of data from the environmental load studies and the (alternative) sanitation surveys for possible future situations. The analysis of data according to the provisional IEZ system will produce plotted maps, on which the areas under multiple loads are drawn, and overviews with the consequences for (future) sensitive areas. Of course, the impacts of possible sanitation measures or alternatives can easily be made visible on the map. This all makes this instrument a powerful support in managing the data in an IEZ project and for the decision-maker who wants to evaluate certain options quickly.

MIEZ is built as a shell (application) through which an existing commercial geographical information system (GIS) can be accessed. The advantage of this approach is that the user can use GIS easily, without needing much skill in working with the GIS package itself, through the application. If other GIS functions need to be accessed, then that is possible by bypassing the shell and using the GIS package directly. The user must then, however, have more specific knowledge. MIEZ consists of a number of modules: input of data, output/presentation (plotter, printer, file), storage and data management and analyses of data using the IEZ provisional system. Figure 2 gives a schematic presentation of the functionality of the system. Three different kinds of input are required under the MIEZ application to be stored in the (geographic) database. First, the outcomes of the measured or calculated environmental loads. Second, data on the location of sensitive objects within the impact area of the agents considered. Last the geographic data (maps) of the area under consideration. They are depicted in the figure as "environmental model contours" and "topographic underground".

It is then possible to perform "queries" on the information available in the data base, for instance retrieving data on different calculated loads from the same sources (e.g. a noisy, smelly,hazardous installation). It is also possible to perform "analysis" on the data available, for instance the classification of areas under multiple loads using the provisional IEZ system. Different kinds of comparisons between maps with different load contours are possible.

Alternative sanitation plans can be analysed and judged on their merits. Of course, it is also possible to have the results presented in different ways on screen, plotter as well as on a print-out.

An extension of the system with a financial module, for the calculation of sanitation alternatives, is now being studied. The ease of implementing the system is dependent upon the organizational structure of the different (unique) IEZ pilot project. For this reason, a proposal for the organizational implementation of the system is made available.

4.2.4. Costs of obtaining environmental quality. An important part of IEZ is the question of the costs of obtaining the required integral environmental quality and the issue of who is going to pay for the sanitation measures. In general, the premise that the polluter must pay is followed. This means that, in principle, all the effect oriented sanitation costs for zoning must be paid by the owner of the polluting source. This is certain for the application of the "best practical means". However, in existing situations when only expensive "best technical means" (above normal measures) can be applied to obtain the environmental quality required, compensation costs can be granted. The costs involved with measures for sanitation of undesirable situations of the (existing) sectoral environmental policies fall outside IEZ.

Regarding the state of development of the sectoral policies, with preliminary standards for existing situations for external safety, substances and odours, at the start of the IEZ project there was no general insight into the potential costs. The sectoral sanitation costs needed to be assessed first in order to obtain an overview of possible integral sanitation costs. An investigation was started into the potential extent of the sanitation of each environmental sector (noise, external safety and odour). Different sanitation cost pictures were calculated depending on the level of required integral quality, the (intermediate) target value or the negligible (risk) level. The estimated and extrapolated integral sanitation costs for the whole of the Netherlands were based on calculations with the cumulation method of the provisional system IEZ. The preliminary criteria of the sectoral policies for existing situations were a point of departure. Future changes in this can influence the outcomes considerably.

The sanitation surveys in the separate IEZ pilot projects could benefit from this investigation. The project has been completed recently [13]. The results, however, have not yet been properly screened. Bearing this in mind, the most important outcome is that the sanitation costs can be attributed to multiple load situations only for a relatively small part. Most of the costs is needed for measures to reduce the environmental loads of separate sectoral compartments.

The information obtained will be of key importance in the further discussion on the feasibility of an IEZ-Bill.

5. CONCLUDING REMARKS

The Dutch Ministry of Housing, Physical Planning and Environment has adopted the "quantitative approach" of risk management to be able to understand and deal with the environmental loads from sources of pollution in the context of the existing restrictions on available space in the Netherlands. Experience with sectoral environmental policies, such as the external safety policy for hazardous chemical installations, shows that with today's assessment techniques the geographic spread of the environmental impact of agents can be made visible. Standards are established for the acceptability of load levels in the residential areas which are used for zoning. The high density of the population and of chemical industry requires a policy which takes account of the interests of the different parties involved. The policy must therefore include zoning and siting of sources and the establishing of spatial consequences for (land use) planning of (future) sensitive objects, such as housing. In addition, the policy must recognize that the acceptance of certain considerable loads is unavoidable.

For situation with multiple loads from different agents, the IEZ project tries to work out a solution which will satisfy the parties involved. As far as the technical aspects are concerned, such as the assessment of the cumulation of sources, the management of huge amounts of different kinds of data and the formulation of (spatial) consequences, acceptable solutions were developed and are currently being tested in IEZ pilot projects. Matters, such as the sectoral developments for standards setting, the assigning of sanitation costs and the existing legislative framework are now of influence in the (political) debate on a possible acceptance of an IEZ Bill.

REFERENCES

[1] Environmental Program of the Netherlands 1986-1990, Second Chamber of the States General, session 1985-86, 19204, No. 1-2, Ministry of Housing, Physical Planning and Environment, The Hague.

[2] National Environmental Policy Plan, Second Chamber of the States General, session 1988-1989, 21137, No. 1-2,The Hague.

[3] Besluit risico's zware ongevallen (eng. Decree containing regulations on the notification of hazards of major accidents), Staatsblad 1988, 432, The Hague.

[4] Council directive of 24 June 1982 on the major-accident hazards of certain industrial activities, 82/501/EEC.

[5] Report on the COVO study to the Rijnmond Authority, Reidel,1979.

[6] SAFETI, computer package for quantitative risk analysis,Technica Ltd, London.

[7] Manual for a Provisional System of Integral Environmenta Zoning, Ministry of Housing, Physical Planning and Environment, IEZ report No. 14, Distribution centre public authority publications, The Hague, (Also published in the French languageas IEZ report No 15),1990.

[8] Claassen-Dales W.J.B., Samkalden D., Inventarisatie wetgeving en richtlijnen op het gebied van Milieuzonering, Ministry of Housing, Physical Planning and Environment, IEZ report No.4, Distribution centre public authority publications, The Hague,1988.

[9] Hauwert P.C.M., Keulen R.W., Inventarisatie omvangzoneerbare milieubelasting (MILZON), Ministry of Housing, Physical Planning and Environment, IEZ report No. 3, Distribution centre public authority publications, The Hague,1990.

[10] Kreijenbroek S., Tussentijdse evaluatie proefprojekten IMZ, Ministry of Housing, Physical Planning and Environment, IEZ report No. 10, Distribution centre public authority publications,

The Hague,1990.

[11] Feenstra J.F., Aiking H., de Boer J., Sol V.M., Lammers P.E.M., Haalbaarheidsstudie Milieubelastingsindex, Ministry of Housing, Physical Planning and Environment, IEZ report No. 8, Distribution centre public authority publications, The Hague,1990.

[12] Kreuwel G., Bartels J.,Gies Th., BIMZ, geautomatiseerd gegevensbeheer voor integr milieuzonering (a computer application developed for IEZ), buro GEOPS, Wageningen.

[13] Inventory of sanitation costs IEZ, to be published , Sema Metra Ltd, London.

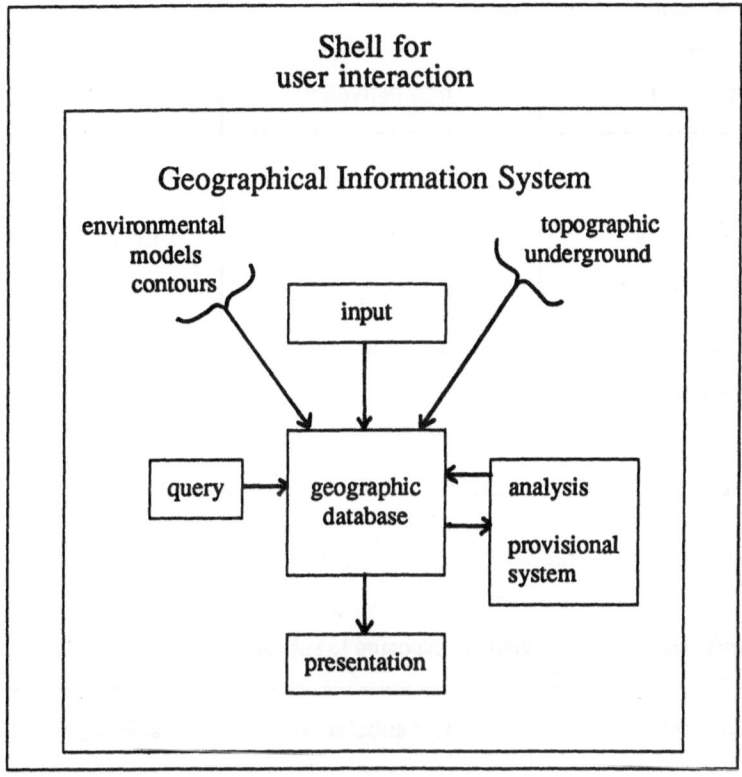

Figure 1. Relationship between source-oriented and effect-oriented policy.

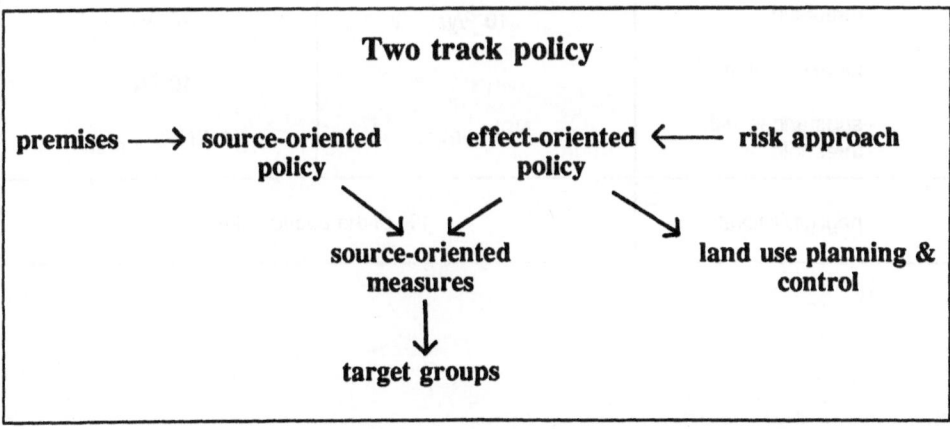

Figure 2. Functionality of the automated system for the management of IEZ projects (MIEZ).

Table 1. Limit values for major hazards.

type of risk	new activities	existing activities
max risk for humans - individual risk - individual risk (cumulative)	10^{-6}/yr 10^{-5}/yr (maximum n*)	10^{-5}/yr (preliminary) –
negligible level	1% of the above values	

*) With a maximum of n=10 activities, the cumulative level is $n \times 10^{-6}$.

Table 2. Limit values for the continuous exposure to substances.

type of risk	new substances	existing substances
max individual risk for humans - substances without threshold - idem cumulatively - substances with threshold	10^{-8}/yr NEL_{human}	10^{-6}/yr 10^{-5}/yr 1% NEL_{human}
negligible level	1% of the above values	

Table 3. Acceptability limits for odour.

type of limit	new activity	existing activity
max allowable for humans (ou/m^3) - odour concentration (hour average) - cumulative	1 at 99.5-th perc. 1 at 98-th perc.	1 at 98-th perc. 1 at 98-th perc.
negligible level	1% at 99.5 percentile value	

Table 4. Acceptability limits for noise.

type of limit	new activity	existing activity
max allowable for humans (dB(A)) - noise load - preferred limit	 60 50	 65 50

Agent	Class E	Class D	Class C	Class B	Class A
Existing situations (Class A has no significance)					
Risk	$>10^{-5}$	10^{-5}–10^{-6}	10^{-6}–10^{-7}	$<10^{-7}$	$(<10^{-8})$
Noise	>65	65-60	60-55	<55	(<50)
Odour	>10	10-3	3-1	<1	(<1 [99.5%])
Carc. subs	$>10^{-5}$	10^{-5}–10^{-6}	10^{-6}–10^{-7}	$<10^{-7}$	$(<10^{-8})$
Tox. subs	> 100	100-10	10-3	<3	(<1)
New situations (Class E has no significance)					
Risk	$(>10^{-5})$	$>10^{-6}$	10^{-6}–10^{-7}	10^{-7}–10^{-8}	$<10^{-8}$
Noise	(>65)	>60	60-55	55-50	<50
Odour	(>10)	>3	3-1	<1	<1 [99.5%]
Carc. subs	$(>10^{-5})$	$>10^{-6}$	10^{-6}–10^{-7}	10^{-7}–10^{-8}	$<10^{-8}$
Tox. subs	(> 100)	>10	10-3	3-1	<1

Table 5. Classification of the sectoral environmental loads based on existing (draft) criteria for the maximum permissible and the negligible levels for existing and new situations. The (draft) levels for existing situations are mostly relaxed by a factor of ten. The in-between values are chosen for IEZ. (Measures are as in tables 1 to 4).

Agent	Effect		
	nuisance	toxic effects	death
1. Noise		##############	########
road		##############	########
rail		##############	########
air		##############	########
industry		##############	########
2 Odour		##############	########
3 Air pollution	##########		########
toxic	##########		########
substances	##########		########
carcinogenic	##########	##############	
substances	##########	##############	
4 External Safety	##########	##############	
Total Effect Cat. Index	Nuisance	Toxic Effects	Death	Env. Load Index

Table 6. Method A of the feasibility study into an environmental load index. Cumulation of the possible effects down the columns of all the agents before aggregation into an environmental load index.

The marked (#) cells represent effects which are not considered because either the effect does not occur or other effects occur at lower load levels (noise, odour, toxic substances) or possible effects are not considered because of the probabilistic nature of the occurrence of effects (carcinogenic substances, external safety).

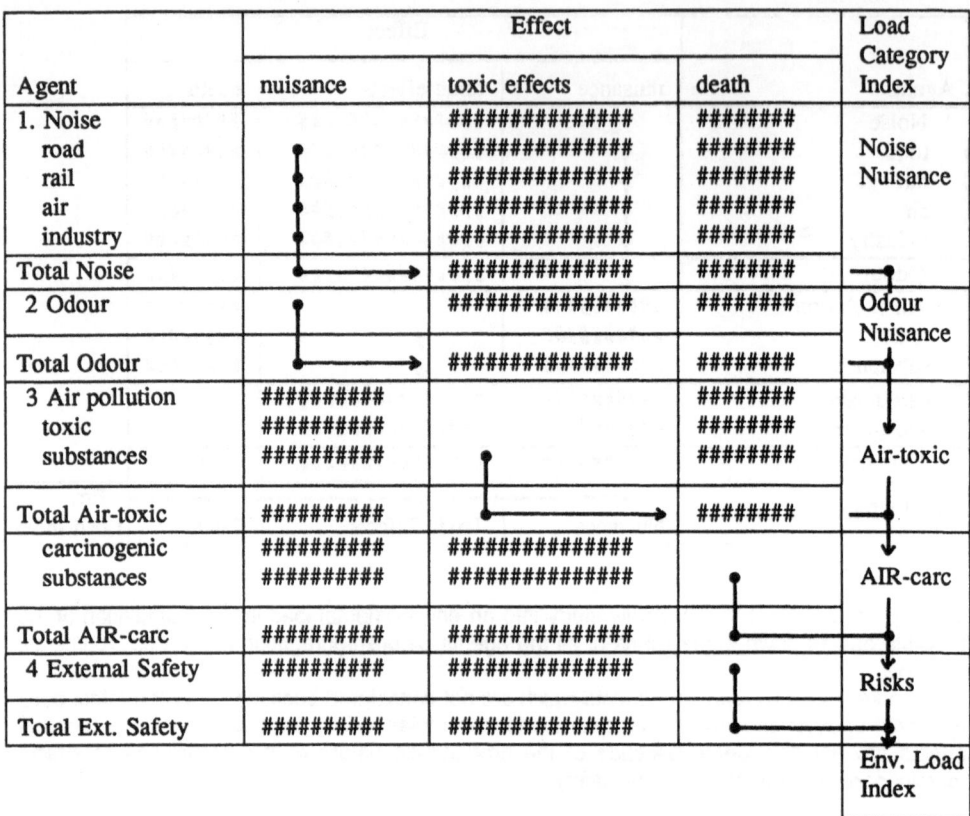

Agent	Effect			Load Category Index
	nuisance	toxic effects	death	
1. Noise		##############	########	
road		##############	########	Noise
rail		##############	########	Nuisance
air		##############	########	
industry		##############	########	
Total Noise		##############	########	
2 Odour		##############	########	Odour Nuisance
Total Odour		##############	########	
3 Air pollution	##########		########	
toxic	##########		########	
substances	##########		########	Air-toxic
Total Air-toxic	##########		########	
carcinogenic	##########	##############		
substances	##########	##############		AIR-carc
Total AIR-carc	##########	##############		
4 External Safety	##########	##############		Risks
Total Ext. Safety	##########	##############		
				Env. Load Index

Table 7. Method B of the feasibility study into an environmental load index. Cumulation of effects of the agents belonging to one environmental compartment along the rows, before aggregation into an environmental load index.

The marked (#) cells represent effects which are not considered because either the effect does not occur or other effects occur at lower load levels (noise, odour, toxic substances) or possible effects are not considered because of the probabilistic nature of the occurrence of effects (carcinogenic substances, external safety).

Agent	Norm Standard				Norm Index
	Negligible level		Max. permissible		
	new	existing	new	existing	
1. Noise					Noise
road	●	●	●	●	Criteria
rail	●	●	●	●	
air	●	●	●	●	
industry	●	●	●	●	
2 Odour	●	●	●	●	Odour Criteria
3 Air pollution toxic substances	●	●	●	●	Air Criteria
----------					
carcinogenic substances	●	●	●	●	
4 External Safety	●	●	●	●	Risk Criteria
					Environmental Load Index

Table 8. Method C of the feasibility study into an environmental load index. The effects of each agent is calculated separately according to the methods of the sectoral policies and is related to the relevant norm. The resulting numbers are aggregated into an environmental load index.

CONTENTS AND PHASES OF AN EIA STUDY

M. Masera[+] and A. G. Colombo
CEC, Joint Research Centre
Institute for Systems Engineering and Informatics
I - 21020 Ispra (Va)
Italy

ABSTRACT. This paper describes critically the phases which must be accomplished to carry out a Environmental Impact Assessment (EIA) procedure. The main legal references which are taken into account are the Directive on EIA of the European Communities and the Italian Decree of 27/12/1988. The objective of the paper is twofold: firstly, to review the phases of an EIA study, pointing out principles and characteristics (with emphasis on the participation of the different "actors", such as developer, authority, public and technical advisers) and, secondly, to argue the comprehensive process that should support the relating procedure. Finally, features which computer aids for EIA studies should have are discussed. A knowledge-based systems approach has been adopted .

1. INTRODUCTION

An Environmental Impact Assessment (EIA) is an activity designed to identify, interpret, predict and communicate information about the impact of a human action (Alberti et al., 1988; Gisotti and Bruschi, 1990; Munn,1975). As an activity it has several facets: analytical, legal, procedural, and decisional. Human actions subject to an EIA are those which because of their size or foreseeable consequences could be hazardous for the well-being of an ecosystem (including human beings and their material and non-material assets).

The fulfilment of an EIA is determined by legislative acts. These legal norms usually define, more or less precisely, the conditions, contents and administrative consequences of an EIA procedure. We will consider as main legal references the Council Directive on EIA of the European Communities (Directive 85/337/EEC) and the Italian Decree of 27 December 1988 (DPCM 27/12/88). We will also refer to the Italian Decree of 10 August 1988 (DPCM 10/8/88) and to the Circular of the Italian Minister of the Environment of 11 August 1989 (CIRC 11/8/89).

These norms are the first legal instruments in the European Community (considered as a whole) and in Italy that deal with the assessment of the environmental implications of major projects. Although these laws form the established normative body which has been used for carrying out EIAs in recent years, not all their principles and characteristics should be conceived as unchangeable. Moreover, the experience gained could mark the present EIAs' weak points.

(+) Visiting Scientist from the University of Buenos Aires, Argentina.

A. G. Colombo (ed.), Environmental Impact Assessment, 53–77.

Not only might the provisions made several years ago prove to be ill-suited for their declared purposes, but new requirements and needs may arise, such as more severe requirements in the quality and quantity of factors that should be considered, a certain level of demonstrable quality, different interactions between the participants, or early provisions for the participation of the public to the consent procedure.

The main objective of this paper is to review critically the phases of an EIA, taking into consideration the relationships between the legal framework, the decision-making process that implicitly accompanies or follows it, and the scientific and technical backgrounds that support the understanding of the present and future conditions of the environment. We especially want to mark out the particular roles that the participants in an EIA (which we will call 'actors') perform, and the need for a common basis for their mutual comprehension (a basis that must include, for example, the same vocabulary and the same methodology, in order to ensure valid communication between the actors) .

The EEC Directive aims at setting out common bases for the evaluation of public and private projects which are likely to have significant effects on the environment, in order to supplement and coordinate the development consent procedures. It fixes a group of projects which are mandatorily subject to EIA, and specifies the types of projects that should be regulated by the Member States.

In its Article 3, the Directive lists the environmental factors that must be considered: "human beings, fauna and flora; soil, water, air, climate and the landscape; the inter-action between ... [them]; material assets and the cultural heritage". It also establishes that an EIA should "identify, describe and assess in an appropriate manner... the direct and indirect effects of a project" on these factors. Obviously, this list is not a definition of "environment", which in any case would not be necessary. But, as a mere list of names, it is vague, and a possible source of contention. For instance: "cultural heritage" and "landscape" can have several meanings depending on the scope and viewpoint of the analyst. Furthermore, it does not specify what must be understood by "significant effect" (Articles 1 and 2), or by "impact"; and so what an EIA properly should be remains only sketched.

The Italian Decree DPCM 27/12/88 states the technical norms for drawing up the Environmental Impact studies, and for formulating the judgements of Compatibility, in consonance with the previous Italian Decree DPCM 10/8/88 and the EEC Directive. This norms should act in correspondence with other laws (for instance, on environmental compatibility, geothermal resources, energy plans, etc.), shaping an interlaced and somewhat complex legal system .

The DPCM 27/12/88 specifies the documentation that must be submitted by the developer, sets the role of a special Commission with the charge of producing the Environmental Compatibility judgement, and states three frames of reference for the study:

i) the **"programmatic"** (linked to the plans and programmes that regulate the territory or the sector of activity),

ii) the **"project"** (linked to the technical solutions and the definition of the installations, equipment and/or works of the project), and

iii) the **"environmental"** (linked to the analysis of the effects on the environment).

At the same time, the environmental components and factors are defined in some detail in Annex I.

In addition to this, each Italian region should regulate the relationships between their own plans and programmes and these norms. Up to now, only Regione Veneto (Legge regionale 23 Aprile 1990, No. 28) and Provincia Autonoma di Trento (Legge provinciale 29 Agosto 1988, No. 28) have issued special laws. Lombardia and Emilia-Romagna have outlined draft versions of EIA norms. Other regions (e.g. Piemonte, Umbria) have anticipated EIA norms for particular objectives.

As an EIA tries to embrace very different present and future situations (human health, technical solutions, economic aspects, consequences on the air, water and soil, etc.), with very diverse purposes (fulfilment of a procedure, legal considerations, technical comprehension, governmental agreement, etc.), it will inevitably be a fairly complex entity. Structurally, it can be broken down into three layers:

    i) a formal legal and administrative procedure (i.e., the "bureaucratic" steps linked to the legal context outlined previously),

    ii) a descriptive and analytical process (i.e., the sequence of "evaluation tasks" developed in order to support defensible decision-making), and

    iii) a set of technical and scientific models and methods (i.e., the deepest "knowledge" and "know-how" that supports the reasoning of the assessment process).

This division into layers can be understood in the following way: the EIA as a procedure is legally determined (and so there is a legal layer which acts as entry and frame for the study), but to fulfil its steps special tasks are needed (a second layer that embraces the assessment process as joint understanding by the group of EIA actors, and so including the agreement between them), and some of these tasks need more thorough analyses to support the comprehension process (the 'deepest' scientific layer).

These three layers are intrinsically and logically independent; i.e., none of them can replace any of the others. For instance: a legal statement cannot replace a toxicological study, and the application of a statistical method for treating data is not properly the evaluation of an environmental effect, but only a computation of numbers which can support the understanding of an effect. In any case, all the tasks performed in an EIA must share a common purpose, viz., to support the assessment of certain environmental effects, and the evaluation of their global impact; and all the other possible goals must be seen as secondary.

The second objective of this paper is to discuss the features that a software aid to EIA should have. As a first impression, any valuable aid would have to reduce the effort needed for the assessment and improve the results obtained. I.e., it should be useful in the double sense of quality and efficiency of the procedure, and, still more, be helpful during the whole set of phases needed to carry out the assessment (from the scoping of the analysis, until the final decisions). Other interesting traits of such computer software should be:

    - to assist in the interactions among the three layers mentioned previously, and

    - to facilitate the communication between the actors (developer, authority, public, technicians) and to aid their mutual understanding.

Our approach takes a "knowledge-based systems" point of view, since it seems obvious that any performance of an EIA would be mainly supported by skills and expertise of a "knowledge" nature, which cannot be modelled by means of algorithms. In section 11 we will point out the chances for knowledge-based system supports to the EIA phases.

Briefly, we view an EIA as a cognitive activity developed inside a legal framework, with the objective of supporting a political decision, based on scientific and technical knowledge, and ideally guided by common understanding and consensus among the participants or 'actors'. Thus, the main keywords of our approach are: knowledge-based, legal-framed, consensus-led and decision-oriented.

2. FRAMEWORK OF AN EIA STUDY

Let us review the characteristics of an EIA study. The main framework for an EIA is set by ad-hoc legal norms. Within this framework the study must be done taking into consideration the project, the environment and their potential interactions. The main objective is to verify (and the legal norms say that this must mainly be done by the developer) the effects and consequences of the project, and to support a decision about their acceptability and/or defensibility. But, as a general rule, legal norms do not define a method. Unfortunately, this absence of a mandatory method often leads to low level studies. Instead, it should encourage the use of up-to-date techniques, which take into consideration the latest developments of the evolving environmental sciences. At the same time, each new study might have its own way of performing the "assessment" (ranging from qualitative to strictly quantitative methods, among other possible characteristics), upsetting both any attempt to standardize the EIA procedure and the evaluation of the studies (Sorensen and Moss, 1973).

In order to outline the problem, we make some remarks about the nature of an EIA:

i) its aim is anticipatory by nature: it talks of future plausible events that might occur if the proposed development is carried out;

ii) several actors play a role in it:
- the developer (the applicant for the authorization of the project),
- the authority (whether the government officials responsible for the application of the law (for example: the Ministry for the Environment) or the body which judges the EIA (in Italy, an ad-hoc Commission of Environmental Compatibility) or other government authorities which have responsibilities in the consent procedure),
- the public (at various levels: individuals, social or political organizations, scientific or professional institutions, etc.),
- technical advisers (experts who act as consultants or assessors of the developer or of the authorities, giving the main scientific background);

iii) other possible actors which could play a role in an EIA are:
- reviewers or auditors (independent board or government agencies with the responsibility of ensuring compliance with published guide-lines, and correctness in the evaluations),
- other countries (neighbouring countries that could suffer the effects of the project), and
- international bodies (inter-governmental organisations at different levels);

iv) this multiple participation forces the adoption of a common language and a common method to ensure communication between the participants and the meaningfulness of the statements made in the study (i.e.: a unique vocabulary, a common semantic model (viz., what is "landscape"), an agreement about the scope, targets, models and techniques, and desirably, a common information system);

v) this multiple participation also implies the need for agreements and consensus about the results and conclusions of the analyses, the question of the link between this consensus and the decision-making process remaining open (mainly because this decision is of a political nature, and is taken outside the EIA study proper);

vi) it must be clear who the decision-maker is, and what his/her responsibilities are;

vii) there must be objective goals and criteria (political, social, etc.) for the assessment (set by laws, regional or sectorial plannings, or local census);

viii) the goals can be partially conflicting, and the criteria may be difficult to express numerically;

ix) the approach must be systematic and comprehensive, aimed at considering all the factors and conditions that characterize the project and the environment in an appropriate manner, with coherent treatment of basic data and antecedents, consistent computations and deductions when applying methods, and congruent derivations of potential future states;

x) it is a multidisciplinary study, involving physical, chemical, biological, economic, social, etc., elements; and the appropriate way of treating this whole system of disciplines is by the right use of specialized "state-of-the-art" knowledge;

xi) beyond the estimates and computations, the assessment implies evaluations based on subjective judgments and criteria, not only because of the non-

quantitative nature of several items (e.g., the landscape), but also because the
final appraisal will always be a matter of personal or social values and
subjectiveness;

xii) the validity of the study is bounded by the limitations of the models and
techniques used, and the inherent weaknesses of a comprehensive
and wide-reaching analysis: it is impossible to know everything (there
is an unavoidable margin of ignorance in the input data and in the previsions),
and what is known is only known approximately (i.e., with uncertainty and
imprecision).

When we talk of computer aid, we think of software which can support a wide range of EIA
studies. All the possible methods that could be used despite the particular ways adopted by each
one should employ, in general terms, the same basic knowledge, i.e., the same scientific,
technical, legal, etc., knowledge. Besides the specific methods applied to make data treatments or
computations, it is important to realize that the basic physical, chemical, biological, etc., knowledge
should be the same for all analyses. And, of course, it is a matter of knowledge, and not of
particular techniques, methods, formulae or equations. Furthermore, this computer aid must be
consistent with the general characteristics of an EIA (such as those mentioned in section 1). So,
for us, the two reference marks for a general EIA software should be:

- accordance with the "nature" of EIA , and
- knowledge-based without a prefixed procedural structure besides that legally
imposed (i.e., flexible enough to be used in diverse circumstances).

Moreover, as an EIA is not based only on numerical data, but also on criteria and judgments,
computer aid should help in producing clear expressions for subjective qualitative values.

Figure 1 shows the phases of an EIA study which are to be treated. They are:

i) **description of the project** (considering its possible linkage with plans and
programmes, and the alternative possible technical solutions),
ii) **description of the environment** (i.e., the place where the project will be installed),
iii) **identification of the environmental effects** (with consideration of potential and
important effects),
iv) **evaluation of the environmental effects** (with discussion of the information needed
and the methods that can be used),
v) **management and control of the environmental effects** (i.e., the mitigation,
compensation and monitoring measures that could modify, minimize, etc., the
effects),
vi) **presentation of the study** (considering how to communicate the results of an EIA),
vii) **participation of the public** (i.e., the problem of how to discuss an EIA study
with the public), and
viii) **judgement of authorities** (i.e., the comparison of the results with criteria and
values (i.e., the proper "assessment"), and the final political decision).

3. DESCRIPTION OF THE PROJECT

For the Directive 85/337/EEC, "project" means (Art. 1):
" - the execution of construction works or of other installations or schemes,
- other interventions in the natural surroundings and landscape including those
involving the extraction of mineral resources",
and the developer must provide (Art. 5):
" - a description of the project comprising information on the site, design and
size of the project,

- a description of the measures envisaged in order to avoid, reduce and, if possible, remedy significant adverse effects,
- the data required to identify and assess the main effects which the project is likely to have on the environment,
- a non-technical summary...".

So, the developer is responsible for furnishing the information, although it is rather indefinite and a clear source of possible controversies which is the range of "data required to identify and assess the main effects".

One critical point is the consideration of alternatives, which is mentioned neither in the Directive, nor in the Italian norms. In any case, the alternative "zero" (i.e., the non-execution of the project) always exists, and it should be the main basis for making comparisons. A logical way of thinking is to compare the future of an environment with and without the development of a project, but it could also be very useful (and fruitful for developing agreements among the participants in the study) to contemplate options of location, size, time, etc. If these options could be treated and analysed in early stages of the EIA (and possibly of the project itself) by the whole set of participants in an interactive way, delays and troubled discussions in the procedure would be avoided.

The Italian DPCM 27/12/1988 states two Reference Frameworks (called programmatic and project) to describe the activity subject to the EIA study. Though these frames are questionable, because of their descriptive more than analytical character, they point out the direct relationships of a project with more general policies, and the importance of the technical solutions adopted. Furthermore, in Annex 3 it is established that for industrial plants, waste and radioactive treatment plants, and electric energy production units, the risks and consequences of failures and malfunctions must be assessed. This fact forms an important bridge between "Risk analysis" and EIA.

3.1. Relationships with Plans and Programmes

As a general rule, the territory and the activity that constitute a project are subject to general pre-established policies, given in the form of official plans and programmes. Whether a project conforms to these policies is not the object of an EIA, but it is obviously an important point in order to coordinate the consent procedure. Furthermore, plans and programmes are more general instruments, oriented towards synoptic views of large areas. As EIAs are focused on specific projects, there must be a trade-off between their scope and the global (and generally more abstract) perspective of a technical-political plan.

The participation of different governmental authorities could be a source of potential conflict in asserting the conformity of a project to generic plans, and its environmental compatibility. For that reason, EIAs must be approached as a unique decision procedure, having a clear separation from other consent procedures. And, at the same time, to be consistent, planning lines should be set in accordance with the values and criteria used for taking decisions and judging the project.

The norms that deal with the territory (for instance, with the land use) and with the environmental compartments (for instance, quality standards for water and air), usually determine directly or indirectly the possible use of the natural resources, and stipulate the nature and characteristics of discharges and wastes. At this level, the temporary work that must be done to construct the final installations might also have a great weight in an EIA. But, generally, there is much more uncertainty about the size and practical consequences of these phases of preparation and construction, than for the final activity, which in turn will have a considerable imprecision in the determination of the long term effects and their compliance with the norms.

The Italian DPCM 27/12/1988 states that the project must be drawn up and described according to the applicable official plans, and that the coherence of objectives, possible discordances between different programmes, and the time evolution of the work must be shown. But the question about the extent of the territory that must be considered is open. This point is related to the extent of the studies and their intersection with regional plans and at the same time it influences the definition of remedies and controls imposed on an activity (for example; the extension of a monitoring network).

3.2. Technical Solutions

A project must be characterized by as precise data and information as possible. Of course, the limits of that set of information can vary considerably. Which is the right framework? Small differences in its definition can mean great differences in the quantity and quality of information.

The Italian decree mentioned specifies a Process Reference frame. First of all, for each project the type of goods or services offered, eventual satisfaction of the demand, and envisaged evolution of the demand/offer rate must be stated. As can be seen, these data are not directly related to the environmental impacts, but are only contextual information which could be useful as general support to the evaluation. Then, there must be a description of the considerations taken into account for the phases of construction and operation, and the criteria used to assess the territorial transformations in the short and long terms, with reference to the infrastructure services required. If this information is not bounded by certain indications, these general descriptions could become either only vaguely interesting or even irrelevant for an EIA.

The outlining of the technical and physical features and the conditions which constrained the project choices are more specific. These points must be referred to the particular norms that rule the sector and the zone, the use of resources, the discharge of wastes, etc. (i.e., the conditions discussed in the previous section 3.1). These references could be made in either a detailed or a vague generic way. It seems reasonable to us that a systematic approach with pre-established schemes would help to avoid arbitrariness and useless narrative documents. For instance, the Italian DPCM 27/12/1988 states that there must be a description of (among other things) the landscape, natural, and historical/cultural ties, without any other type of specification on how to write this description.

A project must also contain information about the measures to be taken to prevent, control and remedy any impact caused either during its normal operation or functioning or in abnormal conditions. Again, if a methodological way of referring these measures to the impacts that are being analysed is not defined, it would be difficult to assess their significance.

Of course, a technical description must be given mainly in numerical terms. Nevertheless, which variables must be defined, with what type of precision, and with what support (equations, physical-chemical formulae, etc.), are points that could be interpreted in diverse ways, and with different levels of detail. This type of information will, finally, constrain the validity of the study.

The technical alternatives should be described defining (for example) various options for location, general dimensions, forms, scheduling or timing of the project development, and the production processes, structures and/or regimes.

4. DESCRIPTION OF THE ENVIRONMENT

Describing an environment for an EIA means giving all the data that could be helpful in identifying it as a system, its components, its condition and trends of development (for instance, information on topology, hydrology, geology, air, biota, etc.), and all the information that would be important in understanding the consequences of the project.

First of all it must specify which local and general environments will be analysed, and the background that supports the choice of certain aspects (for example, regional problems and policies). The spatial scales that must be considered are: immediate surroundings (i.e., the very space of the project), region (i.e., the local area affected), and world (as the rest of the things that could be influenced).

The description of an ecosystem should consist in the account of the categories, factors and attributes (considering magnitudes, interactions, dynamics, and potential future evolutions) that are necessary to establish its state.

An ecosystem is a complex web or network of inter-relationships between components. A preliminary list of elements that must be considered is made up of:
- physical characteristics (soil, water, air, climate, land use, land character, etc.),
- ecological characteristics (habitats, communities, species, etc.),
- human activity patterns (demography, employment structure, transport, etc.),
- infrastructure (electricity, gas, sewerage, solid waste disposal, housing, telecommunications, etc.),
- social and community services (health service facilities, emergency services, etc.), and
- existing pollution (air pollution, water pollution, noise and vibration, radioactivity, etc.).

For an appropriate description, suitable variables are needed; but as not all the intrinsic values can be directly measured or assessed, indicators and indices are used. The key points that must be regarded are:
- selection of the parameters (which can be based on the quality standards most usually employed),
- data sources consulted,
- type and quality of direct measurements, and
- techniques for predicting possible future scenarios.

As numbers make sense only in their context, most of the values must be commented on, with explanations about their basis, scope and significance (for instance, comparing actual values against quality standards established by law).

It is obvious that trying to develop a comprehensive view of the present state of a complex dynamic and multi-level system like an environment is not an easily manageable task. A correct approach could consist in not trying to reach the "essence" of the environment, but to relate its description to the objectives of the EIA. In practical terms, it means that it is important to appreciate the actual status of an environment as a basis to which to refer the possible changes induced by the proposed project. The Italian DPCM 27/12/88 talks of "descriptive, analytical and forecasting" criteria; and we can add that the three criteria must be considered jointly.

A very efficient way could be to give a general panorama, describing in depth only what is going to be affected. But, of course, the impacts will be known after the analysis. So, at least, the factors that at first sight could be directly or indirectly implicated, other factors that could be in a critical state, the present use of natural resources, and the on-going degrading processes must be shown.

Open questions are the extent of the territory to consider (i.e., the limits for the spatial scales referred previously), and the structure which could be given to the statements of qualitative and non-numerical nature (e.g., judgments about the landscape, but also about the imprecisions in numerical values).

5. IDENTIFICATION OF THE ENVIRONMENTAL EFFECTS

Even if the general concept of environmental effect is almost intuitive, it could be interpreted in several ways. Trying to define it in a simple way, we can say that an environmental effect is any alteration of the environmental components. For an EIA, an effect is a process that affects the ecosystem and that is set in motion or accelerated by the project under study (Gisotti and Bruschi, 1990; Munn, 1975). The objective of their identification in an EIA is the recognition of the links between the sources of the perturbations and the final consequences, with the goal of assessing any variations induced.

An effect can be defined at different levels: for instance, changes in the soil (quality, stability), air (quality) and water (quantity, quality, seasonality) are a first layer since they are the media that would immediately receive the outcomes (matter and energy) from the project. Then, the biota (abundance/scarcity, extent) and the human ambits (health, society, economy) can be aggregated in various modes. The definition of each effect (what, when, etc. i.e., the actual targets of the analysis) is not obvious, and it has a great, and sometimes hidden, influence in an EIA for the simple reason that fixing them is fixing what is going to be evaluated. Working with pre-established check-lists is a help, but not a solution of this problem.

Effects are usually divided into direct (induced by the project itself) and indirect. As the latter might be more significant than the immediate consequences of a project, a thorough method must ensure the completeness of the survey of the possible web of effects and their interactions. One important point is the consideration of the irreversible or irretrievable commitment of resources, and of the possible restraints in the range of potential uses of natural elements.

This classification into direct or indirect effects, meaning the closeness between source and target in the cause/consequence chain, can be very useful in structuring the evaluation process. It is possible to make an almost sharp distinction between the straightforward effects of an activity on certain components (when, for instance, there is an obvious discharge of material to water, air and soil), and the effects of one component on another (for example, effects on human health due to polluted water).

An effect could be "good" or "bad", but its judgment (which will depend on points-of-view and criteria) must be made in a subsequent stage, i.e. when assessing the impact. In an attempt to distinguish "effects" from "impacts", it can be stated that the last are appraisals based on values, and so impacts are essentially human (subjective or social) evaluations of net changes. This topic will be discussed in section 10.

It is important to remark that, although this phase constitutes only a first step (mainly based on common sense, and in generally with no deep technical/scientific analyses) towards the final evaluation, it is frequently used as a major source of judgments and values. But the assignment of a number to a qualitative index is neither a real estimate, nor an assessment.

5.1. Methods for Identifying Potential Effects

A project can be viewed as being composed of activities, which at their time are accomplished through actions, which can end damaging effects to the environment. For the first identification of all the possible effects, several methods have been developed that try to act systematically, covering the entire project--environment nexus step by step. The various methods present considerable variety in their conceptual framework, approaches to data presentation, and technical sophistication (Clark, 1979).

Among these methods we mention: check-lists, overlays, impact matrices and cause-effect diagrams. However, we should remember that there are many ad-hoc methods which have been employed for particular cases.

A check-list consists of a simple catalogue of environmental factors, which are compared with the activities to be developed. Check-lists can be comprehensive listings of environmental parameters, actions associated with development proposals, or typically impacts resulting from a specific kind of development. They are simple to use, give a clear structure to the analysis, can establish a direct relationship with policy guide-lines, and can act as an aide-memoire for avoiding the omission of important factors. One example of a check-list is:
- atmosphere (air movements and climate, air quality, visibility, smells),
- water (hydrology, water quality and quantity),
- soil and subsoil (morphology, composition, quality),
- noise (intensity, frequency),
- biota (species, habitats, ecosystems),
- landscape (visual quality, historical and cultural elements),
- human health and welfare (safety, particular elements of health),

- uses and interests (agriculture, fishing, hydrological resources, mineral resources, land use, etc.).

It is worthwhile noting that a check-list includes both environmental components as water and environmental effects as noise.

The **overlays** approach (McHargh, 1968) comes from the field of land use planning and landscape architecture and consists of a series of overlaid map transparencies which show the geographical characteristics of the site, and different graphical presentation of the other traits. Their major advantage is in the presentation of the results related to spatial patterns, but the overlays are weak in providing other type of linkages, more complex interactions, or the uncertainties of an estimation. Even if the evolution of software technology could help considerably in its application, this type of method is not ideal, and should be limited to special cases or aspects of a problem (Munn, 1975).

An **impact matrix** (Sorensen,1972) consists of a pre-established list of project actions and of environmental characteristics or parameters. Thus, the possible (m x n) combinations between both lists are presented as opportunities of interactions. Each list forms an axis, and each cell defined in the matrix must be marked as a probable impact on the environmental characteristic from the correspondent action. Major criticisms of the matrix approach centre on its complexity, the broadness of certain interactions, the focus on direct, first-order effects, and the more or less arbitrary assignment of numerical values. This method of ranking provides subjective estimates, and this could be misleading as the scores do not have a standardised weighting method.

The **Leopold matrix** was the pioneering approach in impact matrices (Leopold et al.,1971) and it was the first comprehensive listing of environmental and socio-economic factors. It was designed to apply to a range of schemes, with the scope of selecting proposals that would be both environmentally and technically acceptable. The number of actions listed is 100, and the number of characteristics 88. Each cell that corresponds to possible effects (from the set of 8800) will have two values: an evaluation of the absolute magnitude, and an evaluation of the relative importance of the interaction (using indexes in a 1 to 10 scale; a score 10 representing the greatest magnitude, and 1 the least). A sum of the last values would allow a very simple qualitative computation of the whole effect of a project on an environment, given a first impression of the magnitude of the general "impact". In practical cases the quantity of interactions would range from 25 to 50.

A **coaxial matrix** (Canter, 1977) attempts to reproduce and evaluate the chain:
  i)   project (activities and potential sources of impact),
  ii) environmental conditions (affected factors),
  iii) potential environmental alterations), and
  iv) effects on human activities
using matrices and indices that try to assess the magnitude of each interaction. The principle is the same as in the Leopold matrix: lists of elements (that try to be exhaustive), cells for describing the crossings, and qualitative evaluations by ad-hoc scales.

A **cause-effect diagram** (Sorensen, 1972) also works with lists, and tries to be intensive in the consideration of the sources, specifying the levels of the possible impacts and linking each possible source of perturbation with its consequence step by step. This type of control must be performed for the complete set of activities that covers the life of a project: i.e., construction, operation/use, and end/discontinuance.

5.2. Criteria for the Determination of Important Effects

To perform the assessment, the most important effects must be identified from among the catalogue of possible ones. Now the question is what must be understood by "important".

The most general criteria for determining the "importance" of the effects are:
- their frequency, duration and geographical extent,
- their reversibility or recoverability,

- the possibility of mitigation,
- their social or political acceptance, and their relation to the community affected and governmental policies,
- the pre-established legal limits (e.g., environmental quality standards), and other precedents and future developments in the same or similar environments and/or projects.

6. EVALUATION OF THE ENVIRONMENTAL EFFECTS

Knowing that an effect could exist, and then making a qualitative or roughly quantitative estimate of its magnitude and importance in ad-hoc scales, is not really equivalent to evaluating it. Evaluating an effect in an EIA context should mean that:
- the parameters which are useful for describing the effect have been identified,
- their variations are computed or estimated, and
- the effect is stated in understandable terms using these variations.

In systematizing an EIA procedure it would be significant to develop a link between the structures of the project, the environment and the effects; i.e.:
- the sources of perturbations (i.e., the proposed project),
- the targets of the impacts (i.e., environmental factors, considering the nature of each one and the interactions between them), and
- the interface between the sources and the targets (i.e., the ways in which the material/energetic relationships are established).

If we remember that the laws that govern EIA do not specify methods, and that in order to reach appropriate results, a very adjusted and suitable rationale would be preferable to rougher options, the corollary may be that the logic of an EIA should establish what information in the description of the project and the environment is important in evaluating and understanding the effects.

6.1. Indicators and Indices

An indicator is a parameter that provides a measure of the magnitude of the change of an environmental characteristic. Indicators can be numerical or linguistic, and have the purpose of "indicating" or showing a value in the absence of direct measurements. Their selection is another crucial step in the evaluation process, because they determine the way the environment is observed. The importance of the indicators derives from the fact that they can be compared with common standards stated by the authorities or of general use. In that way they can give a first idea of the correspondence of a project with the legal context, and show easily understood customary variables which could help in the understanding of a condition.

An index is the result of a function that combines several measurements, indicators, estimates, etc., giving a compressed figure with the goal of summarizing a complex, abstract or not directly observable condition.

The relationships between indicators/indices and effects can be straightforward if these last are narrowly defined (as can be done in general for physical elements), or more indefinite if their interdependence is loosely established (as generally occurs, for instance, for social and economical concerns).

Environmental indicators and indices are discussed in Thomas, 1972 and in Colombo and Premazzi, 1990.

6.2. Information Necessary for the Evaluation of the Effects

Having chosen the indicators/indices which will be used in the evaluation, the information needed to make the computations must be compiled, with special concern for the quantity and quality of the antecedents required. The final precision of the computations will depend on the data considered.

One example of basic information could be the following:

Effect: effects on human health due to atmospheric pollution;

Information: SO_2 concentration (daily annual means, winter mean, concentration at 10 km. from the source, most likely daily concentration in the prevalent wind direction, etc.)

However, most of the evidence will be incomplete and controversial. The EIA basic data (i.e., indicators, indices, etc.) will be uncertain (because of dealing with stochastic systems, and applying statistical methods for analysing the data), imprecise (because of the lack of correspondence between the measurements and the parameters studied, among other causes), and will contain unavoidable traits of ignorance. This condition of lack of determination limits the validity of any evaluation, and forces a careful survey about the status of the basic data (both numerical and qualitative). This is another area where the criteria applied by the assessor could influence the evaluation greatly, and where some type of consensus should be developed among the participants in the EIA.

Another point that merits careful treatment is the manipulation of numbers. Not all the data come from measurements, and usually the lack of evidence forces the use of subjective "best estimates". These approximations to unknown values can be made in a more or less methodical way (for instance, with statistical techniques) over the scale of variation of the parameters. But, as some of these parameters will be indicators defined ad-hoc without links to measurable quantities, the values would be attributed using specially defined metrics, which will obviously be of a qualitative nature even if the judgements are set in terms of "numbers" (say: 1,2,3,4). The problem arises when it is necessary to combine quantitative and qualitative data in some way. If it is not understood what each of the various values means, the result will have no comprehensible value and meaning. And even in the case when a particular meaning is given to a specific value within the analysis, it would not be meaningful for persons outside it (viz., political decision-makers and public).

6.3. Evaluation Methods

The evaluation methods can be classified as formal and informal. The former are based on scientific knowledge (physical, chemical, biological, etc.), are specified by the use of models, and frequently have a mathematical expression. The informal ones yield results based on similarities or expert judgments. Of course, both the extrapolations by likeness and the formalization of the opinions of specialists must be done using rigourous methods, and should have a clear reference to standards.

The main differences between formal and informal methods is the strictness in the formulation of the model and the possibility of producing forecasts of potential dynamic evolutions using mathematical computations. The formal models can be classified according to

i) their representation of the environmental phenomena:
- kind and size of the system, and general conditions of application (constraints, precision, etc.),
- nature of the processes inside the system,
- assumptions and hypotheses (simplifications, default values, etc.);

ii) the outputs that can be computed (results, etc.):
- type of effects that can be estimated (change in concentrations, dispersions, etc.),

- temporal and statistical characteristics of the results (short/long term, media, mode, etc.),
- the formal characteristics of the results (tabular/graphical output, precision, etc.).

The formal methods, besides mathematical expressions, can also be based on experiments (performed in the field or in the laboratory), physical representations (with the purpose of explanation or scaled description), and analogue representations (e.g., using mechanical or electrical apparatus). The mathematical models can be empirical (black box type), or internally descriptive (based on a theory and explicitly describing the relationships among variables).

Sometimes, because of the lack of evidence or antecedents, or the very nature of the effects (for example: "landscape", "ecological value of an area", and in general the measures of diverse "qualities"), the effect is estimated using straight value judgements. These estimates, even if very weak, being obtained by "informal methods", can help in the comparison of alternatives, or in giving a first impression of a perturbation. These subjective evaluations could be made with:
- indicators of value: an objective measurable trait is used as representative of a value of a system (e.g., monetary value, quantity of species, etc.),
- indices of value: a function of several parameters represents the value (for example: noise noxiousness as a function of the noise level and the percentages of various types of vehicles), and
- direct judgments (asking experts, social groups, sample of inhabitants, etc.).

The methods for computing the effects on the air must consider: nature of the emissions, substances, temporal variation and points of release, quantity of pollution per time, quantity of pollution versus the activity (of the project), and the regimes of activity. The effects of air pollution have consequences on the atmosphere (changes in pressure, temperature and humidity, also smells), climate (local effects, and contribution to global effects), and other factors (by deposition or exposure of/to the pollutant substances e.g., water, flora, human beings, and, mainly, human health). Some common sources are: chimneys, deposits and stores, combustion engines, etc.

The methods for computing effects on water must consider more or less the same characteristics as for air, plus the temperature of the discharge, which may affect the models, and, of course, the type of water body. The main consequences of water pollution are: changes in the territory (coasts, landscape, etc.), changes in hydrology (velocity, flow, temperature, direction, regimes, etc.), salinity, biological status (microorganisms, etc.), and use by human beings. Some common sources are: water treatment plants (effluents), direct outputs (drainages) and soil pollution.

The methods for computing effects on the soil are usually considered in relation to underground waters, since the main forms of pollution are by permeation and/or leaching. Soil degradation could be the result of water erosion, salt excess, chemical degradation (like acidification), physical degradation (e.g., loss of permeability), and biological degradation (mineralisation, loss of microorganisms, etc.). There are many methods depending on the specific problem to be treated and the geological framework (depth, zones, etc.).

The biological environment (i.e.: flora, vegetation and fauna) can be studied considering single individuals, populations, species, communities, habitats (for reproduction, shelter, nourishment, transit, etc.) and whole ecosystems. This simple list gives an idea of the complexity of the problems involved. In general, the points that must be considered are: catalogues (which biological elements will be taken into account), functions (the role that biota accomplish in the territory), their dynamics and mutual influence, relationships with geography, geology and climate, etc. The main effects could be the removal or destruction of biota, and the main sources of changes in the air, water and soil qualities (in this way, the biological effects are in general indirect consequences).

The main effects on human beings concern: noise and vibration, landscape and territory, and health.

Noise and vibration can be objectively measured and compared with standards. They could be stationary, variable (fluctuating or intermittent) or impulsive. There are classifications of possible targets (industrial areas, residential areas, etc.) and types of effects (physiological, comfort, etc.). The principal sources are: rotating industrial equipment (pumps, compressors, engines, etc.), explosives, works (building, etc.) and the different types of traffic.

By landscape is meant the whole set of physical and non-physical elements that characterize a territory (from the view to its history). A landscape is usually a very composite system, containing: rural areas, urban areas, historical/cultural elements, roads, etc. Some traits are: morphology, land use, general view, and components, like power lines, buildings, forests, etc. As each perturbation will have its own way of interfering with the landscape, very different methods of analysis and evaluation must be applied.

The evaluation of effects on human health has its weak point in the lack of basic knowledge (mainly in epidemiology, doses, and long term consequences). Not all the elements that should be considered are causes of sickness, because there are many agents that could affect health factors without directly provoking illness. The first difficulty arises in the attempt to characterize the health status of a population, which in any case must be related to the risks expected. The factors that must be considered are: substances, toxicity (acute, subchronic, chronic, etc.), types of effects (cancer, genetic mutation, etc.), bio-accumulation factor, environmental persistence, etc. For each substance one must know: quantity released, temporal variations, products where it is included, impurity and final contents, other substances involved in the process, probable derivations (due to chemical and physical processes).

7. MANAGEMENT AND CONTROL OF THE ENVIRONMENTAL EFFECTS

As every project always has certain negative, even if acceptable, effects, they must be managed or controlled to ensure that:
- the magnitude of the effects are (and will continue to be) inside the limits of acceptance, and
- the conditions that made the project acceptable are not violated.

The measures that could be taken can be classified as:
- mitigation measures (when the objective is to attenuate or to limit the impact),
- compensation measures (when the objective is to repair or make up for effective or potential losses or damage), and
- monitoring measures (when the objective is to check-up on parameters).

These measures should also counteract:
- any possible incorrectness or incompleteness in the EIA,
- rare environmental events or episodes (earthquakes, drought, etc.),
- rare accidents (e.g., generalised fires), or failures (e.g., multiple equipment faults), and
- human errors (e.g., erroneous opening of a valve).

7.1. Mitigation Measures

Mitigation means minimisation of the effects; i.e.: to suppress, subdue or reduce those consequences which might be intolerable in certain phases of the project. The whole project must be analysed with the aim of mitigating the more important effects, once these have been identified.

Mitigation measures can be classified into three types:
- measures inherent to the project, such as the choice of technologies, modifications in the construction or production processes, or waste production /deposit control,
- measures that tend to remedy the potential damages (e.g., oxygenation of waters, repopulating of species, etc.), and

- measures that try to prevent or diminish the effects (e.g., with barriers against noise, dust, etc.).

7.2. Compensation Measures

Compensation measures must be directed mainly towards the community exposed to the effects, with the goal of counterbalancing or indemnifying any "bad" effects. These measures may be part of the conditions laid down by the population for the project to be accepted.

These compensations can be:
- economic (i.e., indemnities for damage, harm, injuries, etc. to economic or social assets, and in general for effects on health),
- social (such as reforestation, approaches to highways, etc.),
- sanitary (any compensation for probable health troubles that might occur in accidents or critical conditions).

The human concerns that must be considered in order to define these measures are: economic and occupational status, social pattern and life style, social and physical amenities, psychological features (e.g., mobility, personal security, molestation), safety and hazards, cultural and aesthetic aspects, etc.

7.3. Monitoring Measures

Monitoring means the periodic or continuous control of environmental variables with the objective of :
- following the status and evolution of pre-established environmental components and systems (also to check that the assumptions made in the study are fulfilled), and
- identifying the modifications and actions that must be carried out to maintain or recover specific environmental conditions.

The actual monitoring of an environment must begin when an area is being surveyed with the scope of determining its suitability for a certain project. Afterwards, it must continue during the construction and operation phases as a routine procedure, and even during the dismantling of the installations.

So, the monitoring activities can be classified as:
- basic monitoring (carried out before the development of the project, and oriented towards the actual environment),
- construction monitoring (carried out during the construction, and oriented towards the special works for this phase),
- operation monitoring (carried out during the active life of a project, and oriented either to the emissions and discharges of the installation, or to the status of the environmental variables), and
- after-use monitoring (to be carried out when the installation is left or has finished its operative period, and oriented to residual emissions, e.g. radiation, and the long term evolution of the environmental conditions).

Monitoring can take the form of spot checks, random site inspections and fully integrated continuous schemes, and in general must be completed with other local activities (for example for social and health problems).

An important point is that any monitoring activity tends essentially to follow the parameters or conditions that have been pre-established. Thus, it must be supported by very clear planning, with definitions of targets, instruments, frequencies and type of samples, reference values, levels for warning and alarm, etc.

Obviously, it is more difficult (due to their vagueness) to monitor the evolution of global, social and human related variables. For that purpose it is obvious that a previous agreed definition of suitable indicators with appropriate scales would be very useful.

8. PRESENTATION OF THE STUDY

The objective of this phase is to prepare and present the results of the EIA in a way that must be comprehensible for anyone who analyses the report, for the public, and in general for people without a technical background. The final report is a public document which will be used by political decision makers, government officers, professionals and lay people, and so the main question is to develop a smoothly readable, not cryptically technical, and complete description of the EIA performed. As can be imagined, the whole study can fail if the results are not communicated in the right way. Two problems can be envisaged:
   i) the reduction of the mass of information generated in the assessment
      process, and
   ii) the belief of the listeners/readers (decision-makers, public) in the models,
      hypotheses and criteria used, and in the results obtained.
   To overcome these difficulties, the report must employ:
      - commonly used terms, in a form that must fit the interpretative capabilities of a
        layman,
      - models, algorithms and computations converted to meaningful expressions in the
        most simple way,
      - graphs, tables, maps, etc.
   The EEC Directive and the Italian Decree DPCM 27/12/88 require the provision of a non-technical summary by the developer, which must include a description of the project, a description of the measures envisaged "in order to avoid, reduce, and, if possible, remedy significant adverse effects" (Article 5 of the Directive). This report should contain: a description of the methods and data used; a description of the environment, the project and the effects studied; an identification of the people, assets and interests concerned; and an explanation of the nature and magnitude of the impacts.

9. PUBLIC PARTICIPATION

The EEC Directive states that the Member States "shall take the measures necessary to ensure that the authorities likely to be concerned by the project by reason of their specific environmental responsibilities are given an opportunity to express their opinion...", and that "The information gathered pursuant to Article 5 shall be forwarded to these authorities." Also, It is stated that "Member States shall ensure that... any information gathered pursuant to Article 5 (is) made available to the public." Furthermore, "the public concerned" must be given "the opportunity to express opinion before the project is initiated."
   As the same Directive observes, the real problems in this stage are:
      - which is the "public concerned",
      - how the information can be consulted, and
      - how they can express their opinions.
   One way that seems efficient for managing these relationships is, as was said above, the early consultation of the "actors" who would play an active role in the EIA.
   However, it would be difficult to reach agreements based only on non-technical summaries. If an EIA is supported by a thorough rationale (i.e., logical foundations that back the choice of the effects to be analysed, the definition of the information and basic data required, the selection of the

evaluation methods, the computations and reasonings made, etc., but also the awareness about the weak points, the technical or scientific gaps and/or lacks, and the zones of ignorance), a sound consensus will be only possible if the rationale is understood. Even if this type of multiple participation could be cumbersome if wrongly guided, it could be systematized with the purpose of facilitating the information flow, minimizing conflicts, and allowing the development of agreed assessments. In that way, the participation of the public will not only be of an "informative" type, but also active and effective in the scrutiny and negotiation steps.

10. JUDGEMENT OF AUTHORITIES

Once the study is finished, the authorities must evaluate the EIA, and give their consent to the project. The concepts that play a role here are: impact, decision and consent. We can make a clear distinction between "effects", as variations of certain environmental conditions, and "impact", as their evaluation in a context where their merit, worth, conditions, pros and cons, etc., can be accounted. In other words, an impact is not a number, but the qualitative appraisal of certain numbers (derived from measures, computations, etc.) that describe a net change (good or bad) in an ecosystem, judged using an objective frame of social and personal values.

The consent procedure concludes when a political decision is taken (either by governmental authorities or by specially designated persons, like the Italian Commission for Environmental Compatibility) about the acceptability of the set of impacts that would cause a project. I.e., it is the positive or negative answer to the request for a project, conditioned to certain special circumstances that make it acceptable or not acceptable.

In that line of reasoning, an impact will not be the destruction of a site, the contamination of a water course, or the changes provoked in the type of employment, but the comparison of an environmental effect with the set of values that supports the perception of the environmental condition that would change (in the sense of environment). I.e., the significance of an impact must be made obvious against explicit criteria, the uncertainties that embrace its evaluation, the associated risks, the comparison of alternatives, the long term evolution of the environment and public involvement and opinion.

The magnitude of the expected impacts, and the penalties to be paid for preserving an environmental quality, will set the differences between decisions with more or less political contents (Munn, 1975). These points will be magnified if the decision might affect neighbouring countries, or, obviously, if the conflicting objectives include political issues.

The result of the environmental decision-making process depends on the comparison of the changes that have been envisaged for an environment, and the environmental values that the decision-maker takes into consideration. There is a possible source of conflicts between society and the government on this topic. As environmental values are not obvious, change continuously with time (as a result of experience, economic and social needs, etc.), and cannot be taken as stated by law (quality standards are only the written, objective and quantifiable part of them), each environmental decision needs as counterpart the formulation of respective values. By whom? By the authorities, but also by the people "concerned". Environmental values are goals that must be satisfied, and concerns that must be observed. In other words, the participants in an EIA (and first of all the public) should not only be involved in the assessment, but should also be considered in the final decision-making.

Finally, the result of the governmental judgement must be communicated to all the parties affected, with an appropriate description of the decision taken, the objections made, the parties involved, and the recommendations and conditions of the decision.

11. CONCLUSIONS

The phases of an EIA procedure have been critically described and discussed, with reference to the EEC directives and the pertinent Italian regulations.

Emphasis is given to the multidisciplinary nature of the problems to be faced and to the need for a common language and valid communication between the actors involved in the study. In fact, a frank and constructive dialogue between developer, authority, public and technical advisers is essential to reach a consensus, which, obviously, must be a compromise.

The main methods for the identification and evaluation of the effects on the environment of a project are described briefly. Mitigation, compensation and monitoring measures to manage and control the possible negative effect are also illustrated.

In the appendix, the features of an informatic system to be used as a tool in an EIA procedure are discussed. A knowledge-based system approach has been adopted, as it allows us to store and integrate the abilities necessary in an EIA, such as very specific scientific knowledge and technical skill, and making them operational to non experts also.

REFERENCES

Alberti M. et al. (1988), La Valutazione di Impatto Ambientale, Franco Angeli, Italy.

Canter L.W. (1977), Environmental Impact Assessment, McGraw-Hill, New York, USA.

CIRC 11/8/89. Circolare 11 agosto 1989 del Ministero dell'Ambiente della Repubblica Italiana. Pubblicita' degli atti riguardanti la richiesta di pronuncia di compatibilita' ambientale di cui all'art. 6 della legge 8 luglio 1986, n. 349; modalita' dell'annuncio su quotidiani.

Clark B.D. et al. (1979), Environmental Impact Assessment in the USA: A Critical Review, Research Report 26, Departments of the Environment and Transport, UK.

Colombo A.G. and Premazzi G. (Editors) (1990), Proceedings of the Workshop on "Indicators and Indices for Environmental Impact Assessment and Risk Analysis" held at Ispra (VA), Italy, 15-16 May 1990, EUR 13060 EN.

Directive 85/337/EEC. Council Directive of 27 June 1985 of the European Communities. On the assessment of the effects of certain public and private projects on the environment.

DPCM 10/8/88. Decreto del Presidente del Consiglio dei Ministri della Repubblica Italiana del 10 Agosto 1988, n. 377. Regolamentazione delle pronunce do compatibilita' ambientale di cui all'art. 6 della legge 8 luglio 1986, n. 349, recante istituzione del Ministero dell'ambiente e norme in materia di danno ambientale.

DPCM 27/12/88. Decreto del Presidente del Consiglio dei Ministri della Rebubblica Italiana del 27 Dicembre 1988. Norme tecniche per la redazioni degli studi di impatto ambientale e la formulazione del giudizio di compatibilita' di cui all'art.6 della legge 8 Luglio 1986, n. 349, adottate ai sensi dell'art.3 del DPCM 10 Agosto 1988, n. 377.

Gisotti G. and Bruschi S. (1990), Valutare l'Ambiente - Guida agli studi d'impatto ambientale, La Nuova Italia Scientifica, Roma, Italia.

Hayes-Roth F. (1984), "The knowledge-based expert system: A tutorial", Computer IEEE, 17(3), pp. 11-28.

Leopold L.B., Clark F.E., Hanshaw B.B. and Balsley J.R. (1971), "A procedure for evaluating environmental impacts", US Geological Survey Circular 645, United States Geological Survey N71-36757.

McHargh I.L. (1968), "A comprehensive highway route-selection method", Highway Research Record No. 246, p.1-115.

Munn R.E. (Editor) (1975), International Council of Scientific Unions, Scientific Committee on Problems of the Environment, Environmental Impact Assessment, Principles and Procedures, SCOPE Report 5, Toronto, Canada.

Nichols R. and Hyman E. (1982), "Evaluation of environmental assessment methods",
ASCE Journal of Water Resources Planning and Management, 108, pp. 87-105.

Provincia Autonoma di Trento. Legge provinciale 29 Agosto 1988, n. 28. Disciplina della
valutazione di impatto ambientale e ulteriori norme per la protezione dell'ambiente.

Regione Veneto. Legge regionale 23 Aprile 1990, n. 28. Modifiche ed integrazione alla legge
regionale 16 aprile 1985. n. 33.

Rosenberg D.M. et al. (1981), "Recent trends in EIA", Canadian Journal of Fisheries and Aquatic
Sciences, 38, pp. 591-624.

Schibuola S. and Byer P.H. (1991), "'Use of knowledge-based systems for the review of
environmental impact assessments", Environmental Impact Assessment Review, 11,
pp. 11-27.

Sorensen J.C. (1972), "Some procedures and programs for environmental impact assessment",
in Ditton R.B. and Goodale T.L. (Editors.), Environmental Impact Assessment: Philosophy
and Methods, University of Wisconsin, Wisconsin, USA.

Sorensen J.C. and Moss M.L. (1973), Procedures and Programmes to Assist in the
Environmental Impact Statement Process, COM-73-11033, University of California,
Berkeley, USA.

Thomas W.A. (Editor) (1972), Proceedings of the Symposium on "Indicators of Environmental
Quality" held in Philadelphia, Pennsylvania, USA, 26-31 December 1971.

Waterman D.A. (1986), A Guide to Expert Systems, Addison-Wesley, USA.

Whitney J.B.R. and MacLaren V.W. (1985), "A framework for the assessment of EIA
methodologies", in Whitney J.B.R. and MacLaren V.W. (Editors), Environmental Impact
Assessment: The Canadian Experience, University of Toronto, Toronto, Canada.

72

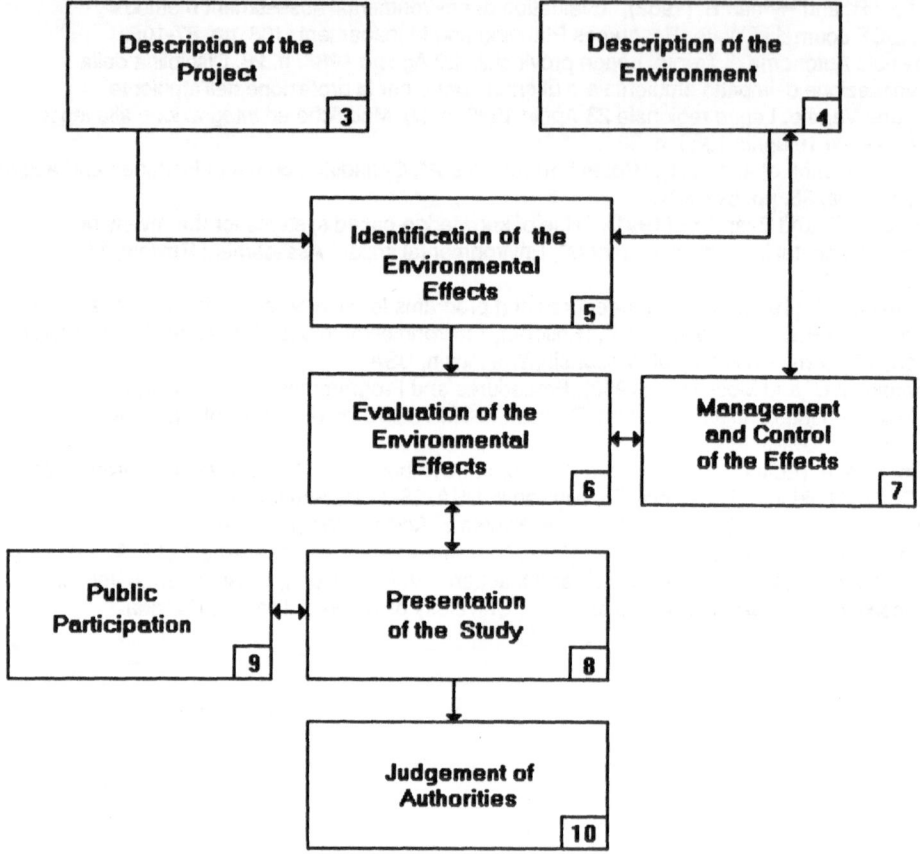

Note. *The numbers in the boxes correspond to the number of the sections in the paper.*

Figure 1. Main Phases of an EIA study.

APPENDIX. A DISCUSSION ON SOFTWARE AIDS FOR EIA

As most of the tasks that must be performed to carry out an EIA are based on very specific scientific knowledge and technical skills, it seems obvious that Knowledge-Based Systems KBS) are a potential way of capturing these abilities and making them operational in a computer (Nichols and Hyman, 1982; Schibuola and Byer,1991). Some of the attributes that an EIA should exhibit are: rationality, comprehensiveness, accountability, and participation and inter-action of various actors. Several reviews have shown how it is very easy to develop an unsound EIA, especially if it is based on ad-hoc techniques and is not supported by a scientific background (Rosenberg et al.,1981; Whitney and MacLaren, 1985). The fulfilling of a detailed and complete appraisal, and communication between the actors, are tasks closely related to the solidity of the reasoning process. How can people discuss things that they cannot understand? And if there are points which cannot be solved by applying "state-of-the-art" knowledge, at least there could be agreement on this ignorance.

An actual EIA should proceed by consulting experts in all the particular fields (or, what is the same in terms of knowledge, by using literature, manuals and handbooks which describe the sound ways of making things happen). It is not only a problem of method (several can be imagined, and a large number have been proposed), but of content. And, whether it be the developer making the studies, the government reviewing them, or the public criticizing the results, the participants cannot produce these aims only by expressing personal beliefs or subjective judgements. They must apply knowledge (specialized, up-to-date knowledge) in a rational manner.

A KBS is a computer program that uses human knowledge and expertise to solve problems ordinarily requiring human intelligence (Waterman, 1986). A KBS should be usable by people who are not specialists in the subject, and a KBS must be able to explain its lines of reasoning to the user. As general guide-lines, it can be said that to construct a KBS, five major tasks can be recognized (Hayes-Roth, 1984):
- **Identification:** definition of the problem, the sources of knowledge and the goals and objectives that the system should meet; the system is designed to a manageable size;
- **Conceptualization:** thorough exploration of the problem to identify key concepts, relations, control mechanisms, subtasks, strategies, data and constraints needed to solve the problems (and the particular subproblems);
- **Formalization:** expressions of the concepts acquired in a formal knowledge representation scheme (for example: rules, frames, etc.);
- **Implementation:** transformation of the formalized knowledge into a functional computer program (using a computer language, in general of a high level);
- **Testing:** systematic assessment of the results produced by the system and the interfaces with the user.

The KBS approach can be compared (in abstract terms) with other possible sources of knowledge. Some characteristics to consider are: cost, time, availability and quality of results. A human specialist is the most expensive solution (and surely the most difficult to arrange). Specialists are scarce, and their time is very precious and in demand. On the other hand, literature is not always practical, and much time and effort is usually needed to obtain the right answer to a problem. A paper manual should be able to provide reasonably good results if it can summarize the knowledge needed, but in any case it will not be operative (in the sense of being "static" and not "dynamic", i.e.: it does not accept questions and give the right answer automatically).

The problems of efficiency and quality for the EIA procedure depend on the efficiency and quality of the knowledge support given. A KBS could give a good result if applied to problems that can be systematised. Rough questions and vague issues are commonly present in human intensive tasks. Nonetheless, it is not the case that the computer program will replace the human (for example, guiding the discussions or giving personal criteria), but that it should aid the people that must develop the study on specific topics with specialist knowledge.

Actually, we are thinking of a set of KBS, each with a specific objective, integrated in an appropriate software structure. This structure must reflect the logic of an EIA procedure, and should help in managing the data and information required. Thus, the ideal complete computer system should provide a joint information and documentation system, with tools for:

- consulting general and specific knowledge (the proper KBS),
- reading, treating and editing texts (possibly with the use of Hypertexts),
- expressing judgements, subjective estimates and criteria (also KBS),
- considering models (physical, chemical, biological, etc.) and applying them (KBS for the appraisal, algorithms for the computations),
- watching and preparing graphics,
- reading and storing numerical data (Databases),
- linking computations, tables, graphics, etc. (spreadsheets plus graphics), etc.

Table 1 summarises the characteristics of each of the phases considered in the paper, and gives the main features to be considered for the software tool.

Table 1. Main features of the phases of an EIA study and relevant software tools

Phase I:	**Description of the project**
Objective:	**identification and characterization of proposed activities**
Input:	project phases, processes, products, materials, risks, etc.
Output:	site, emissions (matter/energy), resources, alternatives, scheduling, technical solutions, relationships with plans & programs
Problems:	indeterminations and uncertainties, technology limitations
Knowledge sources:	developer, technicians
Software hints:	Graphics, KBS, Data Bases, Spreadsheets

Phase II:	**Description of the environment**
Objective:	**identification and characterization of the ecosystem**
Input:	local and global environments, human concerns, standards, existing pollution
Output:	actual state, possible evolutions, critical points
Problems:	sources of data, measurements/metrics/estimates, models
Knowledge sources:	government, public, specialists
Software hints:	Graphics, KBS, Data Bases, Spreadsheets

Table 1. (Continuation)

Phase III:	**Identification of the environmental effects**
Objective:	**identification of the interactions project/ecosystem**
Input:	phases I and II, methods
Output:	list of potentially important effects
Problems:	criteria (importance), qualitative judgements
Knowledge sources:	public, specialists
Software hints:	Graphics, KBS, Data Bases

Phase IV:	**Evaluation of the environmental effects**
Objective:	**estimation of the magnitude and traits of the effects**
Input:	phases I, II and III, data, information and methods
Output:	results of computations and subjective estimations
Problems:	criteria (targets, which model to use), qualitative judgements
Knowledge sources:	specialists, public
Software hints:	Graphics, KBS, Data Bases, Spreadsheets

Phase V:	**Management and control of the environmental effects**
Objective:	**mitigation, compensation and monitoring measures**
Input:	phases I, II and IV, data, information and methods, standards, technical possibilities
Output:	measures (monitoring plans, etc.)
Problems:	criteria (thresholds, which model to use), qualitative judgements
Knowledge sources:	public, government, specialists, developer
Software hints:	Graphics, KBS, Data Bases, Spreadsheets

Table 1. (Continuation)

```
Phase VI:        Presentation of the study

Objective:       report preparation
Input:           phases I, II, III, IV and V
Output:          report, general agreements
Problems:        communication

Knowledge sources:      government, specialists (risk communication),
                        developer
Software hints:         Graphics, KBS
```

```
Phase VII:       Public Participation

Objective:       negotiation with the public
Input:           phases I, II, III, IV, V and VI
Output:          opinions, comments
Problems:        communication, subjective appraisals

Knowledge sources:      public, government, specialists (risk communication),
Software hints:         KBS
```

```
Phase VIII:      Judgement of authorities

Objective:       decision-making and communication
Input:           phase VI
Output:          decision, communication
Problems:        communication, technical/political trade-offs

Knowledge sources:      government, specialists (risk communication)
Software hints:         KBS
```

RISK ANALYSIS IN ENVIRONMENTAL IMPACT STUDIES

S. Contini
CEC, Joint Research Centre
Institute for Systems Engineering
and Informatics
I - 21020 Ispra (Va)
Italy

A. Servida
TECSA
Research and Development Division
Via Caravaggi
I - 24040 Levate (Bg)
Italy

ABSTRACT. This paper deals with the subject of risk analysis in the framework of environmental impact studies of industrial installations. Starting from the fundamental definitions of risk, hazard, and accident, the main phases of the Probabilistic Risk Analysis (PRA) procedure are described. The most important methods for implementing the different phases of a PRA are reviewed. When performed within an environmental impact study, the risk analysis must be extended to consider minor accidents; because of the large number of such accidents, advanced software tools are indispensable.

1. INTRODUCTION

The aim of a Probabilistic Risk Analysis (PRA)[(+)] is to provide a measure of the risk associated with an industrial installation. The term Probabilistic Risk Analysis is commonly used for the process that allows the following questions to be answered:
- what can go wrong in the plant ?
- which are the consequences of malfunctions ?
- how often will they occur ?

Each of the above questions are dealt with by a specific kind of analysis. More precisely, Hazard Identification (HAZID) analysis deals with the first question; models for estimating the damage to man, the environment and property are applied to answer the second question; probabilistic techniques are used to give the answer to the third question.

In risk analysis both qualitative and quantitative approaches are available.

The **qualitative approach** relies on experience expressed in the form of check-lists and codes of practice. Several check-lists have been developed for site selection, process materials, aspects of design, commissioning and operation. However, in some cases, reliance just on check-list cannot be considered sufficient to identify all important hazards and related causes. For instance, for new processes and new plants, a systematic analysis of possible malfunctions may become essential to achieve an acceptable safety level.

[(+)] This paper uses the acronym PRA - used in the nuclear field to refer to risk quantification - even though, in the chemical field, the equivalent acronyms QRA (Quantitative Risk Analysis) or, more specifically, CPQRA (Chemical Process Quantitative Risk Analysis) are frequently used [1].

A. G. Colombo (ed.), Environmental Impact Assessment, 79–103.
© *1992 ECSC, EEC, EAEC, Brussels and Luxembourg.*

Another available technique is that of hazard indexes, e.g, the Dow index [2] and the Mond index [3]. These methods make use of a number of empirical factors which are related in some way to the probable frequency and magnitude of potential accidents, without specifying how these may arise. Frequently, in risk analysis these methods are applied for screening purposes, i.e, to rank the units of a complex plant (e.g, refinery) according to their potential hazard and to define, for each of them, the type of risk analysis procedure to be applied [4].

The **quantitative approach** is obviously more sophisticated, as it aims at producing a consistent and quantified picture of the risk induced by the operation of the plant. It is based on well-established procedures in which systematic techniques for hazard identification, accident frequency and consequence estimation are applied. Because of the uncertainties in data and models, as well as the assumptions the analysts must make to overcome the lack of information, the clarity of the whole evaluation process is considered an essential requirement of any PRA. In spite of the many sources of uncertainty, the real value of a PRA resides in the critical identification and examination of the plant behaviour to verify the adequacy of the safety measures against possible accidents. The risk analysis is in fact an iterative process of "design improvement" leading, at the end, to an acceptable level of plant safety. In other words, the PRA allows analysts to check the validity of the implementation of design rules and safety criteria, and to demonstrate that the design includes adequate safeguard actions for both preventing accidents from occurring and reducing their consequences.

Consider now the problem of performing an Environmental Impact Assessment (Directive 85/337/EEC) for industrial installations, e.g, for those falling under art. 5 of the Directive 82/501/EEC on major hazards. It can easily be realized that risk analysis must form an important part of the study. However, the PRA cannot be limited to major hazards only, but should also consider minor accidents, i.e, so-called high-frequency-low-consequence events. Furthermore, this analysis needs to be complemented with a study of the impact of other associated activities (e.g. waste treatment, transport of dangerous substances, etc.), as well as of all other releases during the normal operation of the plant.

This paper gives an overview of the probabilistic risk analysis procedure, as it is applied to fulfil the requirements of directive 82/501/EEC, and points out the additional analyses to be performed when the EIA is of concern. Section 2 gives the basic concepts and definitions in risk analysis. In section 3 the quantitative risk analysis procedure is briefly described. Sections 4 to 7 contain a synthetic description of the methods for carrying out the different phases of the procedure. In section 8 the list of the main sources of uncertainty is given. Finally, some considerations on the necessary extensions of the analysis to fulfil the objectives of an EIA are given in section 9.

2. BASIC CONCEPTS AND DEFINITIONS

Risk is a very complex concept. Although its meaning might be very easy to realize, it can be very difficult to express. In [1] the risk is defined as a measure of the economic loss or human injury in terms of both the accident likelihood and the magnitude of the loss or injury.

In [5] alternative definitions of risk are given, such as:
  i)   risk is a measure of an uncertain damage,
  ii) risk is a ratio of hazards and safeguards,
  iii) risk is a triplet of "accident - likelihood - consequences".

In definition i) the risk is a combination of **uncertainty** and **damage**, i.e, there would be no risk if one knows the exact time and place of an adverse event; also there would be no risk without damage.

Definition ii) calls for the important concepts of hazard and safeguard. The **hazard** can simply be defined as a "source of danger". For instance, the LPG or the chlorine in a plant represent a hazard, since the release of these substances may lead to a danger to the population and to the environment. But hazards can be controlled by means of suitable **safeguards**, i.e, by adopting technical means to reduce the likelihood of accidents and the damage caused. Consequently, for a given hazard, the more effective the safeguards, the lower the level of damage that may be caused. Therefore, an uncontrolled hazard may give rise to an accident, i.e., a release of mass/energy able to cause, under certain circumstances, significant effects. Such a release may generate different accident scenarios, depending on the possible evolutions, e.g, on the presence of ignition sources, the working/failure of isolation/ mitigation systems and operator intervention.

Finally, in definition iii), the risk is defined as a **set of scenarios** as represented in Tab. 1. This table gives a simple representation of the spectrum of the abnormal events and associated consequences potentially occurring in the plant.

The risk can be **immediate** or **delayed**, depending on the time of the damage. The risk is perceived in different ways by different people, depending on the degree of involvement in the risk acceptance decision process. In this respect, the risk can be either **voluntary** (e.g, smoking, drinking, driving) or **unvoluntary** (e.g, living close to a potentially dangerous installation, fighting in a war). Hence, perceived risk can also be seen as a "feeling of insecurity".

In many common human activities risk can be estimated through the statistical analysis of historical records, and a comparison can be made between the different sources of risk. In some cases, however, statistics cannot be prepared because of the lack of significant data, e.g. a new industrial installation. Thus, the decision on whether to accept or reject an installation needs an a priori quantification of the risk, i.e, a probabilistic risk analysis.

The final results of a probabilistic risk analysis can be represented in two different forms, namely as individual risk and societal risk.

Individual risk is defined as the frequency with which an unprotected person, living permanently in a certain location, is affected by a specific type of damage. Generally, the type of damage one is interested in is death. The individual risk is represented in the form of iso-risk contours, as shown in Fig. 1, as it depends from the distance from the plant. An iso-risk curve delimits the region where the risk is greater then or equal to the associated value. These curves, superimposed on the map of the site, give important information on the areas where the risk may need to be reduced.

The **societal risk** is the relationship between the frequency of an accident and the minimum number of people injured or killed. With respect to the individual risk, the societal risk calculation also requires the population distribution. The societal risk is represented by an inverse cumulative function showing the frequency F of having N or more deaths (F N curve). The basis for constructing the F N curve is more or less the content of Tab. 1, once the triplets are ordered by increasing severity of consequences and the corresponding inverse cumulative frequency is determined. Fig. 2 gives examples of discrete and continuous F N curves.

3. THE RISK ANALYSIS PROCEDURE

The general quantitative risk analysis procedure can be subdivided into four main phases, as shown in Fig. 3, and namely:
1. hazard identification,
2. accident frequency and consequence estimation,
3. risk calculation,
4. risk reduction/acceptability.

The **first phase**, i.e, hazard identification (HAZID), aims at producing a list of potentially hazardous situations (accidents) arising from loss of containment caused by one or more "initiating events" (e.g, component failure, operator error, utility failure, deviation of plant input variables outside the normal range). Hazard identification is fundamental for any risk analysis, since no protection measure can be implemented for unidentified hazards. To be successful, this phase must be performed by a team of experts with thorough knowledge of the plant (e.g, process engineer, mechanical engineer, instrumentation engineer), coordinated by an experienced risk analyst. The team "brainstorms" on each significant deviation of the plant behaviour from normal conditions, at the same time sending suggestions to the designers on modifications needed to improve plant performance from the safety viewpoint. To focus (i.e. limit) the analysis to the significant problems, some screening criteria are applied, based on engineering judgement, on both the occurrence frequency and the damage caused by each identified accident sequence. Those accidents for which the frequency, the damage or a combination of both give an insignificant contribution to the risk are neglected. Therefore, the result of this phase is a set of "conceivable and significant accidents" to be subject to the quantification phase. Accident data bases are also required to enhance the completeness of the analysis.

The **second phase** is composed of two parts: the estimation of the occurrence frequency [6] and of the damage produced by each of the previously identified significant accidents.

The first part is performed by means of a set of probabilistic methods (see section 5) for modelling the different sequences of events leading to the accidents under study. These methods allow the plant to be broken down into its constituent parts up to the detail of "simple item", i.e, of components/subsystems for which the failure probability can be statistically estimated from past experience. The analysis of these models allows analysts to estimate the accident frequency, the different causes and the major contributors. With these results the designers can rationally identify the best cost effective design solutions to reach the plant safety desired.

The term "consequence analysis" refers to:
- the estimate of the physical effects of accidents, and
- the estimate of the damage caused to defined targets.

Typical accidents are fire, explosion and release of toxic substances. The models available allow analysts to determine the physical effects in terms of heat radiation for fires, blast wave intensity and fragment trajectory in the case of an explosion and ground concentration in the case of a toxic release (see section 6). To estimate the damage, dose-effect (dose-damage) relationships are needed for people, structures and the environment. In risk analysis both "on-off criteria" and the probit model are frequently applied (see section 7). The results of these calculations allow the analysts to verify the adequacy of the existing protective devices and, if necessary, to suggest the implementation of other protection measures, whose effectiveness must be verified through additional consequence analyses.

In the **third phase** the results of the previous phase are used to estimate the risk and to represent it in the form of iso-risk contours on the map site and F N curves.

Fourth phase. Finally, decisions can be made on further risk reduction or on its acceptability. In the former case the analysts have all the information necessary (generated during the previous phases) to identify the best cost effective improvements.

4. HAZARD IDENTIFICATION TECHNIQUES

The hazard identification (HAZID) represents a critical step in risk analysis: hazards not identified in this phase will almost certainly remain unknown for a long period of time during which the plant is not adequately protected. As a matter of fact, it is very difficult to discover all the potentially hazardous situations because of the complexity of the problem (think for instance of human factors), the large number of ways in which the plant can behave which forces analysts to define "cut rules" in order to have a manageable number of accident sequences for quantification.

Hazard identification begins with the conceptual design and follows its development until the construction of the plant. For this reason there is no single method applicable for all design stages. The HAZID analysis should also be performed during the life of the plant if there are significant modifications. The importance of performing the analysis starting from the very beginning of the design is that the most suitable design improvements can be made at the lowest cost. Indeed, the cost for removing a hazard is completely different if it is done at the design stage rather than on the pilot plant or on the industrial plant. As HAZID is both time consuming and expensive, the depth and extent of the study are defined according to the inherent hazard level presented by the plant units. Therefore, a screening study is performed to identify the plant areas on which to carry out the HAZID analysis. In many companies the screening is based on index techniques (e.g. Dow, Mond). Hazard identification is performed by an interdisciplinary group of design experts representing the different departments involved in the project. The involvement of these experts guarantees the study's success.Hazard identification is usually preceded by an investigation of past accidents in similar plants, aiming at enhancing the completeness of the results. Accident data bases are required to give information on causes, evolution and consequences of accidents.

The most frequently applied techniques for hazard identification are:
- Process/ System Check-list,
- Safety Review,
- Preliminary Hazard Analysis,
- Failure Mode and Effect Analysis (FMEA),
- Hazard and Operability Analysis (HAZOP), and
- Systematic Identification of Release Points (SIRP).

These techniques are not mutually exclusive: none of them can be used for all circumstances [6]. As a matter of fact, more than a single technique should be applied to give a more complete list of hazardous situations (e.g. the use of HAZOP for the process, FMEA for utilities and SIRP for pipeworks). The choice of the technique depends upon the type of process (well known or new), the intrinsic hazard level, the phase of plant development, etc. For the purposes of this report only some of these methods will be considered, i.e. methods based on the identification of sequences of plant perturbations leading to significant accidents. Hence, methods based on check-lists (design practice), safety review and preliminary hazard analysis are not described.

FMEA and HAZOP are the most frequently applied structured techniques. Plant behaviour is examined starting from an initiating event (i.e. an event which gives rise to a perturbation in the plant) and studying its propagation up to the identification of the final effects. A substantial difference among these methods is linked to the way in which the initiating events are selected. The choice of the technique depends greatly on the analyst's experience and preference, the characteristics of the plant under study, the aim of the analysis and the computer codes available.

The final result of the hazard identification study is the identification of possible accidents, together with a judgement on the suitability of the existing protective/preventive measures. Those hazards for which the plant is considered not to be adequately protected are submitted to the attention of the designers and the best design modifications are discussed and identified. The number of accidents identified is generally high; hence, some criteria are to be applied to select those to be retained for the risk estimation. These criteria are generally based on a subjective judgement of both accident frequency and consequences.

4.1. Failure Mode and Effect Analysis (FMEA)

FMEA is a qualitative procedure to identify potential hazards starting from the failure of single components [7]. As such, FMEA is particularly suitable for application from the first stages of development of a design and can be periodically updated.

The results of the analysis enable the designers to identify the relatively most critical areas on which design modifications may be needed. After defining the boundaries of the system to be analysed, the mission profile and the level of detail, the design is carefully examined to identify the final consequences of single component failure according to the procedure shown in Fig. 4. The information collected by the analysts is laid out on specially designed forms. The FMEA form can be customized so as to be suitable for the level of detail of the design and for the analysis' aims.

Typical questions the group responsible for the analysis must answer, for each component, are of the following type:
- What are the modes of failure of the component?
- What are the causes leading to them?
- What are the consequences of each failure mode?
- What are the means to detect the failure?
- What are the design changes most suitable to reduce possible unacceptable consequences?

For each failure mode of selected components all possible causes are identified, independent of whether they are intrinsic or induced by external events. Each failure mode leads to a different performance of the plant. The consequences of a particular failure mode are identified by studying the plant performance with the assumption of unavailability of protective actions, in order to evaluate the actual criticality of the initiating event. Hence, it is possible to draw conclusions on the convenience of making design changes, i.e. upgrading the control/protective systems or revising the emergency operating procedures.

4.2. Hazard and Operability Analysis (HAZOP)

As for FMEA, HAZOP is applied to identify the potential hazardous situations starting from process variable deviations and to verify qualitatively the adequacy of the existing protective/preventive measures [8, 9]. Fig. 5 describes the main steps of the procedure.

The points in the plant at which the process variables should be examined are selected on the main process lines. The questions the group responsible for the analysis should answer for each process variable that may undergo a variation at the particular node under examination are the following:
- What are the possible deviations of the variable ?
- What are the causes producing such deviations ?
- What are the possible effects of the deviations ?
- What are the means for detecting the deviations ?
- What actions are provided to avoid or reduce the effects of the deviations?
  Are they sufficient ? If not, which new measures can be implemented ?

The answers to these questions are recorded on specially devised forms. Guide words are used which are aimed at guaranteeing the systematic examination of all possible deviations; guide words are meant to prompt the group to carry out a complete and thorough investigation of the deviations. Examples of guide words are: MORE, LESS, NOT, PART OF, BEYOND, EQUAL TO, OPPOSITE OF, ... For each guide word, both causes and consequences of the deviation of the process variable under examination are identified and recorded. Generally, for a given deviation, there can be different consequences depending on the performance of the preventive and protective measures. Once the consequences are identified, the adequacy of the automatic/manual protections already provided in the plant is judged. The inadequacy of such protections, if any, implies the description of the corrective actions that are deemed most suitable.

4.3. Systematic Identification of Release Points

This technique is based on the systematic identification of all conceivable breaks of components and pipes of the plant in a number of "characteristic hole sizes" from which a release of flammable or toxic material can occur. These hole sizes (and related occurrence frequency) are obtained from the statistical analysis of historical failure data, and therefore they represent the mean probable rupture sizes. Examples of characteristic hole sizes for pipelines: full bore rupture (both ends open), 20mm, 10mm, 3mm diameter equivalent hole. In practice the number of such failure cases is large, even for simple plants. For each identified release point the quantity of material released is estimated on the bases of the process conditions, the time delay before isolation, and so on.

To keep the number of failure cases within manageable limits, a screening criterion is applied to neglect those failures which do not present any significant risk at a predefined distance. The screening procedure is based on a release rate threshold q_{min} below which the effects are considered negligible. For example, in the case of a toxic gas release, this value is obtained by assuming a minimal lethal concentration, at a chosen distance d_{min} that would lead to x % fatalities for an assumed exposure time t and atmospheric conditions. Thus, only accidents with release rates greater then or equal to q_{min} (i.e. significant contributors) are considered for risk quantification. A similar method is applied when the accidents are fire and explosions: the minimal quantity of substance on fire or in the explosive range is determined as a threshold level. The HAZID analysis finishes with the list of hazards considered significant for risk quantification. Often, the event tree technique is applied to describe the possible evolution of the releases, leading to different types of accidents in more detail (see example in the next section).

5. FREQUENCY ESTIMATION TECHNIQUES

The selected accident sequences for which the occurrence frequency must be estimated can be either elementary (associated with a single component/pipe failure mode) or complex (associated with a more or less complex system/subsystem) failure mode. In the case of complex events, for which data are not generally available, techniques for system decomposition down to the level of components are applied [10, 11, 12].

5.1. Fault Tree Technique

The fault tree is a deductive analysis tool [13] to describe systematically the possible malfunctions which can cause the plant to reach a particular predefined critical state referred to as "Top event". A top event can be an accident sequence or any plant state considered important for risk quantification. Examples of top events are: explosion in the reactor, release of toxic or flammable substances from a safety valve, spurious shutdown of the plant, unavailability of a safety related system, etc. The procedure for fault tree construction starts with the description of the events which can cause the top event directly. The relationship between the top event and its immediate causes will be a disjunction (AND logical operator) if only one cause is sufficient, it will be a conjunction (OR logical operator) if all causes are necessary. Each cause so identified is then examined in a similar way; the same deductive procedure is applied to develop events up to the level of detail chosen by the analyst. Generally the limit of development of the tree is represented by the modes of failure of the components/ subsystems of known probability.

To give an example of a fault tree, consider the simplified reactor plant shown in Fig. 6. In Fig. 7, the fault tree describing the event "reactor explosion due to runaway" is provided. It can easily be realised that, depending on the system complexity, a fault tree may present hundreds of events. The quantification of a fault tree, to determine the top event occurrence frequency, requires the determination of the minimal cut sets (MCS) i.e. plant failure modes. A cut set is defined as a set of component failures whose occurrence implies the occurrence of the top event. The order of a cut set is determined by the number of events forming it. A cut set is a minimal cut set (MCS) if it contains no other cut sets. For instance, (SV1-F, CS-F) is one of the MCS whose occurrence leads to the reactor explosion. The determination of the MCSs requires the use of computer programmes since, except for extremely simple cases, their manual computation is almost impossible.

Under certain hypotheses it is possible to obtain bounds for:
- unavailability/unreliability of the top event,
- unavailability/unreliability of each MCS,
- importance indexes of each component.

The information obtained with the quantitative analysis allows us to identify the relatively most critical points of the plant, with reference to the top event occurrence; the designer will focus attention on these points to achieve the needed improvements at minimum cost. The analysis of the fault tree of the revised design represents a good way of using the method. In fact, because of different sources of uncertainty, the results of the quantitative analysis should not be strictly considered as absolute, but rather as reference values to compare different design solutions. Naturally, by reducing or controlling the sources of uncertainty, it is possible to obtain reliable results for risk quantification.

5.2. Event Tree Technique

The event tree technique is an inductive analysis tool [1, 5] to determine the states which the plant can reach because of the occurrence of the initiating events identified during the HAZID analysis. For each of the initiating events an event tree is constructed, where each branch represents a combination of states (success/failure) of the safety systems needed to prevent or mitigate the possible consequences of the event. In order to facilitate the use of the method, the construction of an event tree can be subdivided into several stages through which, starting from a very general event tree, one obtains the final event tree corresponding to the level of detail selected. Each physically possible sequence is associated with a description of the physical consequences. The sequences with significant consequences are selected and submitted to a more detailed analysis in order to determine the frequency of occurrence and to identify the possible design changes to be made to make the plant safer. An example of event tree for the simplified reactor system of Fig. 6 is given in Fig. 8.

The event tree technique is also applied to describe possible further developments of the accidents following the loss of containment. For instance (see Fig. 9), the release of a flammable material in the gas phase may be either instantaneous or continuous and may find or not find a source of ignition of sufficient energy to ignite it immediately. If the released gas is ignited, a fireball or a jet fire occur, otherwise a cloud forms (either denser or lighter than air) which can disperse without damage if it does not find an ignition source. Available mitigation actions are also represented in the event tree to complete the description of the possible evolution of the accident.

5.3. Cause Consequence Diagram

The cause consequence diagram (CCD) allows us to describe the possible consequences (critical states) of one or more initiating events and the set of causes of each particular consequence [14]. Therefore this method combines the event tree and fault tree analysis methods. The development of the accident originated by the initiating event(s) is followed in its evolution in time by appropriate graphic symbols, taking into consideration its different consequences (plant states).

The graphic method is based on the use of "decision boxes" with two inputs and two outputs associated with each component, subsystem or system (hereafter generally called "item") of interest (see Fig.10). An input event (A) generates a situation requiring the intervention of the item to which the decision box refers. The item's function is described in the decision box (event description). The possible behaviour of the item generates the output events (B), (C), when the function is satisfied and not satisfied, respectively. To each of these output events is associated the corresponding occurrence probability. Often, to obtain the probability of event (C) it is necessary to construct the corresponding fault tree. Naturally, the probability of the event (B) is the complement to 1. The analysis of a plant by the cause consequence diagram starts with the definition of the initiating event and proceeds with the description of the propagation of the perturbation up to the final accident. The procedure for the construction of the diagram is similar to that seen for an event tree, but, thanks to a richer formalism, it is possible to describe more complex situations in which for example time plays an important role. An example of a cause consequence diagram is shown in Fig. 11 for the simplified reactor system of Fig. 6.

The quantification is carried out by working out, by appropriate algorithms, the fault tree for each significant consequence (top event) and analyzing this tree to determine the minimal cut sets, the accident frequency and the component criticality factors. Therefore, all the considerations made for the fault tree are applicable to cause-consequence diagrams.

5.4. Generic Reliability Data Bases

In order to perform the quantitative analysis of Fault trees and Event trees, the failure parameters of each component must be available. These parameters can be obtained from generic reliability data bases, which are continuously updated to increase the confidence in the estimated data.

6. MODELS FOR CALCULATING THE PHYSICAL EFFECTS OF ACCIDENTS

The second part of phase 2 is the calculation of the physical effects of the accidents identified in the HAZID study and for which the occurrence frequency estimation has been performed. Many models are available to calculate the effects of an accident [1, 15]. Integrated computer packages, also running on personal computers, have been developed by many research institutes and private companies.

Models cover the following phenomena:
- discharge rate,
- evaporation,
- fire,
- explosion, and
- release.

These phenomena may not be independent, e.g a fire may produce, as combustion product, a toxic gas; a run away reaction may produce an explosion followed by a fire. A simple classification of the models to study these phenomena is represented in Fig. 12.

6.1. Discharge Rate

Calculation of the effects of an accident starts with the estimation of the source term (i.e. quantity and physical conditions of the released material). Models for calculating the discharge rate for different flows are available:
- liquid phase flow,
- gas phase flow, and
- two phase flow.

Liquid and gas outflow phenomena have been thoroughly investigated theoretically. The two phase discharge occurs, for example, when pressurised vessels fail, where the liquid outflow partially flashes. These types of releases are more difficult to model than the other two, which, in any case, can be considered as limiting situations.

6.2. Pool Evaporation

The rupture of a vessel containing a liquid substance leads to the formation of a spreading pool which then evaporates more or less rapidly. Depending on the type of material and the physical conditions of the storage, part of the outflow may flash and the remainder forms the pool. The dimensions of the pool depend on the quantity of the escaped material, the shape of the ground, the existence of secondary containments (bounds), etc. Models for estimating the evaporation from a pool on land and water are available [16]. The results of these models are needed to estimate the dispersion of the cloud formed.

6.3. Fire Models

The models available for fire can be classified on the basis of the properties of the substances involved, the release conditions and the time of ignition after release [17, 18]. The types of fire considered in risk analysis are as follows:
- jet fire,
- pool/tank fire,
- flash fire, and
- fireball.

A jet fire occurs when a flammable substance escapes from a puncture in a pressurised vessel or pipe and ignites immediately. A pool/tank fire occurs when the vapours of the flammable material, stored in a tank or spread on the ground, due to the loss of containment, are ignited. A flash fire is caused by the ignition of a vapour cloud and the propagation of the flame occurs at low velocity, i.e. without exploding. When a pressurised vessel is overheated by an external fire an explosion occurs and the contents of the vessel rapidly vaporise forming a cloud which is ignited and burns rapidly.

The models available generally determine:
- the geometry and dimensions of the fire,
- the heat radiation intensity at different distances,
- the flame temperature, and
- the heat transferred to nearby objects.

Knowledge of the heat radiation at different points from the centre of the fire is important for:
- estimating the damage to people,
- estimating the damage to structures with possible domino effect,
- estimating the minimum distance at which the operators are not affected
  and, hence, the type of protective clothing needed,
- identifying, at the design stage, the safe location of particularly critical components
  (tanks, vessels, control room, etc.),

- optimum dimensioning of the mitigation systems.

6.4. Explosion Models

An explosion is a rapid exothermic reaction (i.e rapid decomposition of the substance) with a sudden release of energy: the violence of the explosion depends on the rate of energy released. An explosion can be classified as a detonation or a deflagration depending on the flame front velocity. In a deflagration the flame front velocity is relatively low, whereas in a detonation the velocity is faster than the speed of sound. The shock wave which accompanies a detonation may cause significant damage. Explosions can also be classified as confined (occurring in a closed or semi-closed environment) or unconfined (open air) [19].

Types of explosions of interest in risk analysis are as follows:
- Unconfined Vapour Cloud Explosion (UVCE),
- Boiling Liquid Expanding Vapour Explosion (BLEVE),
- Physical Explosion,
- Runaway Reaction Explosion,
- Dust and Gas-Dust Mixture Explosion.

The UVCE occurs when a cloud of flammable material is formed and ignited when its dilution in air is within the explosive range. Physical explosions are typically those which occur when a pressurised vessel fails catastrophically with the release of high energy (i.e. a vessel containing pressurised vapour). Runaway reactions are chemical explosions caused by the rapid decomposition of the materials. Finally, dust and gas/dust explosions may occur; the ignition energy decreases with decreasing dimensions of the dust particles. Models for explosions give results concerning the blast wave intensity and the trajectory of fragments (missiles) which may lead to the rupture of other parts of the plant with consequent domino effects. These results allow us to estimate the probable damage caused by the explosion and to verify the adequacy of the existing protection devices.

6.5. Cloud Dispersion Models

The modelling of the dispersion of a cloud of a flammable/toxic substance, firstly depends on the density of the cloud, which can be either heavier or lighter than air. This density changes in time due to wind speed, the stability of the air (e.g. Pasquill stability classification system) and the atmospheric turbulence which affects the entrainment of the air.

Dispersion models can be classified according to the duration of the release (either instantaneous or continuous) and the phase of the material released (gas/vapour, liquid, two phases) [20, 21]. As mentioned above, the determination of the phase and the emission rate, i.e. the source term, is one of the most important factors affecting the modelling.

Lighter-than-air releases are generally well modelled by means of the Gaussian model, whereas the process of dispersion of heavier-than-air clouds is more complicated. It is generally the denser-than-air gases that are of concern in risk analysis, since the clouds present concentrations higher than lighter-than-air gases which disperse more easily.

During the first period of the release of a heavier-than-air gas, the dispersion is dominated by the source properties and the initial acceleration of the gas, followed by a rapid transition to regimes in which the dispersion is characterized by the internal buoyancy of the cloud. Thus, from a certain time on the plume should be considered as lighter-than-air and studied by means of a Gaussian model.

A vast literature on gas dispersion modelling is available. For toxic gases these models give the concentration of the cloud (e.g. at ground level) in time; some models determine the concentration at the central line of the plume, others also give the concentration at the borders of the plume. For flammable gases they give the zone in which the concentration is within the explosive range.

7. VULNERABILITY MODELS

In order to determine the consequences to people and structures of major accidents, it is necessary to translate the results of the models mentioned above into the real damage caused. This requires the use of vulnerability models or damage criteria [22, 23]. It should be noticed that the damage considered in risk analysis is of lethal type. Many criteria have been defined concerning the damage to people and structures of explosions (blast wave intensity), fire (heat radiation intensity) and release (toxic concentration).

A technique frequently applied is based on the Probit model. A probit equation relates the dose (intensity effect and duration) to the probable damage caused, i.e, it gives the probability that a human being may suffer a certain damage for a given dose level. Probit can also be applied to structural damages. In the case of a fire, the effect depends on the level and duration of the thermal radiation; in the case of an explosion it depends on the blast wave intensity; in the case of dispersion of a toxic substance the effect depends on the concentration and the time of exposure, also taking into account any evacuation.

8. MAIN SOURCES OF UNCERTAINTY IN RISK ASSESSMENT

Risk analysis is a complex study involving the treatment of a large amount of information by a team of experts in chemical processes, maintenance, systems reliability and consequence modelling. Uncertainties are inevitably introduced in all phases of the analysis. Apart from the fact that it is always more important to prevent than to quantify, an essential feature of a quantitative risk analysis is maximum clarity. The main sources of uncertainty are listed here.

1. Incomplete plant knowledge, which has a considerable effect on the hazard identification phase.
2. Engineering judgement, needed to overcome the problem of missing data and imperfect knowledge of accident evolution, dose-effect relationships, etc. This is an important source of uncertainty, which calls for a multidisciplinary team of experts.
3. Model uncertainty, i.e, inappropriate modelling, inaccurate model parameters, inadequate validation and model limitations which necessarily requires simplifications.
4. Data uncertainty, i.e, source term, reliability parameters, time to operator action, atmospheric data (e.g. wind direction, stability classes) etc.

9. THE ENVIRONMENTAL IMPACT ASSESSMENT
OF INDUSTRIAL INSTALLATIONS

The main concern of a risk analysis, performed to fulfil the requirements of Directive 82/501/EEC, is the identification and estimation of the effects of "major hazards", i.e. of those events which may have catastrophic consequences. The risk is therefore estimated in terms of catastrophic effects on people and structures.The improvement of the preventive and protective measures is the main action undertaken to increase safety. As described in section 2, a risk analysis can be performed in different ways and the results represented in different forms. For instance, the hazard identification can be performed by means of one or more of the techniques mentioned in section 3, the extension of the use of the fault tree technique for frequency quantification can be different, the number of hazards retained for consequence analysis can be different, and so on. It is undoubtedly true however, that, independent of these aspects, the risk analysis represents an important means of contributing to the improvement of the safety level of the installation, provided that the clarity of the whole analysis is guaranteed.

Now, if one must consider the Environment Impact Assessment of a potentially hazardous installation, the risk analysis becomes (an important) part of the study. Generally, the risk analysis is limited to major hazards. However, in environmental impact studies the risk analysis should consider minor accidents as well, i.e. those with high frequency and low consequences, since the effects near the plant may be of great importance from the point of view of the plant acceptability (e.g. think of those unpleasant smelling substances that cause disturbance to the population in the surrounding area). Other aspects to be considered are those associated with plant operation, i.e. waste treatment, transport activity, noise produced, etc.

High-frequency-low-consequence events can be analysed by means of the models currently applied for major accidents. The hazard can, however, simply be identified by means of the technique of systematic ruptures of components and pipes. In fact, as these events occur frequently, generic reliability data bases can give the data (occurrence frequency and dimensions of the rupture) for source term definition.

It should be stressed that in risk analysis the impact to the environment is limited to the very first consequences, i.e. those which are related to the effects in the very short time after the occurrence of the accident. Therefore, the scope of the analysis should be widened to consider the chronic effects of these accidents, i.e. the long term damage. These problems are obviously outside the field of application of risk models, since in PRA the behaviour of the ecosystem is not modelled at all.

Finally, only some PRA models can be applied, i.e. those for calculating the dispersion of pollutants in air and water to estimate the impact of the releases during normal operation. A simple comparison of some properties of the PRA and the analysis of impacts in normal operation is given in Tab. 2 from different points of view. The first part of the table refers to general topics, while the second part outlines the difficulty of carrying out the different phases of the analysis. The table highlights the fact that the inclusion of risk analysis in environmental impact studies is not as straightforward as it may appear, since other techniques must be applied to model the vulnerability of the ecosystem.

10. CONCLUSIONS

An important role in environmental impact studies of industrial installations is played by risk analysis. PRA, however, cannot be limited to the so-called major accidents, but it must extend to the minor ones, i.e. the high-frequency-low-consequence releases. Because such events are frequent, computerised tools are needed. The paper presented a review of the main risk analysis concepts and procedures, as applied to fulfil the requirements of the Directive 82/501/EEC. The main differences between the features of the risk due to major accidents and the risk from normal operation have also been sketched. The complexity of the study requires experts from different scientific disciplines, as well as the use of computerised tools to support the analysis and the different steps of the plant acceptability process.

REFERENCES

[1] AIChE (1989), Guidelines for Chemical Process Quantitative Risk Analysis, New York, USA.
[2] Dow Chemical (1986), Fire and Explosion Index. Hazard Classification Guide, 5th edition.
[3] ICI, The Mond Fire and Toxicity Index.
[4] A. Romano et al. (1992), "Industrial Plant Risk Indices", this book.

[5] Pickard, Lowe and Garrick (1981), Methodology for Probabilistic Risk Assessment of Nuclear Power Plants, Washington, D.C.

[6] AIChE (1985), "Guidelines for Hazard Evaluation Procedures", New York, USA.

[7] MI-STD-1629A (1977), Procedure for Performing a Failure Mode and Effect Analysis.

[8] C.T. Cowie (1976), "Hazard and Operability Studies. A New Safety Technique for Chemical Plants", Prevention of Occupational Risks, Vol. 3.

[9] H.G. Lawley (1974), "Operability Studies and Hazard Analysis", Loss Prevention, Vol. 8.

[10] A. Amendola, A. Saiz de Bustamante (Editors) (1988), Reliability Engineering, Proceedings of the Ispra Course held at the Escuela Superior de Ingenieros Navales, Madrid, Spain, Kluwer Academic Publ.

[11] A.G. Colombo, A. Saiz de Bustamante (Editors) (1990), System Reliability Assessment, Proceedings of the Ispra Course held at the Escuela Superior de Ingenieros Navales, Madrid, Spain, Kluwer Academic Publ.

[12] TNO (1985), Methods for Determining and Processing Probabilities, (Red Book).

[13] W.E. Vesely et al. (1981), Fault Tree Handbook, NUREG-0492.

[14] D. Nielsen (1975), "Use of Cause Consequence Charts in Practical Systems Analysis", in Reliability and Fault Tree Analysis, SIAM.

[15] TNO (1979), Methods for the Calculation of the Physical Effects of the Escape of Dangerous Materials, (Yellow Book).

[16] P. Shaw, F. Briscoe (1978), Evaporation from Spills of Hazardous Liquids on Land and Water, Report SRD R 100, UKAEA, Warrington, UK.

[17] M. Considine (1984), Thermal Radiation Hazard Ranges from Large Hydrocarbon Pool Fires, Report SRD R 297, UKAEA, Warrington, UK.

[18] P.A. Croce, K.S. Mudan (1986), "Calculating Impacts for Large Open Hydrocarbon Fires", Fire Safety Journal, pp. 99-112.

[19] D.K. Pritchard (1989), "A Review of Methods for Predicting Blast Damage from Vapour Cloud Explosions", Journal of Loss Prev. in Process Ind., Vol 2, pp. 187-193.

[20] AIChE (1987), Guidelines for Use of Vapour Cloud Dispersion Models.

[21] S.R. Hanna, D.G. Strimaitis, J.C. Chang (1991), Hazard Response Modelling Uncertainty (A quantitative Method), Vol. II: Evaluation of Commonly-Used Hazardous Gas Dispersion Models, Sigma Research Corporation, Westford, MA 01886.

[22] Dutch Ministry of Housing and Environment (1989), Handbook for Consequence Modelling, (Green Book).

[23] R.F. Griffiths (1991), "The Use of Probit Expression in the Assessment of Acute Population Impact of Toxic Releases", J. Loss Prev. Process Ind., Vol 4.

[24] C.M. Pietersen (1990), "Consequences of Accidental Releases of Hazardous Material", J. Loss Prev. Process Ind., Vol 3, pp. 136-141.

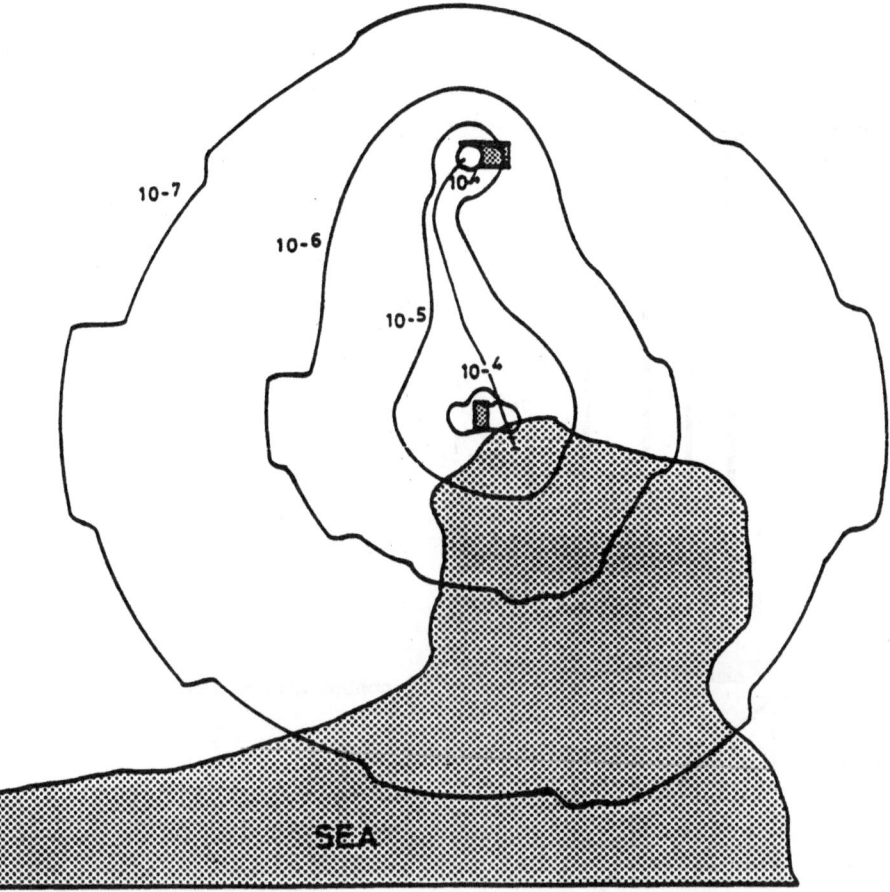

Figure 1. Example of risk contours for a complex major hazard site.

94

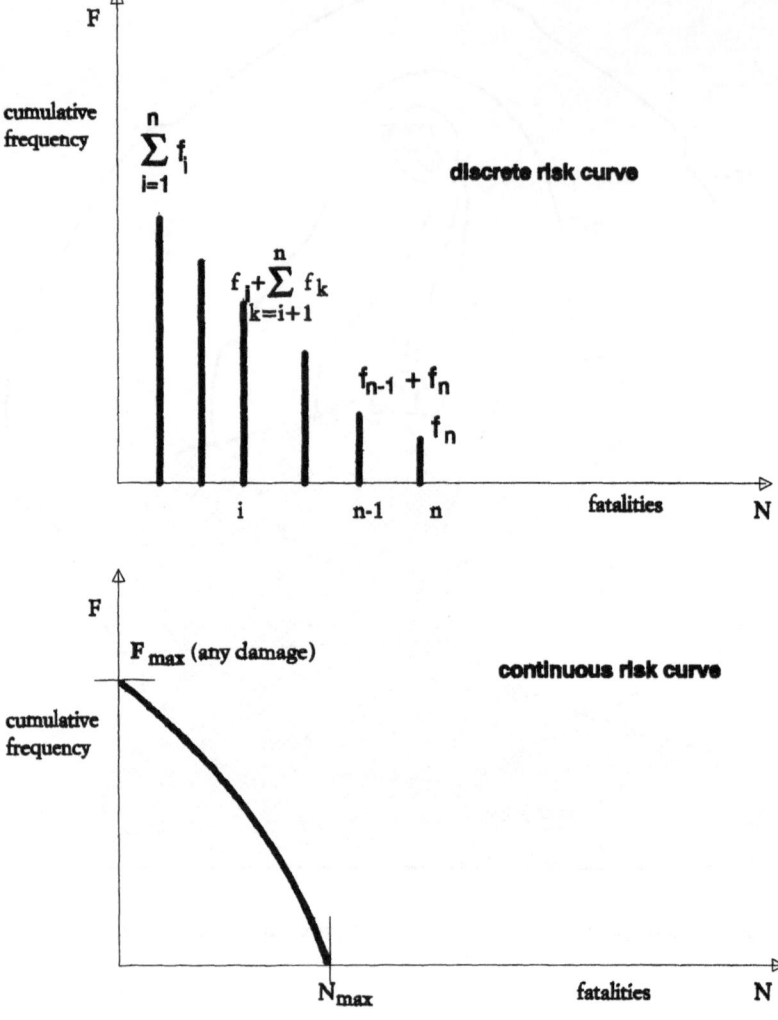

Figure 2. Typical discrete and continuous societal risk curves.

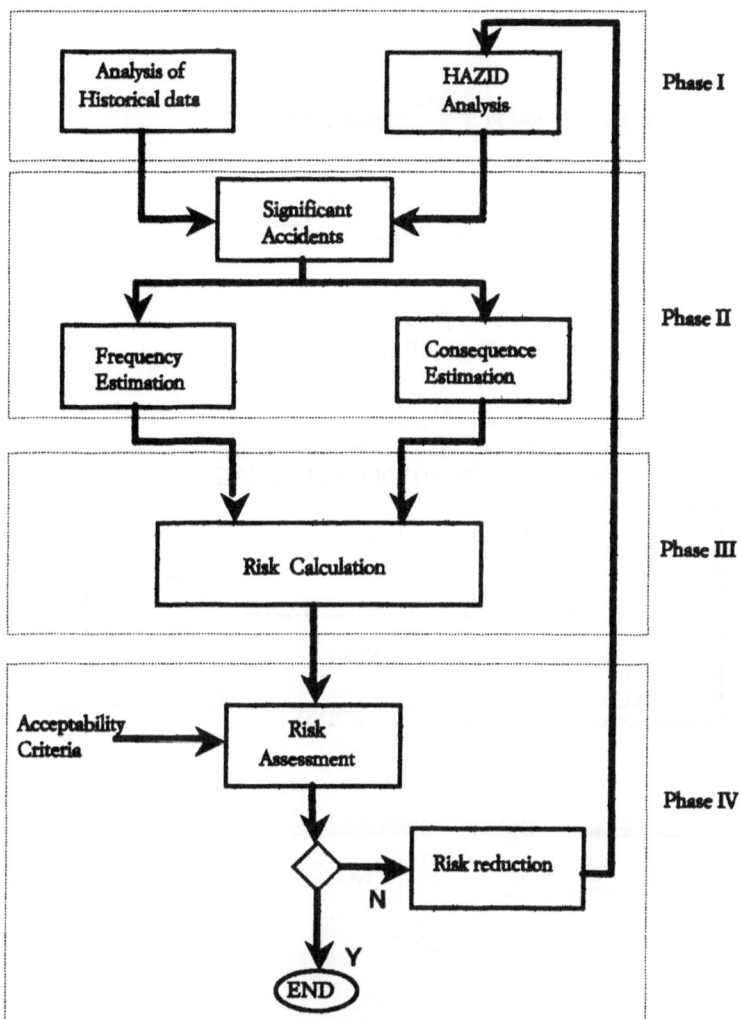

Figure 3. General quantitative risk analysis procedure.

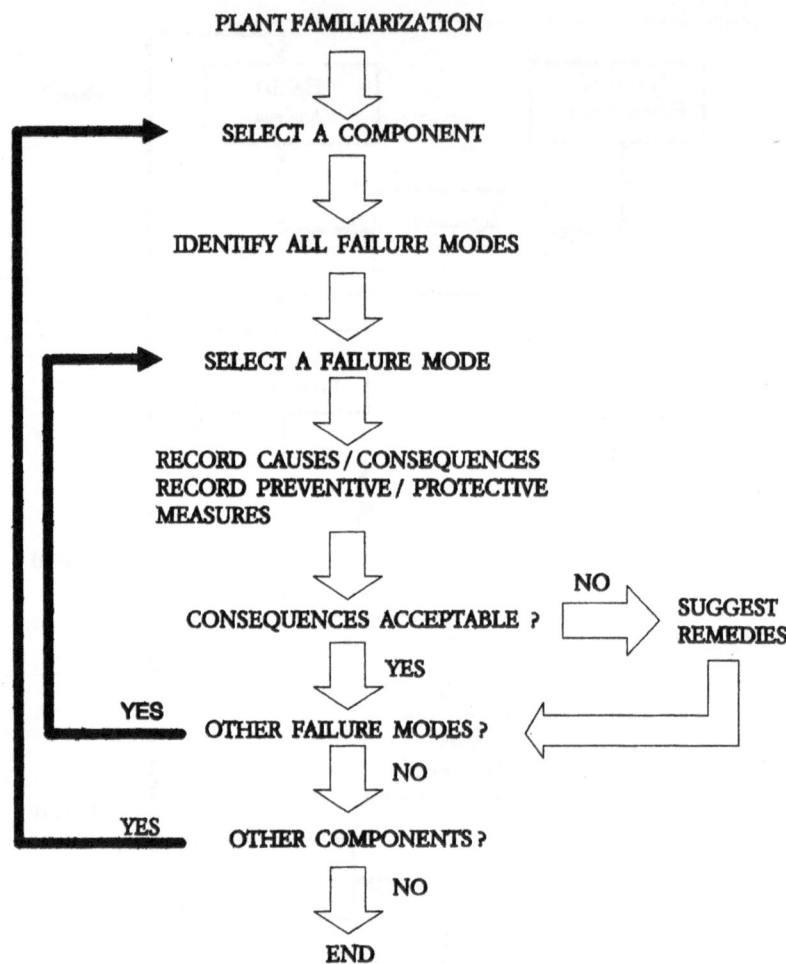

Figure 4. Main steps of the procedure for performing the FMEA.

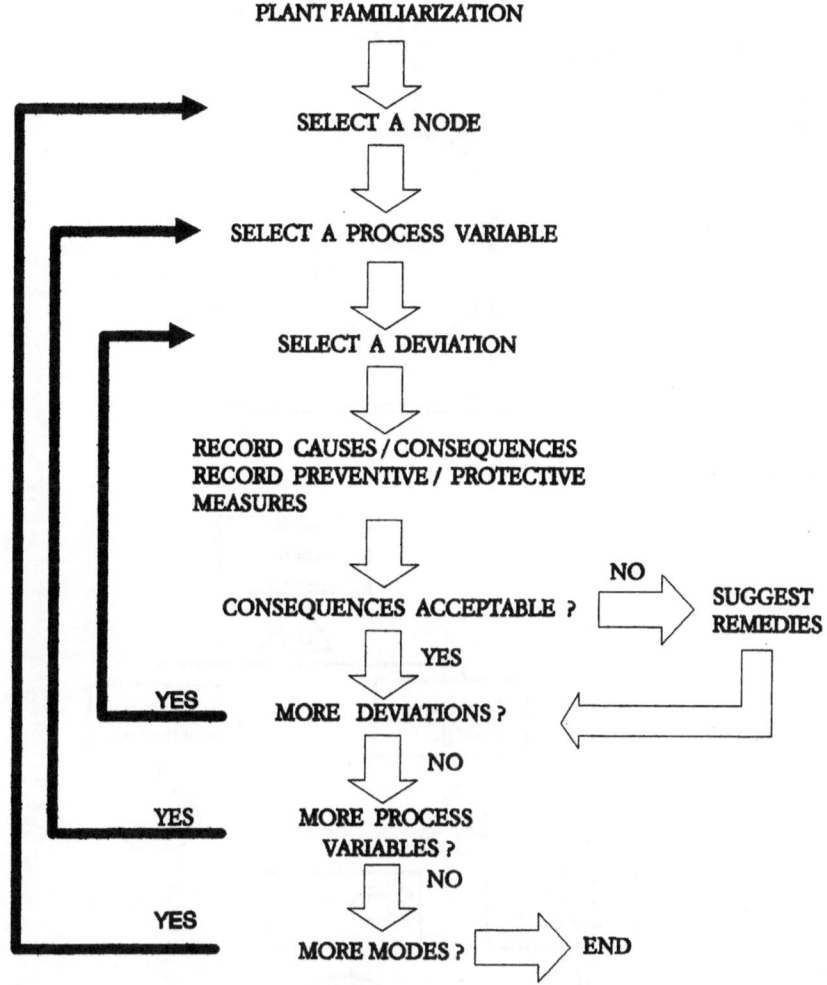

Figure 5. Main steps of the procedure for performing the HAZOP.

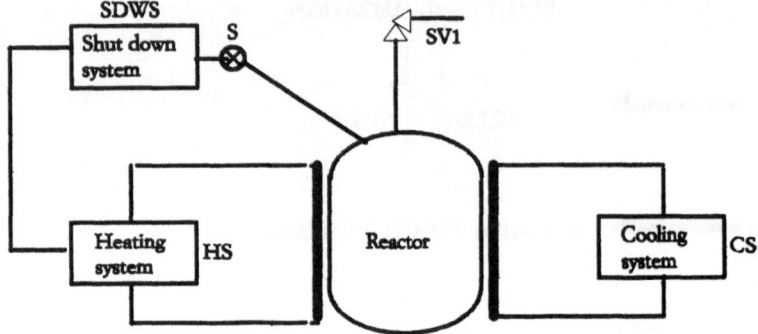

Figure 6. Simplified chemical reactor plant.

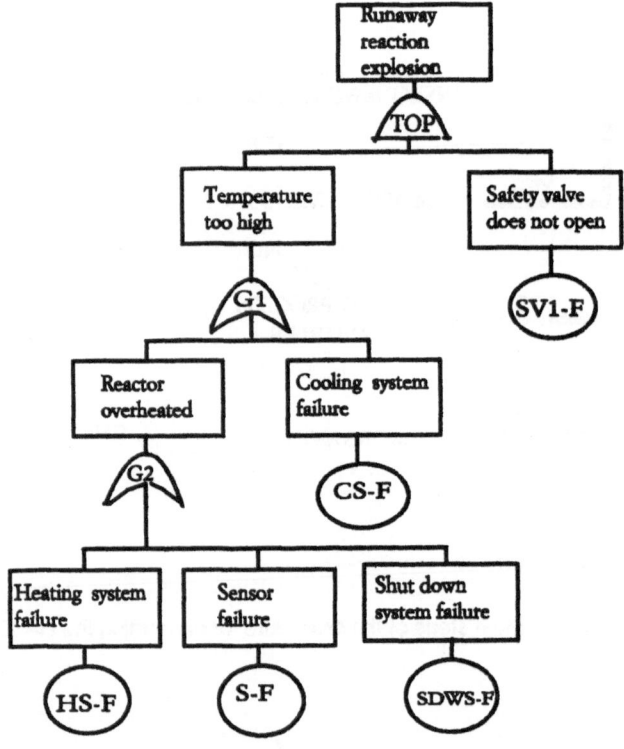

Figure 7. Fault tree for the top event "reactor explosion due to runaway",
(reactor plant of Fig. 6).

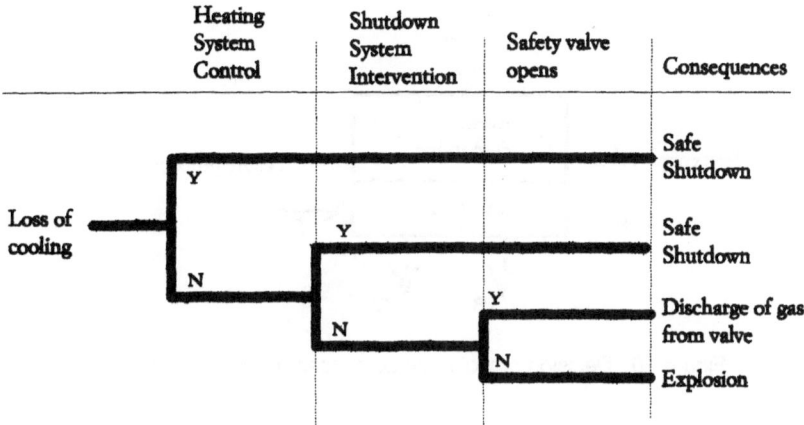

Figure 8. Event tree describing the consequences of the initiating event "loss of cooling", (reactor plant of Fig. 6).

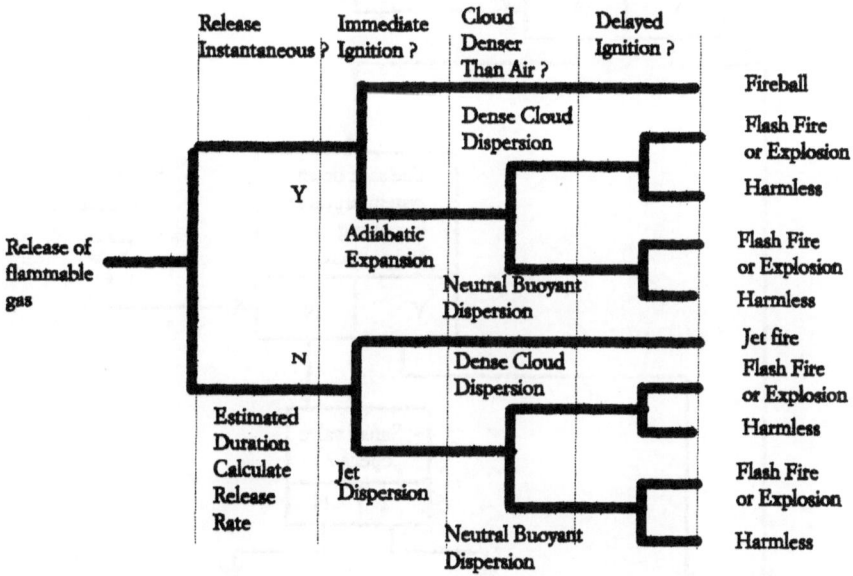

Figure 9. Event tree for the initiating event "release of flammable gases".

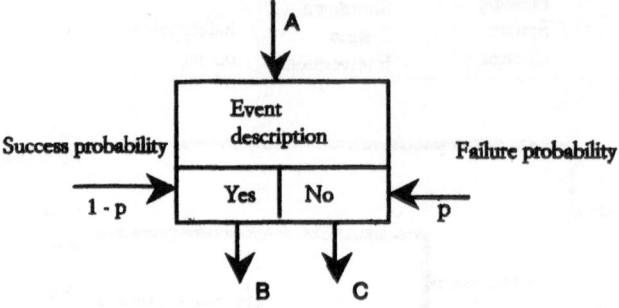

Figure 10. Decision box of a cause consequence diagram.

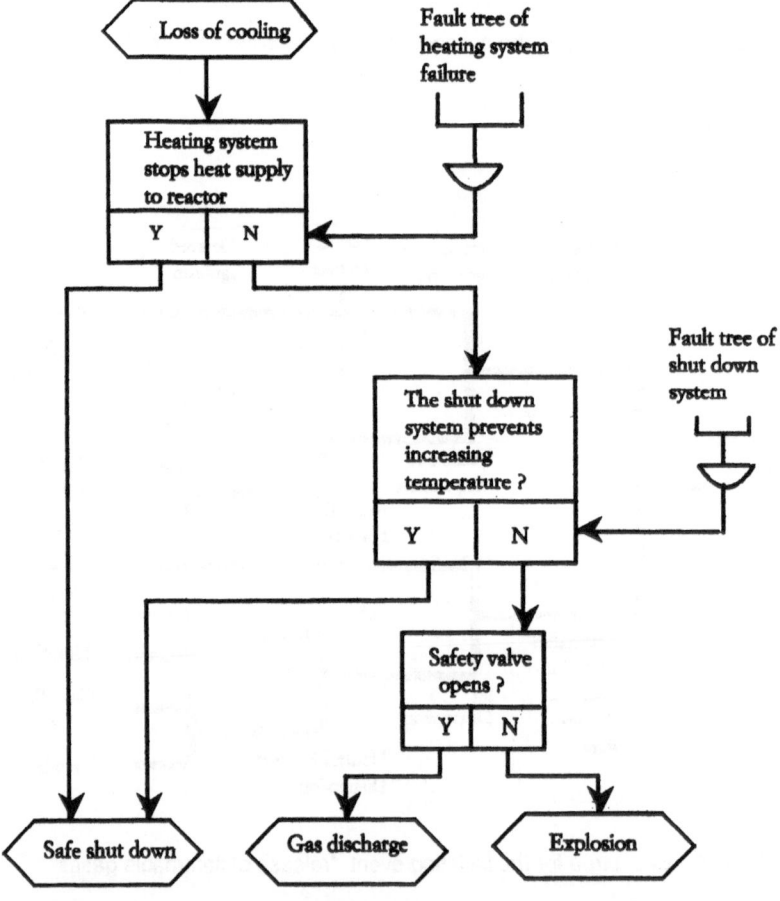

Figure 11. Cause consequence diagram for the event "loss of cooling", (reactor plant of Fig. 6).

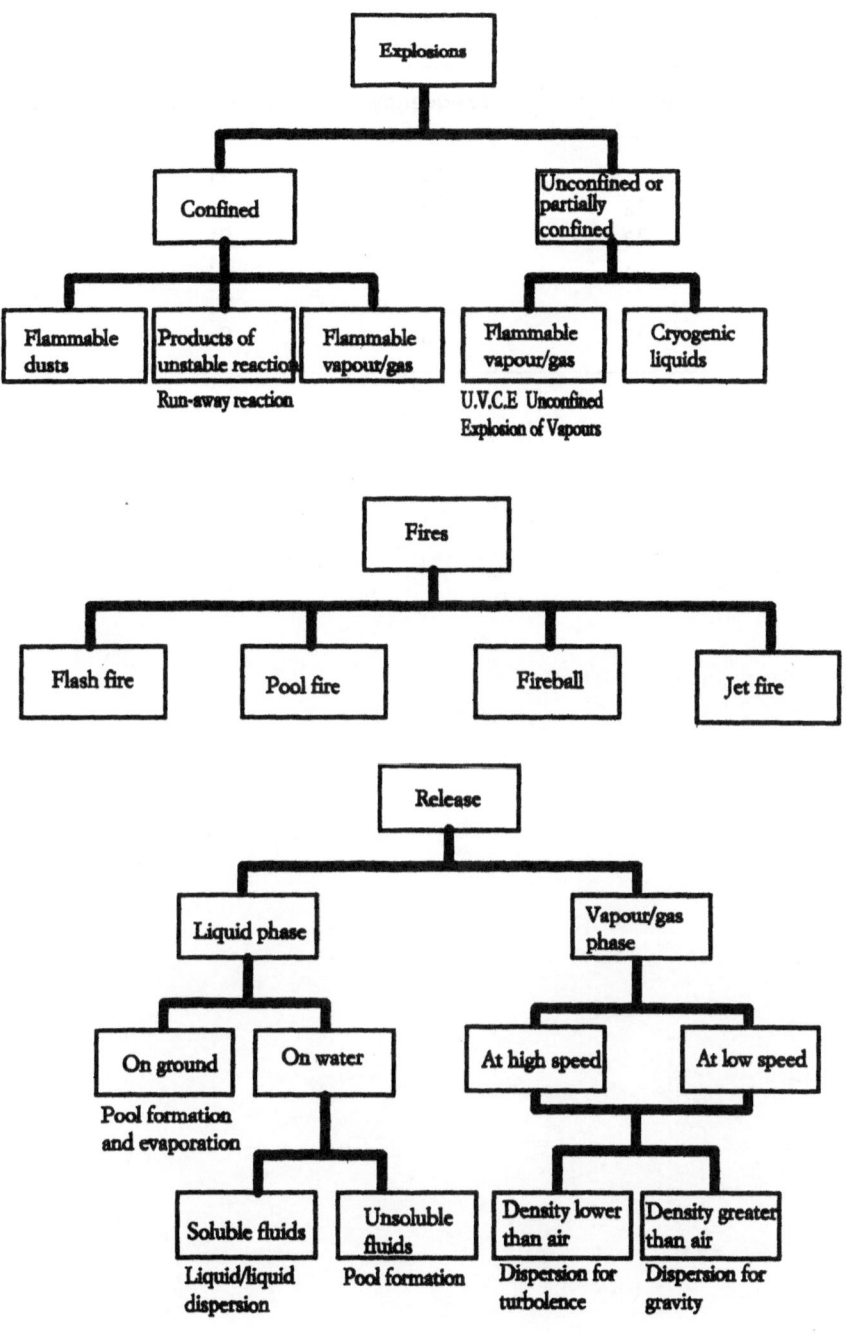

Figure 12. Preliminary classification of accidents.

Table 1. Risk definition as a set of triplets.

accident	frequency	consequence
a_1	f_1	c_1
a_2	f_2	c_2
a_3	f_3	c_3
.	.	.
.	.	.
.	.	.
a_n	f_n	c_n

Table 2. Main features of environmental impact analysis due to accidents and to normal operation.

Item	Analysis of impact due to accidents (non routine releases)	Analysis of impact in normal operation (routine releases)
Analysis focused on ...	Abnormal occurrences	Normal operation
Time span of phenomena	Very short	Very long
Types of impact considered	Sudden/violent; Spot effects	Continuous; Cumulative effects
Types of release	Instantaneous or continuous for short periods of time	Continuous for long periods of time
Damage to ...	People and structures	People; Ecosystem
Nature of damage	Acute toxicity; Catastrophic	Chronic toxicity
Exposure time	Short/very short	Long/very long
Vulnerability models	Available to some extent	Difficult to define
Hazard identification	Completeness cannot be warranted	Easier to carry out
Frequency estimation	Subject to uncertainty	Less uncertain
Source term definition	Subject to uncertainty	Less uncertain
Calculation of physical effects	Fire; Explosion; Release in air and surface water	Release in air, surface water and groundwater; Noise pollution; Thermal pollution; Others

ENVIRONMENTAL IMPACT ASSESSMENT AND RISK ANALYSIS IN DENMARK

F. Bro-Rasmussen
Laboratory of Environmental Science and Ecology
Technical University of Denmark
DK - 2800 Lyngby
Denmark

ABSTRACT. Directives on Environmental Impact Assessment and Risk Analysis have only in recent years been adopted by the Danish Parliament as part of the process of European harmonization. A certain experience, however, has developed over the last decade on the basis of already existing legislation which is illustrated by two practical examples. In one case - a chemical pesticide plant - an ecotoxicological characterization of waste water was developed on the basis of extensive experimental studies on marine fish species and invertebrates. Indicator-organisms were chosen and pollution effect/no-effects were determined on the basis of acute and long-term tests on key species representative of the locality. Legally, requirements imposed on the factory concerning waste water treatment and good manufacturing practice now refer to a risk evaluation accepting zero or only insignificant changes in a dilution zone around the waste water discharge point. The area of the dilution zone is also defined as a result of the risk analysis. In the other case - the major bridge-tunnel constructions across the Danish straits - environmental impact assessments have only gradually been developed, and they are still presented at levels which gives rise to much public concern. Biological indicators are inadequately included in the assessments, and the results presented are uncertain because of the general situation of eutrophication and deterioration of environmental quality of the Danish parts of the Baltic Sea region. An undefined term of "insignificant physical changes" of the benthic zones around the planned constructions, or hydrological parameters, such as the official requirements of "unchanged exchange rate of the water flow through the straits", so-called zero solutions, are, therefore, introduced as the over-all indicators for acceptable impacts, and these terms are therefore replacing the environmentally more appropriate biological and chemical indicators.

1. INTRODUCTION

As far as legislative measures and environmental regulations are concerned, **Environmental Impact Assessment (EIA)** and **Risk Analysis (RA)** are both relatively new concepts.
Specific regulations on these two subjects have only recently been adopted by the Danish Parliament, namely in June 1989 and September 1988, respectively. They are both to be considered Danish adaptations to European Community directives, namely directive 85/337/EEC and 82/501/EEC (amended in 1987), respectively. They thus constitute parts of the general process of harmonisation within the European Communities.

EIA regulation. The first of these two directives, i.e. the EIA directive 85/337/EEC, is in its Danish version called the "VVM-regulation" (No. 446 of 23 June 1989, [1]). In essence, it consists of one paragraph only. In this version it reads, that:

A. G. Colombo (ed.), Environmental Impact Assessment, 105–119.

"any physical planning which includes certain (public and private)
projects shall be presented to the authorities and be followed by
statements concerning the environmental impacts of the projects".

One Annex to the regulation lists which kinds and sizes of projects are to be covered by the regulation, viz. refineries, power plants, airports etc., and another Annex shortly summarises and states, in accordance with the Community directive, that technical information on the installations and information on the impact on the different environmental compartments is required for each project.

Of course, the justification of such a short, Danish version consisting of only 3-4 lines plus a scarcity of explanatory notes, as compared to the 9 pages of the original EEC directive, must be found in the fact that the EIA procedures through this wording are tied to and covered by an already existing piece of Danish legislation, of which the following provisions are the most important:

1) Danish **Law on Physical Planning** through which environmental impact studies
   for the specified projects are given permanent status within the existing Danish
   system of environmental planning, including the important step of public hearings;
2) the EIA is further tied to the general **Environment Protection Act** and its
   requirements on study, assessment and precautionary measures imposed
   on any Danish production plant, installation or project which give - or may
   give rise - to environmental pollution; and, finally,
3) the EIA is also subordinated - or rather coordinated - to requirements and
   criteria which are governed by the existing **Law on Nature Protection
   and Conservation** for the protection of Danish biotopes and natural ecosystems.

RA regulation. The second directive, also known as the SEVESO-directive (82/501/EEC, amended in 1987) is more elaborate, as is also the case for the Danish risk regulation version (No. 545 of 16 September 1988, [2]). This is a regulation in its own right, concerned with the evaluation of industrial installations and measures to be taken in order to ensure safety in connection with "high risk" activities, chemicals and accidents.

Obviously, this risk regulation refers primarily to hazards imposed on persons outside the installations and to the potential consequences resulting from human activities. Thus, the regulation deals with the terms: Risk, Acceptance of Risk and Safety, as these are expressed and interpreted mostly in their human and anthropogenic relations.

It is important, however, that the environmental dimension must be included in the evaluation and in the precautionary schemes of the regulation. Discussion of the risk regulation is therefore fully justified even when reference is made only to the environment as the object of concern, i.e. by dealing with the concepts of risk and acceptance of risk for the environment - or rather: on behalf of the environment - only.

Defining EIA and RA. In comparing the two sets of regulations, it is noteworthy that they have some characteristics in common. Neither the regulations "per se", nor their attached Annexes (or following guidelines) are found to contain any direct explanation or interpretation concerned with their basic terminology: environmental risks or environmental impact. The two terms can be dealt with as parallel concepts:

- which differ physically in targets (existing **or** future installations) and administratively
  in the decision and management processes (e.g. industrial **or** public interests), but
- which **both call for** quantified assessments of possible effects or environmental
  damages **against** potential exposure or stress, and
- which **must both be** assessed in a decision process that involves acceptance
  of certain risks (or impacts) which are often difficult to define or only of a
  qualitative nature.

It is necessary, therefore, in both cases to turn to rules and measures which spring from the existing legislation and regulatory practices, e.g. through a description of methods, by the use of indicators and indices, and also case-by-case decisions, as these have developed over the preceding years of environmental administration and enforcement.

The most important background for this is:
- the Danish **Environment Protection Law** under which the practice of setting emission standards and acceptable exposure limits has been developed, and
- the Danish **Aquatic Environment Quality Planning** which is administered under the responsibility of regional County Councils, and which specifies certain media-related quality measures and objectives for toxic chemicals, and other pollutants, including of course nutrients.

In so far as the Danish regulations on EIA and RA are both relatively new, they have been applied directly only on a very few occasions. However, on the background given here, some experience has been gained, and there are even certain practices which are relevant for the interpretation of the two concepts, as it will be illustrated and exemplified by the following two cases.

2. ENVIRONMENTAL IMPACTS AND RISKS FROM A CHEMICAL PLANT (CASE 1)

The first of these examples deals with the reduction and possible elimination of the environmental stress from toxic chemicals which are discharged with the waste from a chemical production (pesticide) plant into the coastal marine ecosystem. The chemicals being discharged are all related to the generic group of thiophosphate insecticides, either in the form of unchanged, active insecticidal matter, or (in considerably greater amounts) as chemically related by-products or residuals which are lost or released as waste material from the production line.

More than one hundred individual chemical entities were identified and found to be simultaneously present in the waste water. Of these, a high proportion is characterised as esters (mono-, di-, and tri-esters) of thio- or dithiophosphoric acid. Considerable amounts of unchanged chloro- and nitro-phenols, polysulphides and organic solvents are continuously being discharged with the effluent as well.

More detailed information on the chemical characterisation and the ecotoxicological profile of the effluent from the factory has been presented elsewhere [3, 4]. This also includes thorough descriptions of ecological disruptions and environmental damage caused by discharging the waste water into the coastal, marine area during the preceding 2 - 3 decades, namely until restrictions and abatement were ordered by the regional authorities during the years 1985 - 89.

The basis for the official intervention was found in Guidelines published by the National Agency for Environment Protection, the so-called **Danish Quality Planning for Receiving Marine Waters** (1985), in which it is stated as a governing, over-all principle that "flora and fauna shall remain unaffected or **only slightly** affected by human activities".

The word "slightly" is important. From this it is inferred that some (unintended) effects may inevitably be the result of human utilization of the marine environment as a recipient for waste material. This will necessitate a political decision which **either** defines the level of acceptance **and** permits a "certain modification" of the quality level for the specific receiving water, or which calls for an environmental impact assessment and risk analysis in order to ensure, and enforce, that the possible effects are in fact only "slight".

According to the Guidelines any effects shall remain "environmentally insignificant". In the present case this was required by stating that:
- the waste water even under the worst dilution conditions at the point of discharge **should not cause any acute toxic effects** in the receiving water, and

- toxic hazards (defined as the potential for chronic effects under long-term exposure at normal discharge rates and variability) **should not exceed the level of sublethal, long-term effects** when measured in the receiving water "outside the defined dilution zone".

The urgency of such requirements had already been indicated from the results of 8-10 years of preceding, extensive ecotoxicological testing of effluents and individual chemicals [1], **and** from the results of surveys and monitoring programmes which had been initiated in the late 1970s. Through these activities it had been shown that:

- the untreated waste water "per se", was very toxic, and, accordingly,
- acute toxic effects could be expected on certain marine organisms from untreated discharges even at some distance from the point of discharge;
- 21 individual compounds found in the effluent (out of a total of 105 identified) could be classified as "toxic" or "very toxic" to marine organisms;
- 19 out of these 21 chemicals were selected as "environmentally problematic" on the basis of supplementary criteria, such as persistence and potential for bioaccumulation.

Because the "problematic chemicals" are also likely to be selected as **indicator-chemicals in future control programmes**, they were further tested for ecotoxicological long term effects on a series of individual organisms. As for the test organisms, these were in several instances similarly selected as **"key-organisms"** due to their sensitivity, but also because they generally should be representative for the local ecosystem (cfr. figure 1).

These extended tests were performed on the waste water directly from the plant, or in a series of increasing dilutions, **and** eventually also after passage through a biological pilot plant for waste water treatment.

As a result of the biological pilot treatment, the treated effluent was found to be of moderate toxicity only, with a very limited potential for causing effects in the immediate vicinity of the outlet point. It was predicted from these studies that the impact in the dilution zone of properly treated waste water "will be limited to an initial inhibition of the most sensitive species".

These results are illustrated in figures 2 and 3, of which the former shows the proportion of species that are effected in the receiving water close to the discharge point, e.g. about 60% at 200 metres. Some species may still be affected even at 2000 metres, as is the case for 20% of species at an effluent concentration (or: dilution) level of ca. 0.3 ml l^{-1}. In the latter case, the concentration is seen still to be 5-10 times higher than the estimated No-effect level.

Figure 3 shows the results of various biological effluent treatments (cfr. curves 1, 2 and 3). The reduction in the waste water's toxicity will depend on the efficiency of the treatment, and accordingly the risks of environmental effects in the receiving water can be minimised to a degree which depends on the requirements that are legally specified for the waste water treatment before discharge into the receiving water. This is illustrated by the different distances between the operation curves for predictable exposures on the one side, and effects or achieved toxicity reduction on the other.

In figure 3 it is further indicated that a final risk analysis may require the setting of a Safety Factor (SF) in order to ensure that "effects in the dilution zone are insignificant". In the present case this factor has been suggested - and eventually also accepted by all parties - to equal 10 times. In choosing this factor, consideration was given to:

- the quality and relevance of experimental, ecotoxicological tests (accounted for by a factor 2 x), and
- a certain regard for sensitive species and life stages of test organisms (by a factor 5 x).

In summarising the combined exposure and effects assessments, it is the result - as illustrated in figure 3 - that an environmental risk analysis can be performed and a sound basis be created for the important decisions by specifying:

- the degree (or efficiency) of waste water treatment (cfr. curves 1, 2, and 3) which

must be required by regulatory authorities, and on
- the extension of the dilution zone. It can be defined as in this case by the limits of 500, 300, or 200 metres from the point of discharge, with the understanding that "slight effects may still be expected and accepted within the limits, although only as non-acute manifestations on a limited number of sensitive species".

3. RECENT DANISH MOTORWAY - BRIDGE AND TUNNEL PROJECTS (CASE 2)

In recent years, two major projects on traffic installations have developed to become obvious objects for Environmental Impact Assessment. They have both attracted considerable attention among the Danish public, and they are both based on parliamentary decisions with the character of national development projects. They are concerned with the planning and construction of permanent railway and motorway connections between two of the Danish islands, viz. Sjælland and Fyn, and between Denmark and Sweden, respectively. They will be presented here for their potential impact on the aquatic (marine) environment, only.

As combined bridge and tunnel systems, the first one will stretch about 20 kilometres across the Great Belt, and the second about 17 kilometres across Øresund between the cities of Copenhagen and Malmø. Taken together they will pass across the most important seaway passages which connect the Baltic Sea region to the Atlantic Ocean via the North Sea, and they may thus become potential barriers for the free and unhindered water flow through these passages.

The decision on the construction of a "Great Belt Bridge" was taken in 1985/86, i.e. roughly at the same time as the directive 85/337/EEC was adopted by member countries obliging them to perform Environmental Impact Assessments (EIA procedures) on all major construction works and physical installations. Not surprisingly therefore, this decision as well as the more recent 1991decision on a "Permanent Danish-Swedish Øresund connection" gave rise to public discussions and to hectic expressions of concern, both referring to the fact that the environmental consequences of the projects were inadequately assessed in advance **and/or** they were intolerably lenient in their acceptance of environmental risks.

In order to understand this, it is reminded that both decisions were taken during a period in which serious environmental threats to the internal Danish Belt Sea and to open marine waters had become evident. They might, therefore, represent a continuation of an anthropogenic pressure on an already deteriorating Baltic Sea which is a unique brackish-water ecosystem. It is enclosed as a basin within a vast Northern and Eastern European drainage area. Into this basin, human life causes continuous and serious pollution with organic materials, nutrients, heavy metals and industrial chemicals, as the result of the urban, industrial and agricultural activities of about 100 million people.

Quantitatively, this is illustrated in table 1, which shows that especially the yearly load of organic materials (BOD-load), and the pollution of the Baltic Sea region with nitrogen from all surrounding countries both amount to 1 million tonnes or more. Added to these are the contaminating input of phosphorus and mineral oils, both in the order of 35.000 - 50.000 tonnes per year, as well as a great number of individual chemicals and chemical products, especially from the pulp and paper industry, from primary and secondary steel and metal works, and from chemical industries. They all contribute to the pollution of the Baltic Sea and thereby to the serious deterioration of its environmental quality.

This is to-day experienced and confirmed through numerous reports from studies which are predominantly concerned with the **eutrophication process** [6] that has developed and increased over many decades. In this process nitrogen flowing from all the surrounding countries is generally considered to be the primary limiting element:

- the eutrophication leads to an **oxygen depletion** as the result of an increased demand for oxygen used in the degradation of the increased bio-production (cfr. figure 4);
- such oxygen depletion is to-day prevailing and characteristic for **at least 25% of the total benthic area** of the Baltic Sea, and it is extending rapidly into the Danish straits and Kattegat, i.e. beyond the natural thresholds of the Great Belt and Øresund, and now stretching into the more open zones (cfr. figure 5);
- in turn, this also provokes **avoidance behaviour** among mobile benthic species which is to-day frequently observed in response to the lowered availability of oxygen;
- **reproduction changes** and **reduction in productivity and diversity** are regularly reported for coastal flora and fauna in most of the Baltic Sea region, including the Danish Belt Sea and Kattegat;
- these observations are further closely connected to **disturbances and losses for commercial fisheries**, which on the one side may benefit temporarily from increased harvests of pelagic fish species thriving in the open waters as a result of eutrophicated productivity, but on the other side have suffered gravely in recent years from a simultaneous break-down of fish reproduction and productivity in the near-shore, coastal zones (cfr. figure 6).

Table 1. Total Pollution Load in the Baltic Sea (reported to the Ministerial conference on Baltic Sea pollution in Helsinki, 1988 [5]).

Pollutant		Input from land	Atmospheric input
Nitrogen	TOT-N	530,000 t/y	413,000 t/y
Phosphorus	TOT-P	42,000 t/y	6,000 t/y
B.O.D.		1,640,000 t/y	
Mercury	Hg	5 t/y *)	
Cadmium	Cd	60 t/y	80 t/y
Zinc	Zn	9,000 t/y	3,200 t/y
Lead	Pb	300 t/y *)	2,900 t/y
Copper	Cu	4,200 t/y	380 t/y
Oil		36,000 t/y *)	
Arsenic	As	180 t/y *)	
Nickel	Ni	110 t/y *)	
Vanadium	V	290 t/y *)	
Chromium	Cr	0.2 t/y *)	

*)The dataset of these substances is not complete.
Therefore, the values are very preliminary and not representative.

The basic processes behind such development of the polluted aquatic system are reasonably well understood, and the over-all principles which determine the ecosystem's functioning (and its deterioration) are frequently described [6]. This includes the general cause-effect chain-of-events and its influencing factors, such as:
- availability of nutrients,
- conditioning environmental factors including temperature or light, and, of course,

- special hydrological conditions which are determined by the exchange of water
between the Baltic Sea and the North Sea through the straits and Kattegat.

However, in its details and in the development over time of the pollution, the situation becomes highly complex and the environmental impact of the nutrients becomes difficult to assess as part of a risk analysis. The sea passages of the Belt Sea assume the character of thresholds over which the outflowing brackish Baltic Sea water drains as a surface current. This current meets a bottom-going inflow of oceanic salt water which will pass the thresholds regularly, and eventually may penetrate as unpredictable intrusions into the Baltic Sea basin (cfr. figure 7).

A definite halocline is normally formed between the two oppositely-directed streams, and a situation is created in which many pollutants are redistributed between the two layers and transported back into the Baltic. This "nutrient trap" will inevitably become active in situations of increased eutrophication and biomass production in the surface layers, resulting in sedimentation of excessive organic matter through the halocline into the deeper water and the benthic zone. In turn, oxygen depletion develops due to the excess amounts of planctonic biomass, algae etc., which is followed by an increase of anoxic degradation processes in the benthic zones, a release of nutrients and eventually a re-transportation of these into the free water mass of the basin.

The overall processes and functional relations described here can be followed and analytically determined, but they can hardly be controlled. The impact on the marine systems may be profound and the resulting changes in community structures can be dramatic. In the deeper parts of the Baltic Sea basins, the situation is characterised by the increased production of phytoplancton, while in the photic, near-shore benthic zones macroalgae and periphyton are increasing to the point that other aquatic vegetation (e.g. eel grass) disappear. An accumulation on the bottom of increasing amounts of unattached algae will follow, covering the bottom in thick layers with slowly degrading biomass. The already mentioned gradual lowering of the concentration of dissolved oxygen will be the cause of frequently observed "bottom inversions" and episodes of acute fish kills.

Normally, the shifts in marine communities caused by nutrient loading are gradual processes that span over decades. They represent the result of increasing human activity, and of a cumulative impact over years of development. However, the changes are variable and diverse, and they have been found to develop at increasing rates over recent years.

No accurate prediction, therefore, is available for the risks from already existing pollution of the Baltic Sea, and no single ecological indicators or methods of assessment exist for the quantitative evaluation of impacts from any future installations which may interfere with the water flow over the narrow thresholds in the Belt Sea.

4. CONCLUDING REMARKS

It is, of course, the lack of sufficient data which primarily determines the situation that neither any single parameter, nor any defined and reliable set of indicators has been selected for an impact assessment and, therefore, that no satisfactory assessment has presently been made or can be made.

In the parliamentary decision process concerned with the first of the two traffic installations, namely the Great Belt bridge-tunnel system, a basic demand was phrased "that no changes in the existing rate of water exchange must be permitted" - the so-called zero-solution! - **by which it was understood, but not specifically investigated or validated,** that the chemical and biological qualities of the Baltic Sea would also be unaffected.

It was, however, also realised that a zero change in water flow can only be achieved by huge excavations and changes in the sea bottom profile, in order to compensate for the changes which are forced upon the free water flow by the bridge construction. By accepting this, huge amounts of sediment and sea bottom material will be distributed and lost, and an unsatisfactory situation is bound to develop for the local environment, which will be seriously affected.

The limitation is realised that a definite risk analysis concerned with the general development of the environmental quality of the Baltic Sea cannot be carried out. It can also be judged that a conclusive environmental impact assessment for the planned constructions in the Great Belt cannot presently be achieved, because the impact must be defined and accepted for an already severely disturbed ecosystem which is loaded by an existing, still uncontrolled pollution.

Against this background, it is considered that a critical situation has been developed by the second and most recent of the parliamentary decisions, namely concerning the permanent bridge-tunnel construction to connect Sweden and Denmark across Øresund. In this case the demand for "zero change in water flow" has hazardly been disregarded and substituted by more lenient , and still undefined, requests for "minimisation of environmental changes".

Pollution surveys, continued monitoring programmes and acceptance of future impacts and risks now seem to be left as the alternative for following the consequent development. The character and the order of magnitude of environmental disorders will then have to be uncovered as a succeeding - instead of preceding - obligation.

REFERENCES

[1] Miljøministeriet. Bekendtgørelse nr. 446 af 23. juni 1989 om Vurdering af større anlægs virkning på miljøet (VVM).

[2] Miljøministeriet. Bekendtgørelse nr. 54 af 16. september 1988 om Vurdering af sikkerheden i forbindelse med risikobetonede aktiviteter, der kan medføre et størreuheld.

[3] F. Bro-Rasmussen and K. Warnþe (1983), "Harboør Tange - A coastal area polluted with toxic organic chemicals and mercury", in ECOACCIDENTS, edited by John Cairns, Jr., NATO Conference Series, Series 1, Ecology, Volume 11.

[4] Ringkjøbing Amtskommune (1988), Country Council report on Industrial waste water, Vol. II, 1 (In Danish).

[5] Anon. (1987), First periodic assessment of the state of the marine environment of the Baltic Sea area 1980-85, Background document. Baltic Sea Environment Proceedings No. 17 B, 352 p.

[6] AMBIO (1990), Marine Eutrophication, Special Issue, Volume XIC, No. 3.

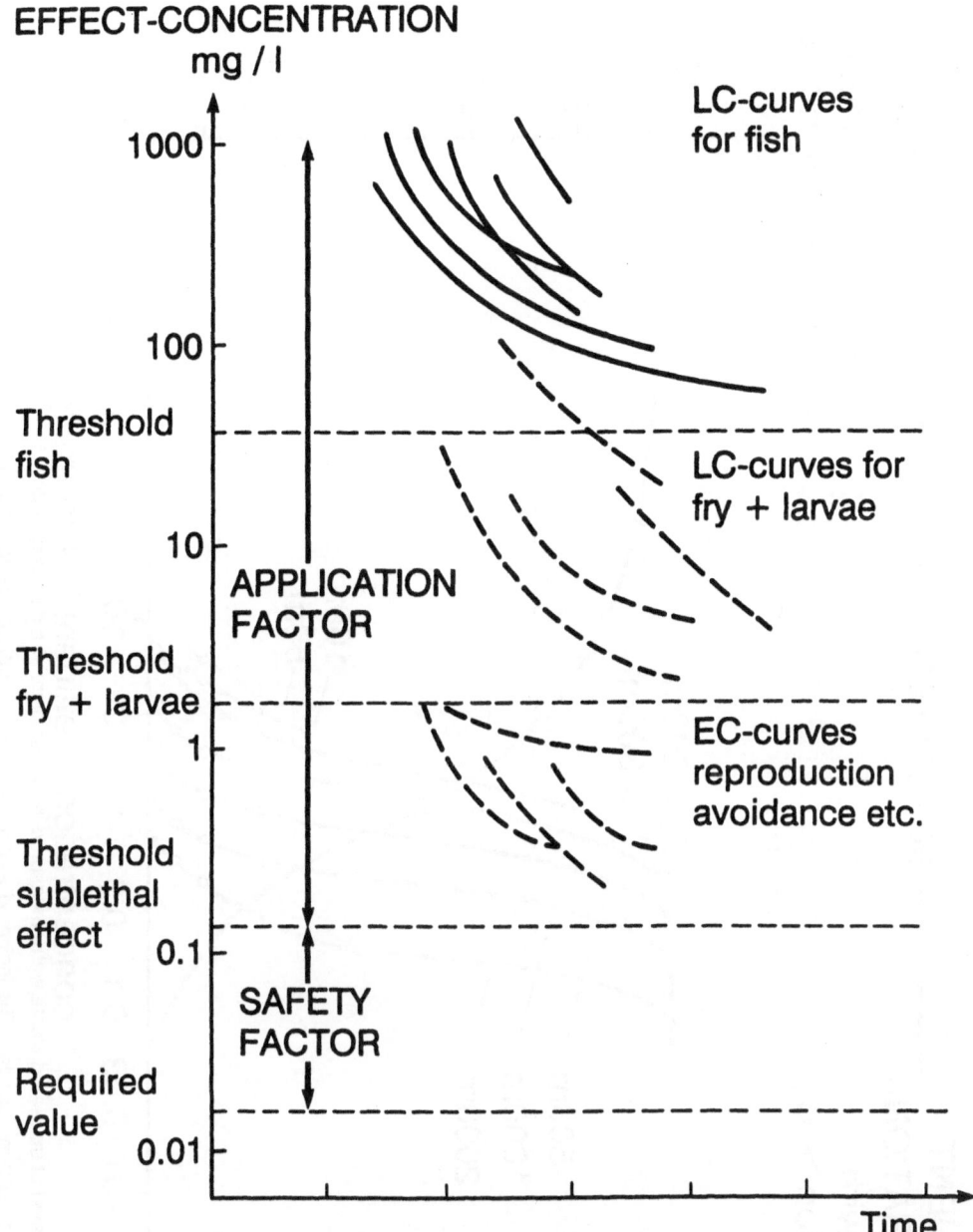

Figure 1. Toxicity of chemicals measured in different test systems.
Effects measured in individual tests against exposure time.

114

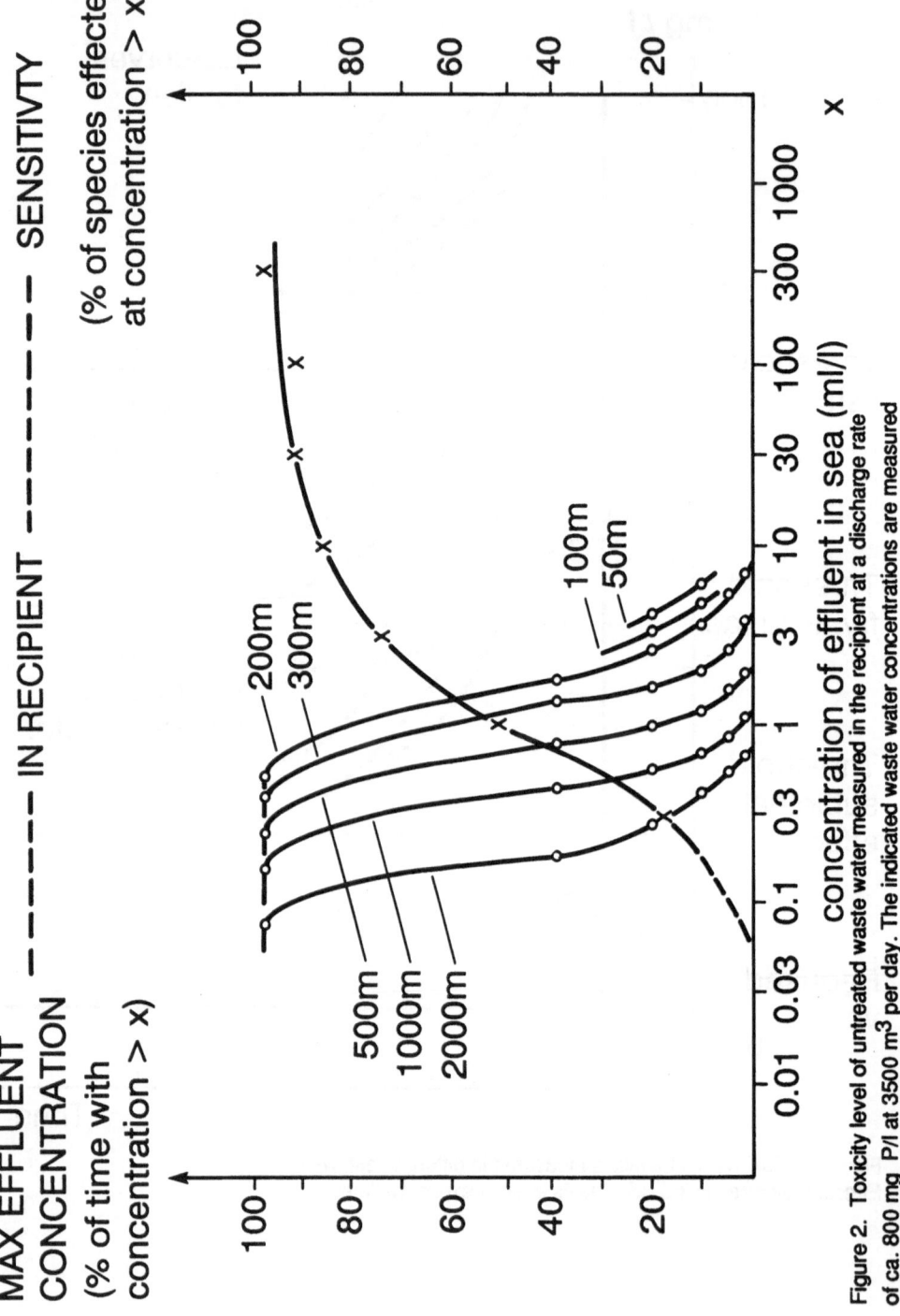

Figure 2. Toxicity level of untreated waste water measured in the recipient at a discharge rate of ca. 800 mg P/l at 3500 m³ per day. The indicated waste water concentrations are measured down-stream of prevailing coastal current.

115

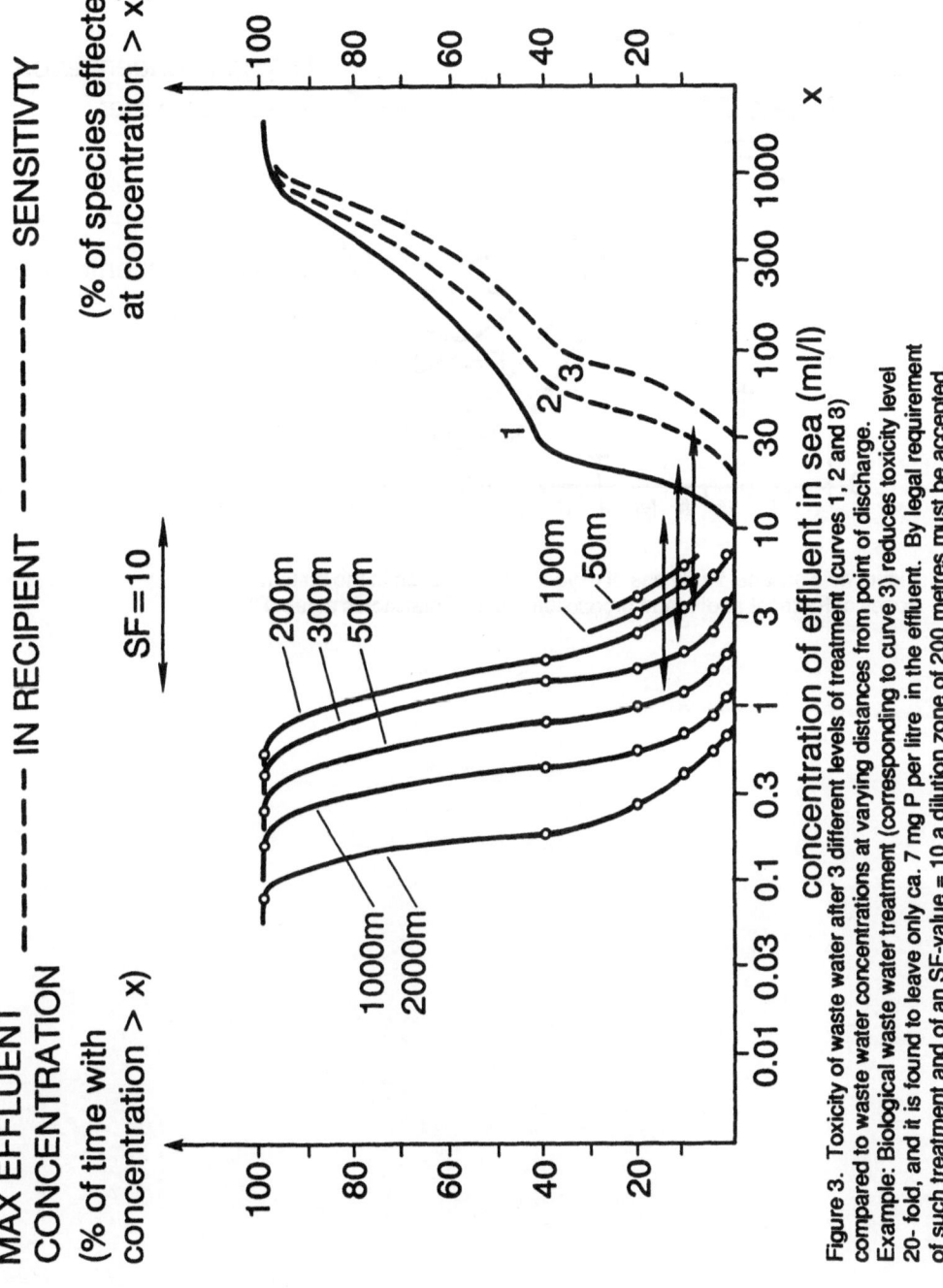

Figure 3. Toxicity of waste water after 3 different levels of treatment (curves 1, 2 and 3) compared to waste water concentrations at varying distances from point of discharge.
Example: Biological waste water treatment (corresponding to curve 3) reduces toxicity level 20- fold, and it is found to leave only ca. 7 mg P per litre in the effluent. By legal requirement of such treatment and of an SF-value = 10 a dilution zone of 200 metres must be accepted.

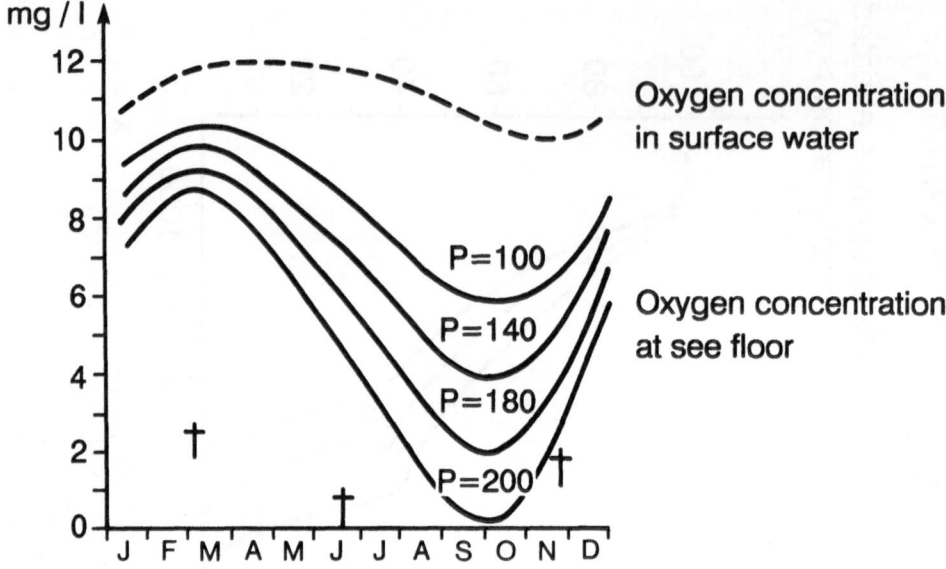

Figure 4. Seasonal variations of oxygen concentration in the bottom zones of Kattegat at different production rates of planktonic algae (P).

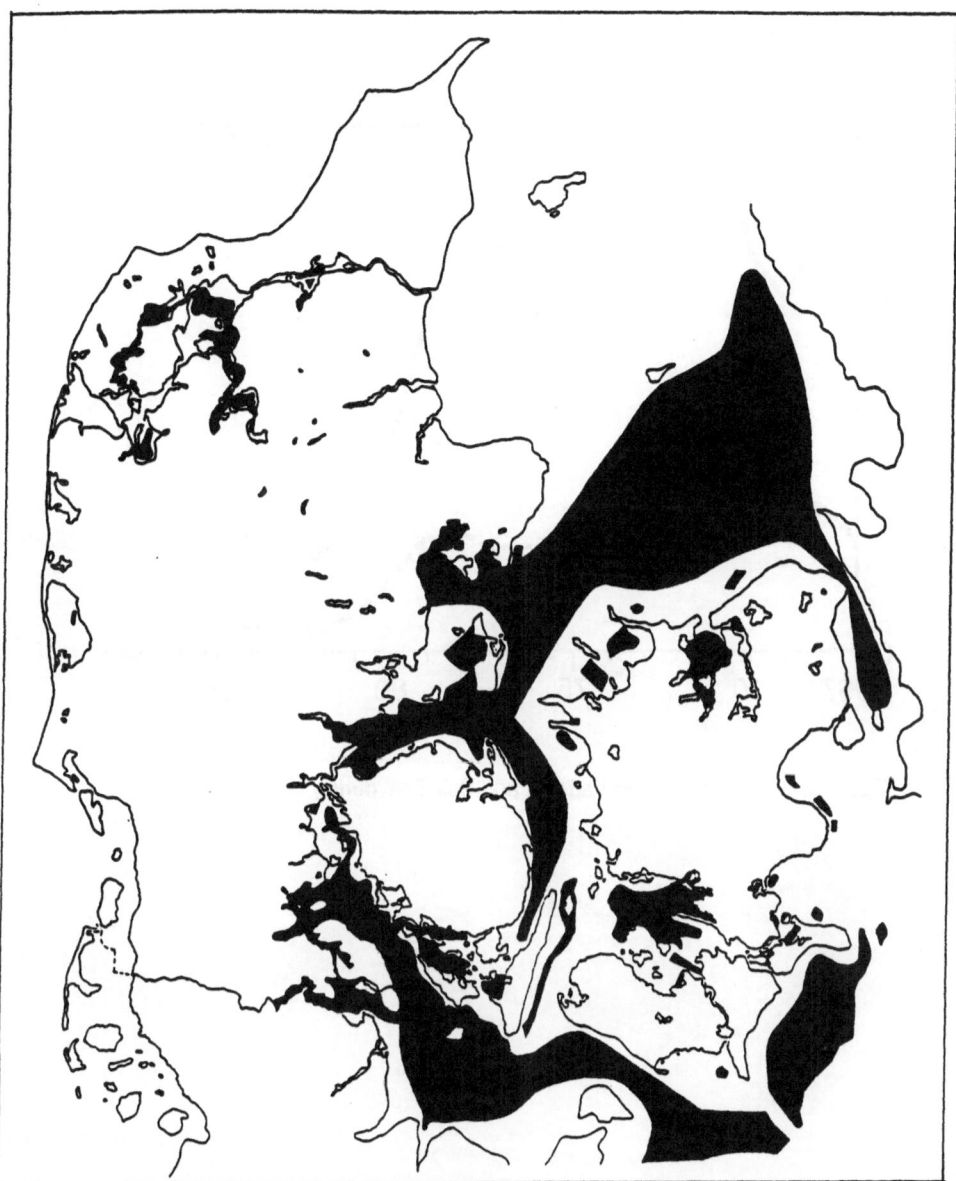

Figure 5. Illustration of oxygen depletion zones (black)
in summer months in recent years through the Danish Belt Sea.

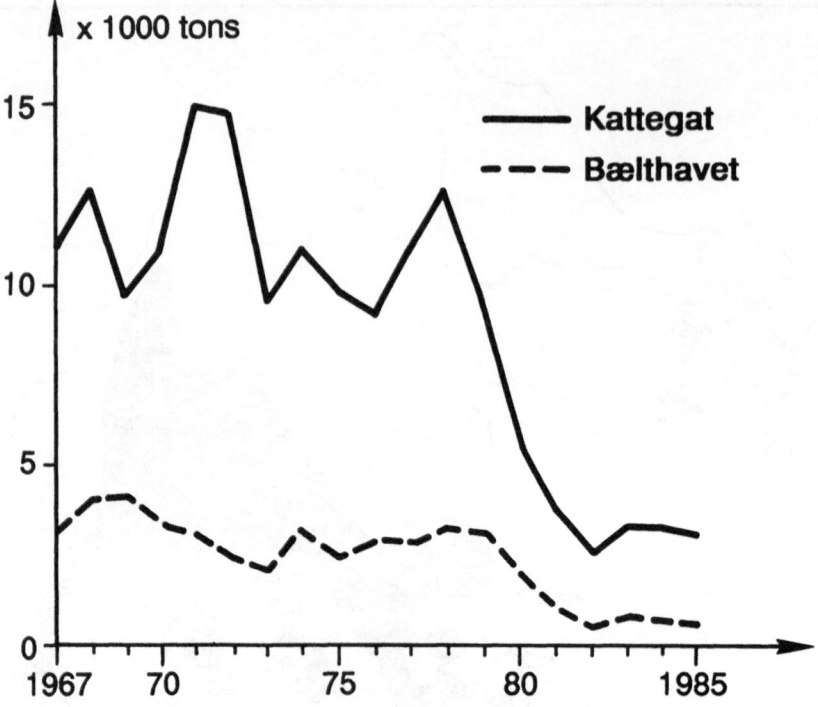

Figure 6. Yields of plaice in Kattegat and Belt Sea during 1967-85.

119

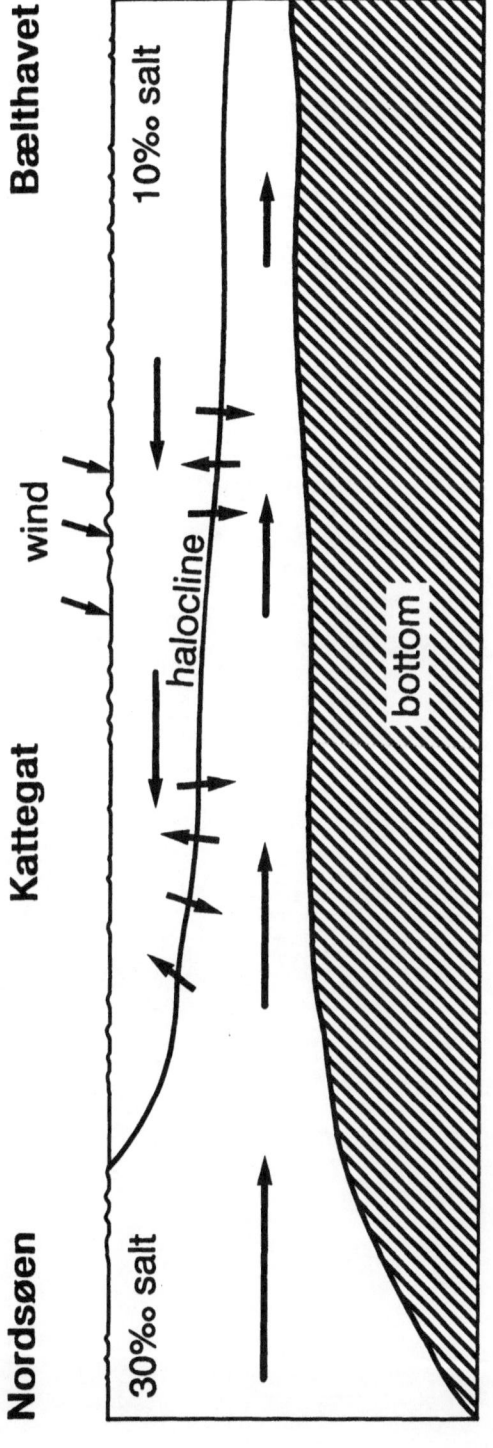

Figure 7. Exchange of brackish Baltic Sea water with salt water from North Sea through Kattegat.

ENVIRONMENTAL IMPACT ASSESSMENT IN GERMANY WITH A SPECIAL FOCUS ON ENVIRONMENTAL PLANNING IN THE RUHR AREA

V. Kleinschmidt
EIA Research Centre
University of Dortmund
D - 4600 Dortmund
Federal Republic of Germany

Abstract. The first part of the implementation of directive 85/337/EEC in the Federal Republic of Germany (FRG) was completed with the EIA law in February 1990. The transformation into the laws of the 16 federal states in Germany is not finished yet, especially for the projects from annex II of the EC Directive; only three states have their own EIA law as a draft (June 1991). The EC directive is valid for projects. In Germany, it is in force on the project level, if not much is known about the project. On this level, the project is tested in a so-called spatial planning proceeding. But many environmental problems are determined by the previous level: a plan or a programme. By the guidelines of the FRG from 1975, the EIA became effective for plans, programmes or the development of guidelines themselves. But this guideline is hardly applied. With the draft of an EC directive about the EIA of plans, programmes and policies there is some movement. As an example, the federal ministry of technology and research authorised investigation of a method of assessing the development of technologies concerning their environmental effects by the Research Centre Jülich, the EIA Research Centre, PRO TERRA TEAM Dortmund and other institutions. At this earliest stage, one can choose a development which includes mitigation measures or better effects on the environment. This method should be obligatory for the development of all technologies with financial support by the state. The EIA of the higher levels of action and the assessment of the projects should cooperate when we reach an environmental policy on a consolidated basis. At least the EIA needs environmental quality goals. The estimation of projects must be based on reference values. Legal values for toxic substances are in general oriented to avoid danger. But the EC directive requires the assessment of cumulative, synergistic and long term effects. One therefore needs precautionary values for the environmental media. On the other hand, environmental quality planning includes spatial developments such as reclamation and development of landscape resources. The international building exhibition Emscher-Park (IBA) aims at this regeneration of landscape parks. Nature protection areas should be increased by up to 10 percent more than today. Areas of constructions (commercial, industrial and residential zones) should consider ecological principles. Roof vegetation and infiltration of the rain to mitigate the effects of the paved surfaces are examples of ecological planning.The main ecological interest is the development and enforcement of natural resources. A research project in the Ruhr Area investigated, computer-aided, the possibility of opening paved surfaces and the effects on the ground water. Beyond this it is tested whether a creek, which is technically enlarged and used for waste water, could be regenerated in its historical course.

A. G. Colombo (ed.), Environmental Impact Assessment, 121–133.
© 1992 *ECSC, EEC, EAEC, Brussels and Luxembourg.*

1. INTRODUCTION

The first stage of implementing the directive 85/337/EEC in the Federal Republic of Germany (FRG) was completed with the EIA law in February 1990. It included the amending of 11 laws and their planning procedures for many projects of annex I and annex II of the EC directive. The transformation into the legislation of the 16 federal states in Germany is not yet completed, especially for some projects from annex II of the EC directive; only three states have their own EIA law as a draft (June 1991), see Tab. 1.

Many environmental problems are determined by the previous level: a plan or a programme. With the draft of an EC directive concerning the EIA of plans, programmes and policies there is a political movement. As an example, the Federal Ministry of Technology and Research authorised investigation of a method of assessing the development of technologies concerning their environmental effects by the Research Centre Jülich, the EIA Research Centre, PRO TERRA TEAM Dortmund and other institutions. The EIA of the higher levels of planning and the assessment of the projects must cooperate, if an environmental policy is to have a consolidated basis.

Environmental quality planning includes spatial developments such as reclamation and development of landscape resources. The International Building Exhibition (IBA) Emscher-Park aims at this regeneration of landscape. Nature protection areas should be increased by to 10 percent. Areas of construction (commercial, industrial and residential zones) should consider ecological principles.

The main ecological interest is the development and reinforcement of natural resources. A computer-aided research project in the Ruhr Area investigated the possibility of opening up paved surfaces and the effects on the ground water. Beyond this it is tested if a creek, which is technically enlarged and used for waste water, could be regenerated in its original course.

2. TRANSFORMATION OF THE EC DIRECTIVE INTO NATIONAL LAW

From the viewpoint of the implementation of directive 85/337/EEC in national law there is a big time-lag in the FRG. There is a national law of EIA, which covers 11 other laws, but, e.g., the Federal Immission Control Law, which includes some of the projects from annex I of the EC directive, sets the EIA in force, when a statutory ordinance to 10 of the Federal Immission Control Law comes into force. This is expected in spring of 1992.

In fact, the environmental impact assessment (EIA) will take place in all cases of annex I of the EC directive and 38 cases of the annex II, if the statutory ordinance to 10, subsection 10 of the Federal Immission Control Law, is given. Up to now, some industrial projects from annex I and annex II do not underlie any legally based EIA and there are some deficiencies concerning projects, which underlie the laws of the Federal States of the FRG (e.g. main roads).

In most cases, the contents of the EC directive are transformed. The time-lag can especially be considered. Considering the obligation to consult all neighbour states (not only the members of the European Community) and the additional obligation to assess genetechnical projects there are two regulations, which go beyond the framework of the EC directive. Viewed collectively, the German EIA legislation is not an extensive transformation. Many items of the directive are implemented without specification.

3. STRATEGIC ENVIRONMENTAL ASSESSMENT (SEA)

3.1. Needs for SEA

The EC directive should not affect the right of the member states to lay down stricter rules (article 13). Many decisions take place at an earlier stage of the planning procedure. For example, the FRG has a law about requirements of roads. Some hundred required roads are named in this law. On this earliest level of planning, an EIA could help to decide on a policy of increasing public transport and reducing the number of roads on the one hand and especially the roads, which cause it - is estimated - the most ecological problems. Only with such a planning level can alternatives be evaluated efficiently.

The EIA law of the FRG only decides that special construction guideline plans connected with EIA obliged projects must be assessed. It is a very small version of plan EIA. Therefore, best wishes accompanies the draft of an EC directive about a Strategic Environmental Impact Assessment of Policies, Plans and Programmes (Doc. No. XII/194/90.DE).

3.2. Strategic Environmental Assessment of Technology

Environmental Impact Assessment aims at the forecasting of environmental impacts, but the project EIA is a reactive instrument. Many problems are based on earlier decisions. Technologies will be developed and later on such technologies will be implemented somewhere in the environment with ecological impacts. The best precaution is to assess the development of the technology. The Federal Ministry of Technology and Research makes extra allowance for research projects to develop technologies. With these grants, the ministry can connect the obligation to assess the estimated environmental impacts.

The example of photovoltaics shows the effects of small technologies on the environment: solar cells, produced based on the toxic heavy metal cadmium, are very efficient in using solar energy. If these cells find extensive distribution in every household, the problems of waste management will increase greatly and will be clear to everybody.

The EIA Research Centre and PRO TERRA TEAM Dortmund developed a method and an organization sheet for such an EIA programme under the leadership of the Research Centre Jülich in a research project. There are some steps and methods, which can be transferred from project EIA experience to SEA (Fig.1):

1. After the application of the research project a screening can take place, which decides about the intensity of examination;
2. Then, a scoping process sets up a framework of questions for the SEA study (Fig.2);
3. Six different ways are possible, from acceptance without any SEA, through several ways of investigation to non-acceptance of the application;
4. The SEA study can take place superimposed (if there is great fear about effects on the environment even in the development phase), parallel (if the project developer does not have enough SEA experience) or integrated in the research project (if the developer is an SEA expert, too);
5 The results of the SEA study must have an influence on the project and the research allowance to similar projects, to modify research activities by duties to the developers and to strengthen or to reduce such activities by the Ministry;
6. A post project analysis must assess the implementation of the technologies to gain experience and better estimate results on the earliest level of planning;
7. The (professional) public should be consulted on the scoping level by discussing the framework of questions for the investigation and a second time by being informed about the SEA study. A yearly report should include the main results of all SEA studies in the year and be available to anyone who has a special interest.

The feedback will bring, on the one hand, more aspects into the investigation and, on the other hand, offers the chance to concentrate the investigations on the main impacts on the environment.

The Federal Ministry of Technology and Research has still to enforce this procedure and make it official practice in every technology development.

4. ENVIRONMENTAL PLANNING IN THE RUHR AREA

4.1. International Building Exhibition (IBA) Emscher-Park

The Ruhr Area is an old industrial region in the North West of Germany, in the Federal State of North Rhine Westphalia. In the centre of this area flows the Emscher, a river which carries all the waste water from the 2.3 million people of this region. This water is purified in one large sewerage plant before it goes into the river Rhine. Now, 6-7 sewerage plants are being planned to decentralize the purification of the water. This should cost about 3 milliard ECU (IBA 1991).

The region is marked by coal mines and the ensuing sinking of the earth's surface. Finally, in such an old industrial region, there is much contaminated soil and because of changing economic structures there are nearly 6,000 derelict unused industrial areas.

The state provides about 5 million ECU per year for the decontamination of soils and the reuse of these areas for new industry, business or trade. The state will invest more than 50 million ECU in the regeneration of the rivers in the Ruhr Area and the development of the river-meadows in a natural way.

All these actions have resulted in the initiation of the International Building Exhibition Emscher Park, which was proclaimed in 1989 and will be completed in the second half of the 1990s.

Many "parks" will be developed; the main ecological interest is in the landscape parks (see Fig. 3). There are six regional green areas of landscape in the Ruhr Area. In the last 40 years, these regional green lines were more and more taken up by industrial, residential and commercial zones. Based on the International Building Exhibition, a so-called "Emscher Landscape Park" will be developed on an area of 320 km^2. An association of the towns in the Ruhr Area - the "Kommunalverband Ruhrgebiet" - worked out a plan about the regeneration of the landscape. The natural resources and the natural potential resources which could be regenerated were added up to a landscape vision for the Ruhr Area. In the year 1991, there are financed measures for the regeneration of landscape amounting to nearly 37 million ECU.

A principle of the International Building Exhibition is to organize planning or design competitions. For most projects, a competition takes place and two to five planning offices make proposals for the development.

In most projects principles of ecology, social aspects and economic interests are integrated. A main objective of the Exhibition is that each project and the whole should make an ecological net profit. Certainly, this is not a simple task!

Some problems should not be suppressed: in the Ruhr Area about 18 waste incinerators and about 23 main roads are in the planning process. It is a difficult task to make an ecological net profit in this region, but there is much positive effort to achieve these aims.

This example shows that the assessment of a single project can not take place without connection to development in the whole region. But one project EIA can not analyse the whole development in the region. Therefore, a plan or programme EIA is necessary.

Up to now, the International Building Exhibition has produced about 70 projects in seven "working fields". The seven working fields refer to the following main projects:

1. Emscher Landscape Park
2. Ecological Regeneration of the River Emscher System

3. Channels as Adventure Fields
4. Industrial Ancient Monuments as Certificates of History
5. Working in Parks
6. Building and Modernization Programmes and Integrated Development of City Areas
7. Offers of Social and Cultural Activities.

4.2. Ecological Resources and Landscape Planning in This Region

Industrial development in the second half of the 20th century has damaged the natural resources in this region. But some man-created biotopes have developed in the damaged areas. E.g., many amphibians live in dig-offs, plants which are on the so-called "red lists", which include the endangered species, grow on waste industrial areas. However, the towns required new residential and commercial zones, Therefore, the areas which include (secondary or primary) nature resources are endangered.

The landscape planning is realized in the Ruhr area only in those areas which are not covered by residential, commercial or industrial zones. In the so-called "outside area", the landscape planning took place primarily in the last ten years. Now, much is known about all natural resources in these outside zones of the Ruhr area. A main task for the 1990s is the recording and improvement of natural resources in the towns. The speed of this task depends mainly on the number of people in the federal landscape protection agencies and their financial budget.

4.3. One Example of Environmental Quality Planning in the Ruhr Area
   (Computer Aided)

In the Ruhr Area, much information is available in digital form. The FRG, their federal states and the bigger towns, are increasing work on using this digital information for planning. The city of Dortmund, in the east of the Ruhr Area, with about half a million inhabitants, set up a computer-based environmental information system in the last six years. Today, there is much information in the system and the next step is to handle this information with a Geo Information System. The town gave a research project to the team of the EIA Research Centre and the Department of Survey and Land Organization. The task was to take all digital information about an area of about 400 hectares, the area round a creek, determined by the highest topographic lines and to digitize the information, which is only available in an analogue form.

The computer takes the information about e.g. soil types, ground water lines, biotopes and transfers them into maps. In addition, in the Ruhr Area, every three years aerial survey photographs are taken around the complete region. These photographs were digitized by a central association of the towns in the Ruhr Area, the "Kommunalverband Ruhrgebiet". The digital information gives answers about real land use. Connected with statistical information about land use types in the Ruhr Area, it is possible to describe the grade of paved surfaces in the region analyzed. Fig. 4 shows real land use in the Dortmund area.

Otherwise, the natural resources were qualified and supplied by values on the basis of a publication which gives instructions on qualifying the individual natural resources such as soil, ground water and biotopes.

On this basis, the whole grade of paved surfaces and their effects on the growth of the ground water were shown. Besides, more is known about the possibility of opening paved surfaces of several types for land use. With this information, the computer can identify the whole potential of opening up these surfaces. One can then estimate the effects on the growth of the ground water.

In the summer, the creek in this area runs dry and in autumn and spring there is flood damage. By optimizing the effects of building areas on the environment the creek could carry water more regularly.

Under these conditions, the technically extended creek could be regenerated. The possibility must be estimated with the aid of the computer. The historical course from 1839 is digitized and the computer tested with the help of information from land owners. The new topographic details give the possibility of showing the creek in its historical course. After this estimation, detailed investigations are necessary, but in most cases politicians ask for feasibility studies at first. If enough actual data about the environment in a city are available in digital form, it is possible, e.g. for the environmental agency of a city, to produce such a feasibility study in a few hours. The next step is to decide about detailed investigations, if the feasibility study shows that the idea is feasible.

The city of Dortmund can now set up a programme of co-financing roof vegetation and opening paved surfaces for the development of natural resources and the realization of a whole environmental concept in one quarter of the city as a brick in an entire concept for the city.

The Geo Information System connected with local databases is a good instrument for environmental quality planning and also for the EIA. Some additional time is needed, but e.g. in the case of two or more planning levels - corresponding to an EIA - and the re-utilization of the information, it can be more efficient than the analogous method. Otherwise, public participation often brings new values into the EIA discussion, the values can be changed in the database and in a few minutes the computer plots out a new valuation map. Therefore, such a system can also be of great help for a clear and flexible EIA.

5. CONCLUSIONS

The EC directive of a project EIA covers a wide field of planning and is a main instrument of environmental policy. More than 1,000 EIA studies are expected yearly in the FRG. However, there are many problems at the early planning stage, which are often due to the lack of an environmental investigation. After deciding on such a programme level (e.g. about the requirements of roads), in the next step one can only decide about the "where" and not about the "if". There are some cases of project EIA in which only a decision about the "how" is possible. Therefore, a strategic environmental assessment (SEA) of plans and programmes is necessary. An investigation at such an early level leads to a better environmental policy by testing more alternatives. This was shown by the example of the SEA technology.

On the other hand, project EIA is a reactive instrument, which in most cases reduces effects on the environment, but over a thousand projects will be performed in landscape per year, with (mitigated) impacts. Therefore, not only compensating and mitigating measures may be performed. The landscape - especially in industrial regions - is often already damaged. An environmental quality planning over the whole area of e.g. a city is necessary to reinforce the natural resources. This can be efficiently supported by a computer-aided Geo Information System, as was shown by the example in the Ruhr Area.

REFERENCES

Bauer, I., Kleinschmidt, V. (1990), "Rahmenbedingungen für die Festsetzung von Ausgleichs - und Ersatzmaß nahmen - dargestellt am Beispiel der Straßenplanung" (Framework conditions for the arrangements of compensation measures - presented by the example of road planning), LÖLF - Mitteilungen 16, pp. 35-39.

Bohne, E. (1990), "Optimale Umsetzung der EG Richtlinie in deutsches Recht?" (Optimal implementation of the EC Directive into national legislation?), Zeitschrift für angewandte

Umweltforschung 3 (1990), pp. 341-348.

Finke, L., Kleinschmidt, V. (1990), "Das Gesetz über die Umweltverträglichkeitsprüfung-eine geglückte Umsetzung der EG-Richtlinie?" (The EIA law - a successful implementation of the EC Directive?), Zeitschrift für angewandte Umweltforschung (3), pp. 348-351.

Hirtz, W., Huber, W., Kleinschmidt, V., Pietzka, F., Heiderich, J. (1991), "Umweltvorsorgeprüfung bei Forschungsvorhaben" (Strategic environmental assessment of research projects), Zeitschrift für Umweltpolitik und Umweltrecht, pp. 179-195.

Internationale Bauausstellung Emscher-Park (1990), Machbarkeitsstudie Emscher Landschafts park (Feasibility Study Emscher Landscape Parc), Emscher Park Planungsgrundlagen Bd.1, Gelsenkirchen.

Internationale Bauausstellung Emscher-Park (1990), Umgestaltung der Wasserläufe im Emschergebiet (Reorganization of the water courses in the Emscher Area), Emscher Park Planungsgrundlagen Bd. 2/1, Gelsenkirchen.

Internationale Bauausstellung Emscher-Park (1991), Abwassertechnische Gutachten zur Umgestaltung des Emschersystems (Waste water managment to reorganize the Emscher system), Emscher Park Planungsgrundlagen Bd. 3/1, Gelsenkirchen.

Kleinschmidt, V., Pietzka, F. et al. (1990), Umweltfolgenabschätzung für Forschungsvorhaben des Bundesministers für Forschung und Technologie (Strategic environmental assessment of research projects of the Federal Ministry of Research and Technology), unpublished research report by EIA Research Centre and PRO TERRA TEAM Dortmund.

Kleinschmidt, V. (1991), "Einbeziehung tierökologischer Inhalte in Gutachten zur Eingriffs reglung und UVP in Nordrhein-Westfalen, Rückblick und Ausblick" (Faunistic investigations in intervention regulation based and EIA studies in North-Rhine-Westfalia, review and outlook), LÖLF - Mitteilungen 16, pp. 46-49.

Minister für Umwelt, Raumordnung und Landwirtschaft des Landes Nordrhein-Westfalen (1990), Ökologieprogramm im Emscher-Lippe-Raum (Ecology-Program of the Emscher Lippe Area), Düsseldorf.

Panteleit, S. (1988), "13 Jahre Landschaftsplanung im Verdichtungsraum" (13 years landscape planning in the Ruhr Area), in Landesbüro der Naturschutzverbände NRW (Hrsg.): Landschaftsplanung - Chance oder Fessel für den Naturschutz?, NZ NRW Seminarberichte 6, Recklinghausen.

Rat von Sachverständigen für Umweltfragen (SRU 1987), Umweltgutachten 1987 (Environmental Expertise 1987), Bundestagsdrucksache 11/1568 vom 21.12.87, Bonn.

Rau, J. (1991), "Werkstatt für eine Industrieregion" (Workshop onan industrial region), Stadtbauwelt 82, pp. 1216-1219.

Schreiber, K. F., Lecke-Lopatta, T. (1990), "Praktische Anwendung der Eingriffs-Aus gleichsregelung aus landschaftsökologischer Sicht" (Actual aspects of the practical application of the regulations concerning the compensation of interventions seen from a landscape-ecological point of view), Landschaft und Stadt 22, pp. 121-129.

Schwarze-Rodrian, M. (1991), "Emscher Landschaftspark - Konzept einer regionalen Entwicklungsstrategie, Emscher" (Emscher Landscape Parc - concept of a regional development strategy), Stadtbauwelt 82, pp. 1230-1237.

United Nations (UN 1991), Convention on Environmental Impact Assessment in a Trans-Boundary Context, Espoo (Finland), 25 February 1991.

United Nations - Economic Commission for Europe (ECE 1990), Post Project Analysis in Environmental Impact Assessment, Environmental Series 3, New York.

UVP Forschungsstelle - FG Vermessungswesen und Bodenordnung (UVP-F/VBO 1991), Rechnergestützte räumliche Konkretisierung von Umweltqualitätszielen am Beispiel von Nettebach und Wideybach (Computer aided spatial definition of environmental quality goals by example of the creeks Nettebach and Wideybach).

Wagner, D. (1991), Implementation of Directive 85/337/EEC in Germany, unpublished paper to the European Commission, Solingen.

Wood, C. (1990), The extension of EIA to policies, plans and programmes: training and research implications, lecture of the EIA Centre of the University of Manchester, Avignon.

129

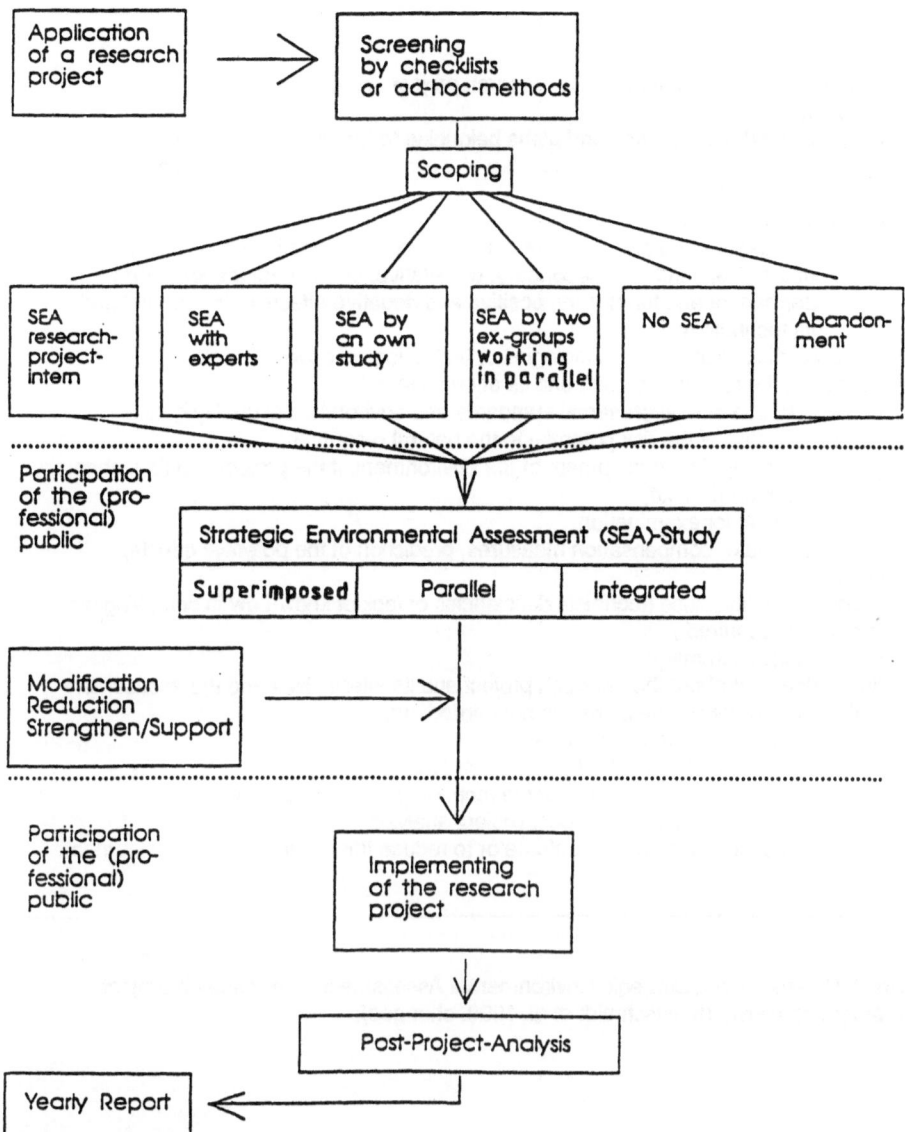

Figure 1. Organization chart for a Strategic Environmental Assessment for a research project on a programme level (Kleinschmidt et al. 1990, changed).

- short description and aims of the research project applied;
- scoping;
- environmental programmes and plans belonging to the research project;
- analysis of the environment (in the case of a pilot project where real impacts can be assessed);
- estimation of effects:
    o scenarios on the development of the technology investigated, estimation of the direct, indirect, secondary, cumulative, short, medium, long term, permanent and temporary, positive and negative effects of the project and its technology,
    o conflicts with other technologies or technological developments,
- assessing the research project and its alternatives:
    o definition of environmental standards and environmental quality goals,
    o estimation of the risk potential to the natural resources,
    o assessing the development of the environment, if the project and its technology is not carried out,
    o proposal for examination;
- mitigation and/or compensation measures, prediction of the potential effects, scenarios;
- indication of difficulties (technical deficiencies or lack of know-how in compiling the information required);
- non-technical summary;
- final judgement about the research project and its alternatives and the assessment of the environment, if the project is not carried out;
- recommendation to the federal agency:
    o additional conditions for the research project,
    o proposals for modification of the technological development,
    o definition of criteria for a post project analysis,
    o proposal for a device to enforce or to reduce this technological development.

Figure 2. Questions of a Strategic Environmental Assessment for a research project on a programme level (Kleinschmidt et al. 1990, changed).

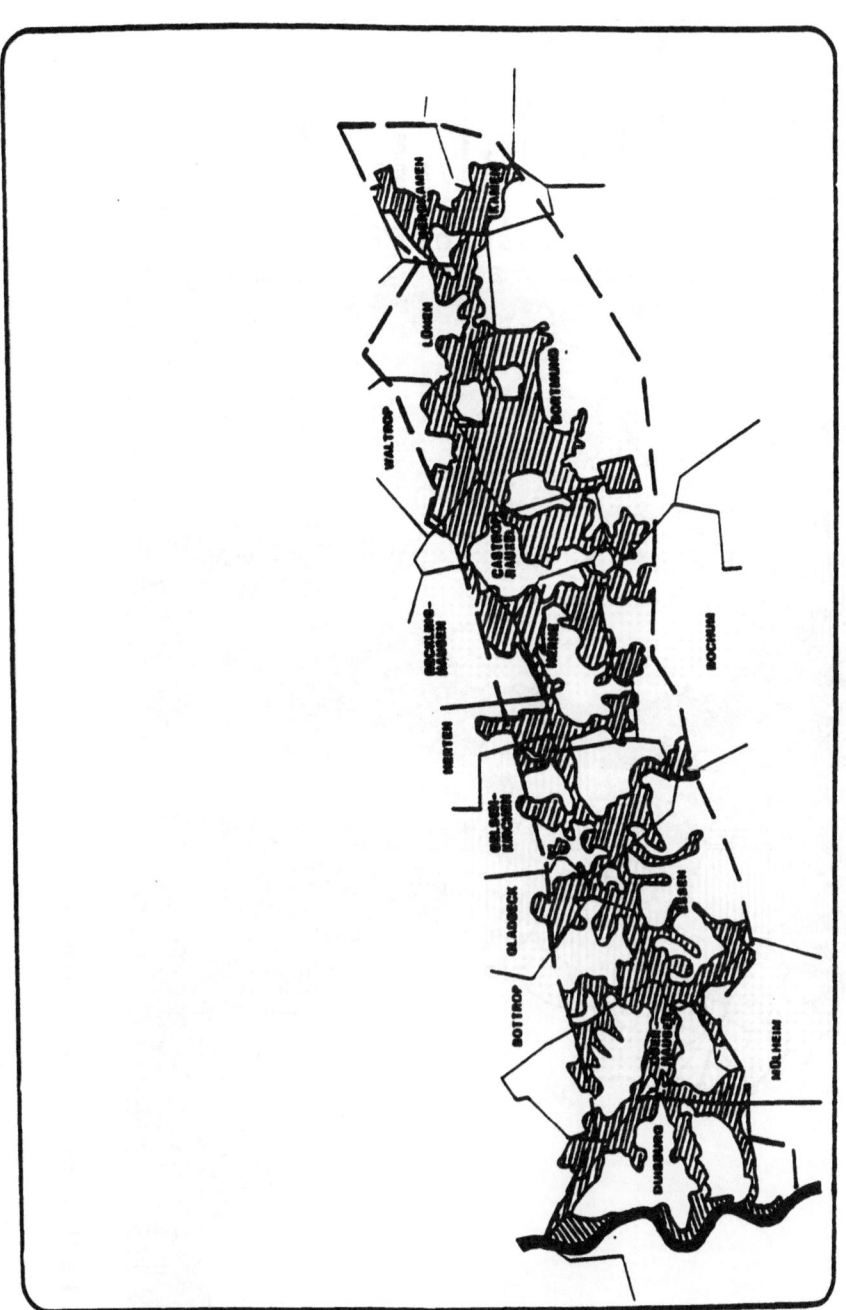

Figure 3. Emscher Landscape Parks.

132

Real land use

LEGEND

LANDUSE TYPES

- residential area / other built-up area
- working area / industry
- public land incl. sport field and playground
- agricultural built-up area
- supply / waste disposal
- park / cemetery / allotment garden
- site vegetation / copse
- waters
- meadow-land
- agricultural useful area
- woodland
- derelict land

INFRASTRUCTURE

- motorway
- residential street
- rail traffic

EIA-Research-Institute
(UVP-Forschungsstelle)

Figure 4. Real land use in an area of Dortmund (UVP-F / VBO , 1991).

Table 1. Implementation of Directive 85/337/EEC in the FRG: a short overview.

EC Directive	Contents	FRG Implementation
Art. 1	defined objectives	transformed in FRG legislation
Art. 2	general assessment obligation for specific projects	see annex I and missing legislation in the federal states of the FRG for some annex II projects, discussion to exempt the first stage EIA for traffic ways in eastern Germany
Art. 3	function and tasks of an EIA	transformed
Art. 4	types of projects, connected with annex I and annex II	types transformed, see annex I and II
Art. 5	authorities should give their information to the developer, developer should give EIA information specified in annex III	mostly transformed, but see annex III
Art. 6	information should be available for consultation	transformed
Art. 7	information of the Member States in case of transfrontier impacts	transformed, information of all neighbour states
Art. 8	developer and consultants information should "be taken into consideration"	transformed without specification
Art. 9	inform the public about decision	transformed
Art. 10	respect of secret information	transformed
Art. 11	information exchange between Commission and Member States	time-lag
Art. 12	application of directive within three years	three years additional time have gone, time-lag for some projects up to now (see annex I and II)
Art. 13	Member States can apply stronger rules	see art. 7 and genetechnical projects
Annex I	nine types of EIA projects requiring EIA	two classes only applied, seven classes will still be adopted in 1991
Annex II	twelve categories and some sub-categories of projects, Member States decide about obligation	only one category and some sub-categories are missing, lack of legislation in the federal states of the FRG
Annex III	contents of EIA information, given by the developer	mostly transformed, but the regulation to assess long termed, synergetic, cumulative and other effects is missing in legislation

POLLUTION ABATEMENT IN OIL REFINERIES AND INORGANIC CHEMICAL INDUSTRIES

G. Drogaris [+]
CEC, Joint Research Centre
Institute for Systems Engineering and Informatics
I - 21020 Ispra (Va)
Italy

ABSTRACT. Essential parts of an Environmental Impact Assessment Study of an industrial activity are: (a) identification of process pollution sources and qualitative/quantitative pollutants emission assessment; (b) evaluation of pollution abatement measures; (c) fate of pollutants in the receivers; (d) environmental quality targets for the receivers. Each of these steps must be tackled both by the manufacturer and the authority in a different manner but in a continuous interaction. The manufacturer must provide the data and adequate support to the conclusions for the three first steps. The authority, however, must verify the submitted data. Sources of data for pollutant identification for various processes, estimation of pollutant emissions and effectiveness of various pollution abatement alternatives are available in the literature (EPA AP-42 provides the best collection for air emissions); the inspector must, however, have experience and be familiar with the plant to make proper judgements. Oil refinery and inorganic acid/fertilizer processes will be used as examples to illustrate the use of AP-42. The manufacturer should also prove compliance with valid environmental regulations (a short review of various pollutant emission and ambient quality standards relevant to the processes referred to above will be given). The authorities, however, must consider the existing environmental condition of the receivers and set ambient quality targets for them before evaluating the compliance of expected pollutant emissions with the environmental strategy. Examples of process modifications that simultaneously reduce pollution and operation cost for oil refineries will be given. Finally, the use of an environmental impact assessment study for pollutant emission and ambient quality monitoring will be discussed.

1. INTRODUCTION

All types of activity performed by man have an impact on the environment. Of all man-made activities, industry mainly attracts public interest with reference to environmental problems. Citing industry as the major contributor to environmental damage is not always justified (vehicles can represent an even greater danger in congested urban development). Nevertheless, environmental regulations of industrial activities became indispensable not only for industrialized countries but also for countries under development.

[+] Present address:
  CEC, Joint Research Centre
  Institute for Advanced Materials
  1755 ZG Petten
  The Netherlands

A. G. Colombo (ed.), Environmental Impact Assessment, 135–165.

Environmental standards and regulations defining the level of pollutants in the ambient environment, which is considered to be "acceptable" in terms of concentration or deposition fluxes of the respective pollutants, are developed or revised under the pressure of environmental problems. Industry is then called upon not only to cope with such environmental regulations but also to prove it by preparing an Environmental Impact Study. Industry is, in fact, the most appropriate source for information regarding the process(es) used, pollutant emissions and fate in the environment and pollution abatement systems employed.

The relevant local or central competent authorities have, however, the task of examining this study and assessing the compatibility of the industrial activity with its surrounding environment in the framework foreseen by the environmental regulation. Moreover, at the latest by the end of 1992, they must provide on request all relevant information to the public (with the exception of data related to public security, under jurisdictional actions or to commercial/industrial confidentiality) [1].

This paper discusses the general problems that arise in the evaluation of an environmental impact study in the overall procedure of an environmental impact assessment. The general characteristics of environmental regulations in some countries are briefly presented. Oil refineries and inorganic chemical processes are used as examples to highlight the problems involved. For new plants, due attention to environmental constraints in the early design phase can be proved to be a very cost-effective method to minimize pollution. For existing installations, a cost/benefit analysis must be used to evaluate various possible alternatives.

Monitoring of pollution emission at its source must be combined with ambient quality monitoring, both to verify compliance with the regulations and to check the effectiveness of the measures adopted. The Environmental Impact Study is very useful in designing and setting up a proper monitoring programme.

2. ENVIRONMENTAL IMPACT STUDY EVALUATION

Environmental impact studies must be supplied for any new industrial activity proposed in most industrialized countries. Existing installations may also be requested to perform this exercise whenever environmental regulations are revised. It is an obligation for the EEC Member Countries to introduce the Environmental Impact Assessment procedure according to the requirements of the Council Directive 85/337/EEC. The relevant competent authority(ies) must evaluate these studies. The first problem area would be the verification of the information on qualitative and quantitative pollutants emission. An independent, scientifically sound information source is required especially if the Environmental Impact Study refers to a new installation to be constructed, since in this latter case verification through a site inspection is not possible. AP-42 of U.S. EPA [2,3] is the most extensive compilation of air emission factors. There is also literature available for specific branches of industry (e.g. [4] for oil refineries). Similarly, there are general (e.g. [5,6]) and specific [7] publications for industrial waste water. The use of these data requires care, however, due to the complexity and variety of the industrial processes.

An even bigger problem arises whenever newly developed processes are employed, which were not in operation when the published data were collected. In such a case an expert judgement is necessary.

Then, the fate of the pollutants in the environment must be assessed. This task must be accomplished by the industry. There are many models and computer codes for air pollution dispersion [8] and the development of more sophisticated tools is in progress (e.g. [9]). Gaussian models are extensively used and they are presented in another paper in this book [91]. Ref. [38] contains a list of ambient air quality and pollutant dispersion models available from NTIS with a brief description of their characteristics. For the fate of water-borne pollutants there are quite sophisticated computerized codes (e.g. WASP, QULAXE, CREAMS, EXAMS) from EPA as well as some simplified models suitable for calculators [10]. The problem of the fate of water-borne pollutants in the soil with reference to ground water is even more difficult to tackle.

The task of the competent authority in this area should be to check whether the code used is appropriate for the specific application and to check the assumptions and constraints. Very often, simple models are very useful for screening purposes and for obtaining an overview. Therefore, constructive interaction between industry and authorities during the compilation of the study can save time and cut costs.

The remaining steps involve checking for compliance with the regulations and for compatibility with the environment. This latter task presupposes knowledge of the existing environmental situation in the area (ambient air, soil, aquatic environment) and a clear definition of environmental quality targets. Ambient quality standards already exist for some pollutants (see paragraph 3.2.1. below); the definition of ambient quality targets considering all possible environment impacts is currently being tested in pilot applications in the Netherlands [87]. Specific cases may even request research work to define ambient quality targets [88,89]. In other cases, sociological or economic factors should also be considered when defining such targets [90]. The final consensus, however, requires the involvement of the public that is affected, is liable to be affected, or feels that it is affected by the industrial activity in question.

The authorities, by nature, have always had the obligation to interact with the public even before the application of the new EEC Directive on the freedom of access to information on the environment [1]. Industry must realize that coexistence with the neighbourhood presupposes sincere communication.

Effective evaluation of an environmental impact assessment requires a dynamic process. For new installations, interaction with the authorities must start during the feasibility study and the conceptual design. Details for pollutant emissions cannot be provided in this phase. The general information, however, is normally enough for a preliminary environmental study. An environmental review at this stage is useful in avoiding the selection of technologies that could hardly conform with the requirements at a specific site or to specify either design constraints or the requirements for the pollution abatement units. The role of the authority in this phase is to prevent the selection of processes that could hardly, or only with an excessive cost, be compatible with site-specific environmental requirements and constraints. As the project advances, in the engineering and construction phase more environmentally relevant information becomes available. Interaction with the authorities on critical points (selection of technologies, finalization of studies), that can be identified from a preliminary environmental study is very useful and will substantially reduce the time needed for evaluation of the final environmental impact study.

In a similar way, environmental impact studies for existing installations should be dynamically developed. The basis for a cost/benefit analysis must be agreed in an early stage with the authorities as well as the selection of pollution abatement techniques with which the industrial installation must be retrofitted.

3. ENVIRONMENTAL REGULATIONS AND STANDARDS

The rate of implementation of the EEC Directive on EIA in most of the Member Countries is rather slow [92]. However, in most of the Member Countries various environmental standards and regulations had been introduced before this Directive came into force (see e.g. [88]). A brief overview of selected examples of such environmental regulations, which ultimately should be integrated in the EIA procedure, will be presented here.

The basis for setting a standard is the scientifically established relationship between a given pollutant concentration in ambient air and water or deposition flux in combination with the related effects. The effects may be effects on human health, animal, vegetation, soil, aquatic ecosystems, materials or climate. There may be acute effects which cause an abrupt change in the condition of the receptor. There can also be chronic effects where an impairment in the condition of the receptor can occur due to exposure over a long period of time. The actual setting of a standard is a political decision, whereby the degree as to which society is willing to tolerate adverse effects is evaluated taking into account economic, technical, political and other aspects. Because the relative importance of aspects may differ from country to country, it is not surprising that different standards for a given substance have evolved in different countries. Some countries set standards which force the development of higher efficiency control technologies, others specify the means by which control is to be achieved, and in other cases reliance is placed on the ability of industry to achieve acceptable emission standards by whatever means are economically feasible.

There is another important aspect which must be considered when setting standards or when evaluating standards applied in different countries. This aspect refers to the assessment system. The assessment system depends on:
- number and location of monitoring stations;
- sampling time;
- sampling frequency;
- instrumentation;
- calibration.

If any one of these is changed, a resultant change is also produced in the definition of the standard, even if it remains numerically constant.

In many respects the level of legislation and particularly its specificity reflect the level of industrialisation and/or the extent of pollution problems within the country. Transboundary pollution problems may, in the future, have an increasing influence on the stringency of standards.

So, standards affect the location of new sources of pollution and the modification of existing sources. Engineers and managers must constantly make decisions on whether to start a new project, expand, modify or even continue operation. These decisions depend on the existing standards in a location.

Regulations applicable to pollution control are of two types: limiting regulations prescribe a quantitative standard that one may not exceed and administrative regulations direct the growth of major sources of pollution or the rate of reduction in pollutant emission at the level of the whole country or a specific area. The standards and regulations are issued in every country by the relevant environmental authorities or group of countries in the course of an international cooperation (e.g. EEC, Rhine Agreement, Mediterranean Action Plan, etc.). Ambient quality standards or regulations, though very important for an overall environmental policy, cannot be directly used for environmental control of a specific industrial source. Emission standards, on the other hand, that can be readily used for environmental control of pollution sources have no general applicability and cannot be adopted in and transferred from other countries without any extensive guide; they are, however, useful as general guides.

The following parameters must be considered in the process of setting environmental emission standards and regulations:
- ambient quality standards that must be met in the receptor in question;
- existing receptor quality from the environmental point of view;
- identification of all pollution sources affecting the receptor in question;
- determination of the fate of pollutants from every single source in the receptor;
- technical possibilities for pollution abatement and the relevant cost.

3.1. Environmental Regulations

The Council Directive 84/360/EEC requires that certain categories of industrial plants are subject to prior authorisation. For these plants, "all appropriate preventive measures against air pollution must be taken, including the application of the best available technology, provided that the application of such measures does not entail excessive costs (BATNEEC)". The Commission is improving the exchange of information among the Member Countries by preparing technical notes defining the best available technologies for various industries [93].

3.1.1. Environmental regulations in the FRG. German environmental regulations are based on the Anti-Pollution Control Act and are normally prescriptive, defining both pollutant emission rates and pollution abatement systems according to the state of the art (Stand der Technik) [79]. This is valid both for atmospheric emissions (TA-Luft [11]) and for waste waters [12-14]. Maximum concentration of atmospheric pollutants is prescribed in most of the cases. Whenever pollutant emission rates exceed some limit values (e.g. 50,000 m^3/h flue gas, 50 kg SO_2/h or 30 kg NO_2/h or 5 kg CO/h, etc. [11,79]) continuous monitoring at the source is required. Emission regulations for water-borne pollutants are becoming more stringent with time [13,14]. In some cases (e.g. heavy metals) not only a certain maximum allowable concentration is prescribed but also a maximum overall pollutant load per year. According to the waste water disposal law (Abwasserabgabegesetz) there is a charge per unit mass of pollutant discharged, which had been revised in 1987 [12]. More stringent regulations are to be applied for pollutants which are dangerous for waters (e.g. chlorinated hydrocarbons, aromatic hydrocarbons, pesticides, etc.). The relevant regulation already became effective in November 1990 in the state (Land) of North Rhineland Westphalia (NRW) [13]. There are also rather detailed guidelines for various industrial branches (e.g. [15] for oil refineries/petrochemical installations in NRW) giving check-lists for environmental impact study evaluations. Finally, there are also guidelines from the German Engineers' Association (VDI-Richtlinien) [16] on emission control practices for specific installations (e.g. VDI Richtlinien No. 2295: nitric acid production; No. 2296: production of hydrogen fluoride, hydrofluoric acid and crylite; No. 2298: sulphuric acid plants; No. 2440: oil refineries; No. 3450: phosphorus and organic compounds containing phosphorus; No. 3454: Claus sulphur recovery plants, etc.). Separation distances between industrial and residential or other activities are also prescribed in relevant regulations [83,84]. The environmental impact study is part of the documentation required for licensing of industrial installations and is made available to the public for comments as the licensing application is processed [17,79]. The last revision of the Federal Air Pollution Control law [79] introduces the manufacturer's obligation to take all the measures necessary to prevent adverse effects to the environment from an installation, even after this installation has been permanently shut down.

Environmental problems and the interest of the public in them lead continuously to further environmental regulations. In the last six months, in the FRG (01.12.90 to 31.05.91) eleven environment-related laws and regulations have been published in the Official Federal Gazette [18]. The most important for industrial installations with the relevant publishing date are as follows:

- 01.12.90: 17th regulation on the implementation of the Federal
Air Pollution Control law of 23.11.90;
- 01.01.91: Environmental liability right;
- 01.01.91: Important changes to the 3rd amendment of the
Waste Water Disposal law of 19.12.86;
- 01.01.91: Regulation on drinking water;
- 01.01.91: Environmental liability law of 10.12.90;
- 01.03.91: 2nd regulation on the implementation of the Federal

Air Pollution Control law of 10.12.90.

3.1.2. Environmental regulations in France. According to the Act No. 76-663 on installations registered for purposes of environmental protection [81], all installations which "may threaten any danger or nuisance, whether in regard to neighbourhood amenity, public health, safety or sanitation, agriculture, protection of nature and environment, or conservation of sites and monuments" shall be subjected to authorization by the Prefect. Plant owners must submit an application for new activities or for modifications, extensions, etc. before they obtain authorization of operation after a public inquiry that should also define land uses around the installation.

Other relevant acts include:
- Act No. 61-842 of August 2, 1961 on the control of atmospheric pollution and odours;
- Act No. 64-1245 of December 16, 1964 on the regime governing waterways and water distribution and control of their pollution (mainly Sections 2 and 6);
- Act No. 75-633 of July 15, 1975 on waste disposal and the recovery of materials;
- Act No. 76-629 of July 10, 1976 on nature protection (mainly Section 2).

Acts and decrees for hydrocarbon storage of the late 20s are also currently under revision (a new decree for the storage of liquefied flammable gases was issued in 1989 [82]). The decree No. 77-1133 of September 21, 1977 [81] defines the procedures for submission of applications for authorization and for carrying out the public inquiry and the authorization of an industrial activity as well as the content of the application (Environmental Impact Study and Safety Assessment). The whole authorization procedure is decentralized and takes place at Prefectural level. The general regulations for the registered installations are issued by Prefectural Orders. The decision whether an installation is covered only by the requirements of the declaration act of December 19, 1917, or whether an authorization is necessary, is also taken by the Prefect.

3.1.3. Environmental regulations in the UK. Traditionally the Best Practicable Mean (BPM) concept had been applied by Her Majesty's Inspectorate of Pollution (HMIP) for environmental control. Environmental legislation started in the UK with the Alkali Act of 1863 (later to become the Alkali Act of 1906), the Public Health Act of 1875 and the Rivers Pollution Prevention Act of 1876. Notes in annual reports on the Alkali Act contained BPM for some branches (e.g. oil refineries [19]). There are also notes published for various processes (see e.g. [20-24]). The new Environmental Protection Act 1990 introduces the application of the Integrated Pollution Control (IPC) for the first time in the UK and aims to push the industry towards new, less polluting technologies. The IPC is based on the Best Available Technique Not Entailing Excessive Cost (BATNEEC) [25]. This new act mainly covers the prescribed processes known as "Part A" processes ("Part B" - all other - processes are under local authority air pollution control and are subject to separate guidance). Prescribed substances (see Table 1) are regarded as the most potentially harmful or polluting when released into the environment. HMIP must decide what is the BATNEEC to prevent their release to specified media or (if that is not practicable) to minimise such releases. The Chief Inspector must issue a final guidance by [25]:
- Jan 1991: for big (more than 50 MW(t)) combustion plants;
- Sep 1991: for other fuel and power industries;
- Jan 1992: for waste disposal industries;
- May 1992: for mineral industries;
- Oct 1992: for petrochemical, organic, chemical pesticide and pharmaceutical industries;
- Apr 1993: for acid manufacturing, halogen, chemical fertiliser and bulk chemical storage industries;
- Oct 1993: for inorganic chemical industries;

- Jun 1994: for metal industries (September 1994 for non-ferrous metal industries); and
- Apr 1995: for other (paper manufacturing, di-isocyanate, tar and bitumen, uranium, coating and coating manufacturing, timber and animal and plant treatment) industries.

The processes mentioned above are the prescribed ("part A") ones. Six months after the issue of the final guidance the process comes under IPC, which means that an authorisation is required. The application for authorisation is sent to the relevant statutory consultants (e.g. the National Rivers Authority) and is made available to the public for comments. Effective 1st April 1991 no new process may operate without an authorisation. Land Use planning was introduced in 1944 and is normally performed at town/county level within general national planning [85]. The HSE may issue an advice based on the safety assessment of industrial installations [86].

3.1.4. Environmental regulations in the USA. Regulations in the USA consider a combination of various ambient quality and emission standards. An overall view of the US Clean Air Act (CAA) regulation scheme [26] is given here as an example that should be helpful in sorting out the various regulations and standards applicable to limit air pollutant emissions. Application of the CAA foresees four levels of action:

a) National Ambient Air Quality Standards (NAAQS) are determined for pollutants that endanger public health and welfare (the criteria pollutants: particulate matter (TSP), sulphur oxides (SO_2), carbon monoxide (CO), hydrocarbons (HC), oxidants, nitrogen oxides (NO_2), lead (Pb) and ozone). Each state adopts emission regulations (SIP regulations) designed to provide ambient air that meets the NAAQS for the various criteria pollutants.

b) There are some large stationary sources of pollution that pollute significantly. These require special attention since the expansion of these categories of sources, owing to industrial growth, will probably cause a violation of the NAAQS. Thus, national emission standards are developed for these sources. These regulations, usually called New Source Performance Standards (NSPS) (e.g. [27]) are generally more restrictive than the SIP regulations. Although the NSPS apply only to new and reconstructed or modified facilities, if the pollutant so regulated is not listed as a criteria pollutant or a hazardous pollutant, each state must adopt a limit applicable to similar, existing facilities.

c) In addition to the more common pollutants that adversely affect health and welfare and those especially large and growing sources of criteria pollutants, there are hazardous pollutants. These result in an increase in mortality or an increase in serious irreversible, or incapacitating reversible, illness. These pollutants are not listed in criteria pollutants but, instead, become the subject of the National Emission Standards for Hazardous Pollutants (NESHAP). The NESHAP regulations generally apply to new and existing emission sources alike.

d) Control the emissions of aircraft and motor vehicles. Owing to the interstate nature of these pollutant sources, regulation is generally preempted by the EPA.

In addition to the published emission limitations applicable to stationary sources, there are also those that may be considered as negotiated limitations. These are often negotiated when seeking a permit for new, modified, or reconstructed sources. They may be identical to the limitations published under the NSPS but more probably they will be that control limit which is considered, for the particular installation in question, best available control technology (BACT) or lowest achievable emission reduction (LAER).

3.2. Environmental Standards

Environmental standards can be mainly divided into two categories: ambient quality and emission standards.

3.2.1. Ambient quality standards. EEC Directives set ambient air quality standards for sulphur oxides (measured as SO_2), nitrogen oxides (measured as NO_2), particulate matter and lead, while a directive proposal for air quality objectives for ozone is under preparation [93]. In Table 2 the first three standards are compared to the USA and WHO standards. There are also ambient air quality standards for carbon monoxide and ozone in the USA. In the FRG a network of ambient air monitoring stations in 80 cities (plus 5 new ones in two of the new federal states) is used for monitoring SO_2, NO_2, particulate, CO and ozone ambient levels. Maximum and average weekly values are published every week in the VDI-Nachrichten and compared to the TA-Luft Standards (however, no such standard has been yet established for ozone). Finally, it is worth mentioning that the EEC Directive allows for more stringent SO_2 and particulate standards for areas of sensitive ecological systems. There are various aqueous effluent receiver quality standards depending both on the nature of the receiver (sea or fresh water, open sea or closed creek, lake or river) and its use (potable water course, fishing, bathing, industrial water intake, etc.). The major concern for aqueous receivers is not limited to organic load (in terms of BOD_5 and/or COD) and phosphates/nitrates, but is extended to toxic substances, also called priority pollutants. Defining priority pollutants or substances, dangerous for waters, is not an easy task. There are various lists of priority pollutants issued by the EEC, the Rhine Agreement, EPA, etc. [29]. Categories of priority pollutants for aqueous receivers are also defined in the Council Directive of 1976 [30]. Aqueous receiver standards for some of these priority pollutants and for various types of receiver have been defined both by the EEC [31] and the EPA [32]. Legislation in the FRG both at federal and at state level on storage and transportation of substances dangerous for waters is rather extensive and complex [33].

3.2.2. Emission standards. Combustion and power plants attract major interest for emission standards for the most common pollutants (SO_2, NO_2, particulates) [11, 26, 28, 34, 35]; there are also process-specific emission standards [11,27]. Furthermore, the EEC tries to achieve a global reduction in the emission of SO_2, NO_2 and particulates from big combustion sources [34,93]. Emission standards for water-borne pollutants are receiver-specific (e.g. [31]).

3.3. Requirements Related to Environmental Accidents

Major accident hazards are regulated by the SEVESO Directive [36] and are discussed in [94]. There are, however, various contingencies of operational upsets that might cause increased pollution compared to normal operation without constituting a hazard for a major environmental accident. These aspects (partial or total failure of pollution abatement systems, operation upsets or operation modes causing a temporary overloading of pollution abatement systems, etc.) should be adequately covered in the environmental impact study. Analysis of such potentially dangerous situations is useful for the development of contingency plans or reactions (e.g. continue operation at reduced capacity) to be agreed upon with the authorities. Early consideration of some general problems is also very useful. The possibility of proper temporary storage of untreated waste water or construction of pollution abatement units in multiple independent trains has a cost impact and must therefore be considered in the final feasibility study before a decision to proceed with a project is taken. Those environmental requirements may even affect the overall project design since alternative cleaner processes might thus become economically more attractive.

4. OIL REFINERIES

The refining industry has already settled down to a new list of priorities. The list includes the following: reduced growth rate for overall throughput capacity, more emphasis on lighter products, more stringent environmental regulations both for refinery operation and on the specification of final petroleum products and reduced energy requirements of individual processes.

The basic purpose of refining has always been to make lighter products. Refiners are spending much to add downstream processes in order to shift total product distribution toward these lighter products.

Improved efficiency of crude oil conversion is fostered by higher crude oil prices. The refiner must be more careful about how and where his own fuels are used within the processing scheme. A popular term for this preoccupation with fuel use is conservation.

The commercial processing and refining of crude oil into salable products began approximately one hundred years ago. The initial refining operations consisted of batch stills for separation of kerosene from crude oil. The so-called hydroskimming type of refinery which can be seen in operation today, though not as popular and rather obsolete, encompasses, besides crude distillation, processes like hydroprocessing, reforming, sweetening and hydrodesulphurization. A diagram of this simple refinery scheme is shown in Figure 1.

The need for other fractions of crude oil (gasoline, fuel oils, lubricating oils, solvents, asphalts and petrochemicals) increased rapidly and the demand for other petroleum products soon developed. To supply the demands, production plants were built to separate the crude oil more effectively, to crack or separate the high molecular weight hydrocarbon molecules, to polymerize or join hydrocarbon molecules into larger hydrocarbons, and to remove impurities from the various fractions effectively. This demand has tremendously increased the sophistication and complexity of the commercial plant to the point where the typical petroleum refinery today is a vast complex of process equipment, as depicted in Figure 2.

The installation of conversion units to European refineries started at high rates in the 1970s. In the EEC member countries, between 1974 and 1981, the total refining capacity increased marginally by 2.3% while the conversion capacity increased by 44.3%. In the early 1980s, though some refineries closed down, there was still an increase in conversion capacity [37].

Major pollution sources and typical pollution abatement systems for oil refineries will be briefly presented to allow a discussion of some interesting problems that might arise in the evaluation of an environmental impact assessment.

4.1. Pollution Sources from Oil Refineries

4.1.1. Air emissions. The major sources of air emissions from oil refineries are summarized in Table 3. Process heaters and fluid catalytic cracker regenerators are the major sources. Fugitive hydrocarbon emissions and sulphur oxides must also be considered.

i) Process heaters. Refinery heaters and power boilers are fired with the most readily available and economic fuel, which is normally a combination of refinery fuel gas and fuel oil. Fuel oil is typically a residual fuel oil, though in some cases commercial fuel oil grades (i.e. a blend of residual fuel oil and distillates) can be employed. Usually, 50% or more of the refinery heating needs will be provided by refinery fuel gas.

Refinery fuel consumption depends mainly on the type of process unit employed and the degree of heat conservation achieved and ranges between 1 and 6% (hydroskimming refineries) or between 1.5 and 9% (conversion refineries) of the processed crude (on a weight basis). Fuel consumption is increased by further 0.5 to 1%wt of the processed crude oil if the electric power requirements are also considered.

Flue gas rates and compositions can be calculated from a typical fuel composition and the percentage of excess air employed. For typical refinery fuels, flue gas flow rate is expected to be in the range of 16,000 to 20,00 Nm^3/t (fuel gas) or 12,000 to 16,000 Nm^3/t (fuel oil). A typical flue gas composition is given in Table 4.

The major pollutants in flue gases are sulphur oxides, nitrogen oxides, carbon monoxide and particulates. Carbon dioxide can also be accounted as a pollutant, due to its contribution to the greenhouse effect.

Sulphur oxide emissions are a function of the fuel sulphur content and can be calculated as 20 S kg SO_2/t (S being the weight percent sulphur content of fuel) [2]. On the average more than 95% of the fuel sulphur is emitted as SO_2, about 1 to 1.5% as SO_3 and almost 3% on the average (but with a strong data scatter) as particulates [39].

The term **nitrogen oxide** is applied to the mixture of NO and NO_2 emitted to the atmosphere but is always expressed as mass of NO_2 equivalent. The reason is that although up to 95% of NO_x emitted by combustion flue gases is NO, NO is converted rapidly to NO_2 in the atmosphere. The nitrogen content in the fuel is more readily transformed to NO_x ("fuel NO_x") [40,41]. Thermal NO_x is produced by the fixation of atmospheric nitrogen in the flame. At low flame temperatures, thermal NO_x production is not significant but emissions increase exponentially in the range of flame temperatures above 1,200 to 1,300°C [40,41]. Typical NO_x emission factors are given in Table 5. An API survey in three refineries [39] showed that on average NO_x emissions were only 60% of that predicted by the EPA factors; however, individual data were widely scattered (values measured varied from 16% up to 129% of the predicted NO_x emissions). Nitrogen oxide emissions from gas turbines are much higher (typically 4700 kg $NO_2/10^6$ m^3 [2]).

Carbon monoxide emissions are normally negligible provided that due attention is paid to proper heater operation (i.e. excess air) and maintenance. Typical emission factors for various fuels and heater types are given in [2].

Particulate emission is almost negligible from fuel gas combustion (16-80 kg/10^6 m^3 [2]) and increases as heavier liquid fuels are used (0.24 kg/m^3 for distillates, 1.25 kg/m^3 for residual oil [2]). Particulate emissions also increase with the sulphur content of liquid fuels. For heavy residual oil, EPA suggests an emission factor in kg/m^3 equal to 0.38 + 1.25 S (S being the sulphur % wt content of the fuel). The API survey [39] concluded that particulate emissions measured were:
- 7-65% of predicted (gas fired units);
- 52-87% of predicted (fuel oil fired units);
- 9-125% of predicted (mixed fuel fired units).

Unburned hydrocarbons or Volatile Organic Compounds (VOC) emissions from combustion sources are normally negligible with proper operation and maintenance (see also CO emission).

Comparison of the EPA emission factors with the results of the API survey indicates the limitations for proper application of the literature emission factors. Such factors are mean average values over long-term operation and cannot be compared to a spot measurement of the expected performance under certain operating conditions given by vendors. They are, however, good estimates of order of magnitude and any big deviation in the EIS from such figures should be adequately supported by the manufacturer. Furthermore, such estimate can be used in an early design phase to decide on possible environmental terms and constraints (e.g. use of certain fuel types only, installation of gas turbines or not, requirements for flue gas treatment (see point 4.2.1.i below), etc. for the project and hence avoid wrong decisions and delays during execution of the project.

ii) The **Fluid Catalytic Cracking (FCC) regenerator** is a major emission source for particulates, SO_2, CO, VOC, NO_2, aldehydes and NH_3. Modern units with a CO boiler or a High Temperature Regenerator practically do not emit CO, VOC, aldehydes and NH_3 [2,43]. EPA [2] has given emission factors only for two extreme cases (see Table 6). There are, however various possibilities of controlling particulate emissions (see Table 7). SO_2 emissions vary with the sulphur content of the FCC feed.

iii) **Sulphur oxide** emissions from combustion and FCC units have already been mentioned. The sulphur contained in the crude oil is partially converted to hydrogen sulphide in the distillation or in cracking operations and the remainder is transferred to the products with a tendency to be concentrated in the heavier fractions. Additional H_2S is produced in the hydrotreating units. This H_2S ends in the raw fuel gas system and will be part of the SO_2 emissions from the process heaters. However, due to environmental or even process reasons, H_2S is scrubbed in a H_2S washing unit. The H_2S rich stream is normally fed to a sulphur recovery (Claus) unit. Sulphur oxide emissions from the sulphur recovery unit can be estimated from the sulphur recovery efficiency and the load to this unit (an overall sulphur balance can be used to define the average feed to the sulphur recovery unit). Typical sulphur recovery efficiencies for various configurations of the Claus unit are [2]:
- two-stage reactors: 90-95 %;
- three-stage reactors: 95-97.5%;
- tail gas treatment: 98-99.9%.

These values are also representative of overall Claus plant performance and are normally adequate for taking decisions on general environmental constraints (e.g. tail gas treatment or not, configuration of the unit, etc.) at an early design stage. In some cases, decisions such as unit configuration (single or multiple trains in parallel) could be shifted at a defined latter stage of the design of the project without, however, leaving the project owner with illusions as to what the environmental requirements would be.

The flue gases from this unit also contain negligible H_2S (normally below 10 ppmv provided that the proper incineration temperature is maintained). Minor CO, particulate and VOC emissions from the fuel gas used in the incinerators could be also present while NO_x emissions are low because of the incineration temperature (~ 800°C).

The nitrogen in the crude oil follows the same route as sulphur but ammonia combustion contributes to NO_x emission only if high combustion temperatures are employed (which is not the case in a sulphur recovery unit).

iv) **Fugitive hydrocarbon** emissions. The main sources of VOC emission with the relevant references for emission estimation are:
- storage tanks [2,44,45];
- loading/unloading operations [2,46,47];
- oil/water separator [2,48-50];
- plant fugitive emissions (valves, flanges, vessels, pumps, compressors, safety valves, etc.) [2,51-54].

The vents of overhead drums in various distillation units should be normally connected either to the LPG recovery or the fuel gas system. Direct venting in the atmosphere would constitute additional emission sources of VOC, H_2S and NH_3.

Vacuum distillation units have another potential source of VOC and H_2S emission if the uncondensed stream is not vented to a firebox [2,50].

Landfarming of oily sludges are also emission sources of VOC and semi-volatile hydrocarbons - polynuclear aromatics (PNAs) in the form of particulates, especially during application of the waste and subsequent filling of the landfarm plots [80].

Operation of the flare [56], catalyst regeneration [26], furnace decoking and process unit turnaround [50,55] are also sources of intermittent atmospheric emissions.

A recent paper summarizes sources of plant refinery hydrocarbon emissions, emission rate estimation methods, dispersion modelling methods, ambient monitoring and the emission control techniques available [80].

4.1.2. Liquid emissions. The waste water flow from refineries varies in a wide range according to design features and the cooling system applied. Effluent flow rates might range from 0.1 to 0.3 m^3/m^3 of crude oil processed (use of cooling tower) up to 0.5 to 3 m^3/m^3 of crude oil processed (use of an open once-through cooling water system) [7,57,58]. The main pollutants are hydrocarbons, other organic compounds (phenols, sulphonic acids, alcohols), sulphurous compounds (sulphides, mercaptans, thiosulphates), sodium salts, ammonia, heavy metals, dissolved solids and suspended solids. Verification of the expected waste water effluent flow rate by means of a water balance is hardly possible since good estimates of steam losses (steam traps, equipment steam-out operations, heater soot blowing) and evaporation losses (cooling tower or skimming pond) are needed. Waste water effluent in European refineries in the early 1970s ranged from 0.08 to 2.8 m^3/t crude [58].

Process water is normally collected as sour condensate from overhead drums of distillation units containing hydrogen sulphide, ammonia, phenols, etc.

Effluents from specific processes (Cu-sweetening, doctor-sweetening and dimersol process) may contain heavy metals (Cu, Pb and Ni, respectively).

Spent caustics from sweetening units contain various acidic materials such as hydrogen sulphide, mercaptans, phenols, thiophenols and organic acids and usually also have a bad smell.

Desalted water contains hydrocarbons, NaCl and H_2S.

Tank and machinery drainages contain hydrocarbons and probably other pollutants depending on the material stored. Crude oil drainages might be of salty water and will contain sulphides. The total quantity depends on the amount of products imported.

Ballast water appears only for refineries served by marine terminals. The quantity depends on the frequency of product delivery and the size of the tankers [59]. These waters, which are usually highly saline, contain a large amount of hydrocarbons and suspended solids.

Cooling water effluents can only be contaminated accidentally. If cooling towers are employed, they will contain a higher level of dissolved solids as compared to fresh water, depending on the number of concentration cycles used.

Finally, boiler blowdown is clean condensate with some dissolved solid and the rest of the antifouling and reducing chemical additives used (phosphates, SiO_2, some Fe and Cu salts, Na_2SO_3/Na_2SO_4, hydrazine or synthetic non-carcinogenic hydrazine substitutes).

Sewage is produced by the refinery personnel. Finally, stormwater run off from process units and tank storage areas must be considered.

4.1.3. Solid wastes. Oily sludges collected during periodic cleaning of storage tanks or other equipment (e.g. heat exchanger bundles) and spent catalysts including sludge from alkylation and/or polymerization units are the major sources of solid refinery wastes. Sludge can also be a by-product of refinery waste water treatment facilities.

4.1.4. Other. Refinery machinery, especially compressors, are noise sources. Because of the size of refineries, their lay-out (storage area between process units and fence) and their noise control measures (enclosure of compressors) for occupational health purposes, normally, there are no problems with noise from refineries to the surroundings. Hydrocarbon emissions possibly with the presence of hydrogen sulphide, mercaptans, sprung acids, etc. can create smell problems. Air emission abatement techniques should also deal with these problems. Finally, soil contamination may occur because of accidental leakages from storage tanks or cross-country pipelines. It should be noted that CONCAWE members (western European refineries) operate nearly 200 different service pipelines of about 18,900 km combined length (status end 1989). The volume of crude oil and refined products transported through pipelines in 1989 amounts to 535 million m^3. Gross oil spillage reported in 1989 was 2,184 m^3 [60]. Various soil decontamination techniques have been reviewed in a recent publication [61]. Much research and development work in the soil decontamination field is carried out in Europe because of the limited availability of land. A review of the most promising soil decontamination techniques, mostly developed in the Netherlands and the FRG, is given in [62].

4.2. Pollution Abatement Technologies

A brief description of air emission and effluent control technologies will be given herewith. Solid waste disposal problems will also be discussed in brief.

4.2.1. Atmospheric emission control technologies. The major areas of concern for air emissions from refineries are combustion, FCC regenerator flue gases, sulphurous compound emissions and hydrocarbon emissions.

i) Combustion. Carbon monoxide, VOC and shoot emissions can be controlled by proper operation and maintenance. Sulphur oxide emissions can be reduced using a low sulphur fuel (low sulphur fuel oil, H_2S removal from fuel gas). Flue gas desulphurization [63] is widely applied to power plants but not yet widely applied to refineries [26,52]. There are several combustion modifications that can be employed to reduce NO_x emissions [2]:
- limited excess air can reduce NO_x emissions by 5-20%;
- staged combustion by 20-40%;
- low NO_x burners/flue gas recirculation by 20-50%;
- ammonia injection by 40-70%.

Water or steam injection and low NO_x burners can be used to control NO_x emissions from gas turbines [41].

While low NO_x burners and load reduction by firebox oversizing seem to be the most effective NO_x emission control techniques for new installations, the use of catalytic reduction with ammonia injection seems to be the most feasible way of meeting very stringent emission standards [41]. Ammonia injection is rather simple but there are process control problems.

Development of cost-effective flue gas treatment processes continues (e.g. [64]). There have been also quite a few attempts to develop of combined SO_x / NO_x removal processes (see e.g. [65-67]), but the main field of such applications remains thermal power plants.

ii) FCC regenerator. VOC, CO, aldehydes and ammonia are practically eliminated by a CO boiler or a high temperature regenerator [2,43]. Sulphur oxide emissions vary with the content of the sulphur in the FCC feed. Particulates are mainly controlled by two or three stage cyclones [2,26]. Further particulate recovery efficiency may be enhanced by means of electrostatic precipitators [2,26,42] or Venturi scrubbers [2,42,43]. Venturi scrubbers can simultaneously reduce SO_2 emissions producing an aqueous effluent (sulphites should be oxidized).

iii) **Sulphurous compounds.** Much of the sulphur contained in crude oil is transformed to H_2S during refinery processing. The H_2S rich gas must normally be routed to a sulphur recovery (Claus) plant, where most of the sulphur is recovered in the form of elemental sulphur. The Claus process consists in partial H_2S combustion. SO_2 and H_2S then react thermally or catalytically to produce elemental sulphur. Sulphur recovery efficiency can be up to 95% (two-stage reactors) or up to 97.5% (three-stage reactors). The use of pure oxygen instead of air for H_2S partial combustion can substantially increase the capacity of existing Claus plants [68]. The tail gas would then be either incinerated or routed to a tail gas treatment plant to reduce SO_2 emissions further.

The expected sulphur recovery efficiency varies with the process employed. Typical overall sulphur recovery efficiencies are as follows [69]:
- BSR/Selectox I process: 99% (in conjunction with a three-stage Claus plant);
- Sulfrin: 99%;
- Amoco CBA: 98%;
- Maxisulf: 99% (cited by vendor);
- IFP-1: ~ 99%;
- Wellman - Lord process: 99.9% (or 250 ppmv SO_2 in the tail gas flue gases);
- Beavon process: more than 99.9% (off gases typically contain 20-80 ppmv carbonyl sulphide and trace species of other sulphur gases and do not require incineration);
- SCOT process: more than 99.9% (150-250 ppmv SO_2 guarantee in flue gases, normally lower incineration temperature, ~ 540°C is required);
- ARCO process: ~ 99.9% (or 250 ppmv SO_2);
- Beavon/MDEA: same as for SCOT process.

iv) **Hydrocarbons.** There are quite a few hydrocarbon emission sources in a refinery. The main control technologies, classified according to each source, are as follows [52]:
- Combustion: proper combustion control and maintaining of adequate excess air;
- Flaring: use of steam at controlled rate proportional to flared stream flow rate;
- Product storage: (i) store light products (lighter than kerosene) in floating roof tanks or fixed roof tanks with internal floating roof or fixed roof tanks equipped with a vapour recovery system, (ii) use the most effective double seal system for floating roof tanks, (iii) use of pressure/vacuum vents on fixed roof tanks, (iv) avoid storage of unstabilized intermediate products, (v) mixing facilities for crude oil tanks to avoid excessive stratification;
- Safety valves and uncondensed gases: routing their discharge to the flare system. Installation of rupture disks upstream of safety valves or regular monitoring may contribute to additional emission reduction;
- Vacuum unit non-condensed gases: recover and incinerate in a fire box;
- Product loading/unloading: use submerged loading and vapour recovery system for light products;
- API separators: use of floating covers or use of closed type separators (PPI, CPI) provided that any operation and safety problems that might arise could be solved;

- Ballast/slop oil systems: vapour recovery systems for the relevant storage tanks;
- Cooling water: minimisation of hydrocarbon leaks into cooling water system, monitoring of cooling water for hydrocarbons;
- Compressor seals: mechanical seals, dual seals, purged seals, monitoring and maintenance programmes, controlled degassing vents to flare system;
- Pump seals: mechanical seals, dual seals, seal oil system, monitoring and maintenance programmes, controlled degassing vents to flare;
- Open ended pipes and valves: installation of cap or plug and regular monitoring;
- Process drains: installation of traps and covers;
- Valves and flanges: regular monitoring and maintenance programmes. The use of packless valves (diaphragm and bellow valves) has been suggested as an alternative mean of emission control, but their performance characteristics, size limitations and cost make their general use impractical;
- Sampling systems: installation of closed sampling loops for the purging of sampling system lines;
- Waste water collection system: venting of drains and sewers through liquid seals provided that safe operating conditions can be maintained, suitable construction of pump bases to allow rapid and complete drainage to the sewers and segregation of process water from storm water to the maximum possible extent;
- Coking operations: cooling the coke drum as much as practical while venting to a vapour recovery or blowdown system (delayed coking) or CO boiler (fluid coking);
- Vacuum distillation and/or asphalt blowing operations: collection and incineration of non-condensable gaseous stream.

4.2.2. Waste water control. The proper segregation of effluents is essential in the overall design of an effluent handling scheme to avoid dilution of pollutants and the consequent oversizing of complicated waste water treatment units.

Stripping of sour process effluents for H_2S and NH_3 is a necessary pretreatment. H_2S is stripped rather easily even at low temperatures. To remove ammonia effectively, temperatures of at least 110°C are required [5,7,70]. In the case of "fixed ammonia" (created due to the presence of acidic impurities), addition of caustic is also required. Stripping can be achieved by direct steam injection, partial reboiling (which permits condensate recovery and reduces the waste water flow), flue or fuel gas. Overhead condensers reduce the amount of sour gases and permit routing of the sour water stripper (SWS) overhead to the sulphur recovery unit. On the other hand, care is required to avoid hydrocarbon contamination in the Claus unit feed, since they might cause catalyst desactivation by coking. It would be better to incinerate small streams for which the presence of hydrocarbons cannot be excluded; the marginal SO_2 emission reduction by routing them to a Claus unit is offset by the more frequent Claus catalyst regeneration that will be required. Other contaminations (e.g. thiocarbonyl and other organosulphuric compounds) can be handled by a layer of suitable catalysts in the first Claus reactor.

The degree of H_2S and NH_3 removal in the SWS depends mainly on the number of trays or the height of the packing material and ranges from 90 to 99% (H_2S) and 30-98% (NH_3). Other pollutants such as phenols (30-60%) and cyanides (30-90%) can also be removed during stripping [5].

Spent caustics can be treated by neutralization, stripping or oxidation [70].

A refinery waste water treatment plant should include deoiling and biological treatment. The BP process (API separator, sand filters, trickling filter and final polishing lagoon) is adequate for hydroskimming type refineries. Extended aeration as a biological treatment step is advantageous in the presence of high phenol loads. The reuse of aqueous waste is very effective for effluent control.

4.3. Process Modifications for Reducing Pollution

There are various design features that contribute substantially to the reduction in pollutant emissions, still giving a positive economic impact [71,72]. Some typical examples in addition to the trivial energy recovery practice are:

- hydrodesulphurization of the FCC feed reduces SO_2 emission from the FCC regenerator and helps in increasing the capacity and yield of the FCC unit;
- installation of third-stage cyclones for the FCC regenerator flue gases permits use of an expander for electric power production;
- H_2S scrubbing of fuel gas reduces SO_2 emission and gives a revenue from the recovered sulphur. Energy recovery from heater flue gases can also be increased since the acid dewpoint of the flue gases is substantially decreased;
- H_2S scrubbing of reformer overhead gives a hydrogen-rich gas that can replace naphtha for ammonia production; sulphur recovery is also increased cutting SO_2 emissions;
- stripped phenolic sour water can be reused in the desalters reducing fresh water demand; simultaneously, phenols are transferred from the aqueous effluent to the crude oil reducing the pollutant load to the waste water treatment plant;
- treated waste water can be reused as service water, for aqueous solution preparation and as a cooling tower make-up.

Consideration of environmental constraints in an early design phase enables the evaluation of various design features that can reduce pollution so that the most cost-effective configuration can be adopted. Interaction with the authorities in this early design phase is, therefore, very important and beneficial both for the environment and the project.

Evaluation of pollution abatement systems for existing installations is more difficult. Generally, a cost/benefit analysis is very useful for selecting alternatives. Investment and operation cost for pollution abatement systems can be found in the literature (e.g. [69] for Claus plants). The cost must be updated by the use of cost indices [78]. Sometimes vendors must be contacted when cost information is not readily available. However, these figures are related to the delivery and erection of a self-sufficient unit, which requires interconnection to the existing facilities before it is brought on stream. Retrofitting is not always an easy task and the eventual impact on the operation and the safety of the installation must be considered. The cost of interconnection is not only restricted to the installation of new piping, and the eventual impact of shut-down must also be considered if the interconnection work cannot be performed during a scheduled turn-around. This implies that the cost of retrofitting existing installations can only be assessed in collaboration with industry.

5. INORGANIC CHEMICAL INDUSTRIES

Inorganic industries cover a wide range of activities and cannot be covered completely in a general presentation, as has been done for oil refineries. Sulphuric acid, nitric acid and fertilizer production will be briefly used as examples of inorganic industries.

5.1. Sulphuric Acid Plants

The contact process is mainly employed for sulphuric acid production. Melted sulphur combustion, H_2S combustion or sulphide ore roasting to SO_2 is the first processing stage followed by SO_2 catalytic (V_2O_5) oxidation. Water absorption of SO_3 finally gives H_2SO_4.

Double absorption plants achieve a conversion efficiency of 99.7% (or even higher if elementary sulphur is used as a raw material [74]) substantially reducing SO_2 emissions [73]. Conversion efficiencies of 96% to 98.5% are achieved in single absorption plants.

SO_2 emissions can be calculated from conversion efficiency [2]. Scrubbing with an ammonia solution may be required for single absorption plants. In this way, ammonium sulphate which can readily be used in fertilizer production is the by-product. There are also other processes available for SO_2 emission abatement (scrubbing with alkaline solutions, with hydrogen peroxide, molecular sieve absorption, etc.).

Acid mist is another pollutant in this stream as well as from oleum production. Electrostatic precipitators and mist eliminators can be used to control acid mist emission [2].

Particulate emission can also occur in the transfer, loading and unloading of solid materials.

5.2. Nitric Acid Plants

Nitric acid is mainly produced from ammonia in three stages:
- catalytic (Pt/Rh) ammonia oxidation to NO;
- NO oxidation with oxygen to NO_2;
- NO_2 absorption to nitric acid.

The first step is favoured by low pressure and high temperature. However, high pressure and temperature reduce the life of the precious catalyst used. High pressure and low temperature favour the second and third steps.

Either single (medium: 4 - 6.5 bar, or high: > 7 bar) pressure or dual pressure plants (atmospheric ammonia combustion followed by medium or high pressure NO oxidation and NO_2 absorption) are designed.

The second is the preferred approach in Europe [73]. NO_x emissions can be controlled either by extended absorption, catalytic reduction with CH_4 (or other means, e.g. NH_3), molecular sieve absorption or wet scrubbing [[6,75].

Typical uncontrolled emission factors are 22 kg/t acid [2] and can be reduced to 0.2 - 0.5 kg/t acid for catalytic reduction or 0.9 kg/t acid for extended absorption tail gas treatment techniques [2].

Tail gas temperature is rather low, so that the air pollutant dispersion models from stacks based on stack plume rise are not applicable for these NO_x emission sources.

5.3. Fertilizers

Nitrogenous or mixed NPK fertilizers are produced from ammonia, nitric acid, sulphuric acid, phosphoric acid and potassium or ammonium sulphate. The production steps are [73]:
- neutralization in aqueous phase;
- solution concentration by evaporation (for NH_4NO_3);
- solid formation (granulation or prilling);
- drying and bagging.

The most important pollutants are ammonia, nitric acid and dust emissions. Fluoride emissions are also present in the production of fertilizers containing phosphoric compounds. Emission control techniques include the collection of vapours from reactor and evaporator for condensation and filtration of the tail gases from other processes. Venturi scrubbers and packed absorption towers are normally employed for particulate emission control. Bag filters are very efficient but can hardly be used in "wet" ducted gases.

Typical emission factors for various operations (including storage and bulk loading) for various types of equipment are given in [2], mainly for uncontrolled emissions. Pollution abatement facilities and their expected efficiency should be used for estimating controlled emissions.

6. MONITORING

Monitoring enables the plant operator to:
- assess the pollution potential of the plant emissions;
- evaluate the performance of emission abatement systems/operations;
- estimate the effects of major emission sources to the environment;
- comply with control agency requirements.

Monitoring in the source is practically restricted to point sources (stacks). Continuous monitoring requirements exist according to German regulations whenever expected pollutant emission rates exceed threshold values (e.g. 50,000 m^3/h flue gas, 50 kg SO_2/h, 30 kg NO_2/h, etc.) [11,79] or according to US regulations for certain processes (e.g. sulphuric acid production, nitric acid production, big heaters, etc. [27]).

Monitoring of the ambient air quality and/or emission measurements is also important since this allows assessment of the effects of all emission sources. Careful planning and selection of the sampling point(s) is required for reliable results. Pollutant dispersion calculations not only from all establishment sources but from all pollution sources in the neighbourhood of the establishment in question is required.

Based on the results of these calculations, locations which are most affected by emissions from this industry and practically not affected by other pollutant sources in the neighbourhood should be identified and selected for ambient air quality measurements. Local environmental conditions (prevailing wind, frequency of wind occurrence from each direction, and wind intensity) must also be considered in the pollutant dispersion study and must be registered during sampling. In many instances, ambient monitoring is a requirement of the local regulatory agency, which will prescribe the procedures to be followed and equipment to be used [80].

Real-time dispersion models can be very useful for a combined monitoring of the ambient air and at the source [76].

With reference to aqueous effluents, grab sampling at the plant outlet and at the receiver is normally used for monitoring. Recording of the total waste water flow to the receiver is also a usual requirement. Sampling and analysis of near-surface soil has also shown great promise for the investigation of potential soil contamination [77].

7. CONCLUSIONS

The examples that have been presented above show that information is available in the literature which enables the identification of emissions from various processes and even a quantitative estimation of average emission rates. Pollutant emission factors obtained from a survey such as those of AP-42 [2] are statistical average values. They are very useful for estimating the order of magnitude of pollutant emission over a long period of time (e.g. year). They are also applicable to specific processes or specific pollution abatement techniques specified. Extrapolation of such emission factors must be done with great care. Hence, there is an obvious need to develop a commonly acceptable definition of the various terms which are used in compiling and evaluating an EIS as well as reaching an agreement on the proper use of available knowledge.

Furthermore, the stages of the EIS development and updating in the life of a project must be defined. For each stage the targets of the EIS evaluation and the necessary extent of information to be given in the EIS must be precisely described. When these conditions are fulfilled, a meaningful communication with the public can be attempted, working towards achieving a consensus based on the evaluation of the EIS both from a natural scientist's and from a socioeconomic point of view. Such a consensus should be indispensable for all stages of the EIS within the framework of the EIA procedure.

ACKNOWLEDGEMENTS

Most of this presentation is based on experience acquired by the author while performing design engineering and consulting work as an employee of Asprofos Engineering S.A., Athens, Greece, or as a private consultant. The environment-related projects were performed together with Mr. D. Davios, Mr. D. Kardomateas, Dr. S. Konstas, Mr. D. Latas, Mr. Th. Papazoglou, Mr. S. Spiliopoulos, Mr. N.Vlogiaris, Mrs. A. Zavitsanou and Mr. E. Zervakos.

REFERENCES

[1] EEC Council Directive 90/313/EEC (1990), On the freedom of access to information on the environment, Official Journal of the European Communities, L158, 56.

[2] U.S. EPA (1985), Compilation of air pollutant emission factors, Vol.I: Stationary point and area sources, AP-42, 4th edition.

[3] U.S. EPA (1986), Supplement A to compilation of air pollutant emission factors. Vol.I: Stationary point and area sources, AP-42, Supplement A.

[4] U.S. EPA (1977), Revision of emission factors for petroleum refining, EPA-450/3-77-030, PB 275685.

[5] G. Drogaris (1988), Treatment of industrial waste waters, ELKEPA, Athens.

[6] S.E. Jørgensen (1979), Industrial waste water management, Studies in Environmental Sci., 5, Elsevier Scientific Publication Company, Amsterdam - Oxford - New York.

[7] M.R. Beychok (1967), Aqueous wastes from petroleum and petrochemical plants, John Wiley and Sons, London - New York - Sydney.

[8] P. Zannetti (1990), Air pollution modelling - theories, computational methods and available software, Van Nostrand Reinhold, New York.

[9] J.G. Bartzis, C. Housiadas, S. Andronopoulos, C. Cuvelier, R. Nijsing and J. Würtz (1991), "ADREA-HF: a three-dimensional finite volume code for dense vapour cloud dispersion in complex terrain", paper presented at ERCOFTAC Managing Board Meeting, Athens.

[10] M. Bonazountas (1984), "Calculation methods and mathematical models", lecture in Seminar on Fate and Risk of Waterborne Pollutants, NTU-EPA-MIT, Athens.

[11] TA Luft, Revision May 1987.

[12] C. Friedl (1989), Schärfere Bestimmungen sollen Wasser schützen, VDI - Nachrichten, No. 27 (07.07.89).

[13] Editorial (1991), Auf dem Weg zur Nullemission, VDI - Nachrichten, No. 11, 46 (15.03.91).

[14] L. Huber, "Schwermetalle in Abläufen von Erdölraffinerien", Erdöl und Kohle - Erdgas - Petrochemie vereinigt mit Brennstoff-Chemie, 30(6), 276.

[15] N.R.W. (1985), Guidelines of the Minister of Nordrhein - Westfalen for work, health and social affairs for operation permits of crude oil refineries and petrochemical plants,

154

MBI NW.

[16] VDI (1988), Richtlinien Handbücher, Beuth Verlag.
[17] A. Amendola and S. Contini (1991), National approaches to the safety report - a comparison, CEC, JRC - Ispra, SP - I.91.07.
[18] VDI-KUT (1991), "6. Mitglieder-Information" (Juni 1991).
[19] 11th Annual report on alkali etc. works (1974).
[20] Her Majesty's Inspectorate of Pollution U.K. (1985), Notes on best practicable means No.17 - sulphuric acid (class II) works - sulphuric acid manufacture.
[21] Her Majesty's Inspectorate of Pollution U.K. (1980), Notes on best practicable means - HF manufacture.
[22] Her Majesty's Inspectorate of Pollution U.K. (1987), Notes on best practicable means No. 25 - chemical fertilizer works.
[23] Her Majesty's Inspectorate of Pollution U.K. (1987), Notes on best practicable means No. 26 - ammonia works (manufacture of ammonium nitrate and of ammonium phosphate).
[24] Her Majesty's Inspectorate of Pollution U.K. (1987), Notes on best practicable means No. 24 - nitric acid works.
[25] Her Majesty's Inspectorate of Pollution (1991), "Integral pollution control - the timetable", The Chemical Engineer, No. 493, 28.
[26] S. Calvert and H. Englund (1984), Handbook of air pollution technology, John Wiley Sci., New York.
[27] U.S. EPA (1986), Standards of performance for new stationary sources - Vol.I: introduction, summary and standards, EPA-340/1-86-005a.
[28] K. Stamelos (1984), "EIS for planning and development", paper presented in the Seminar of the Greek Ministry of Land Use Planning, Housing and Environment, Athens, October 1984.
[29] K. Krisor (1982), Umwelt (4), 234.
[30] EEC (1976), Directive of the Council from 4 May 1976 for the pollution from discharge of dangerous substances to waters (76/464/EEC), Official Journal of the European Communities, L129, 23.
[31] EEC (1986), Directive of the Council, 12 June 1986, for limit values and quality targets for the discharge of dangerous substances defined in the Annex of Directive 76/464/EEC (86/280/EEC), Official Journal of the European Communities, L158, 35.
[32] G. Karantounias and M. Bonazountas (1984), "Process and fate of toxic pollutants", paper presented at the Seminar on Behaviour and Fate of Hazardous Pollutants in Aqueous Receivers, NTU - EPA - MIT, Athens.
[33] E.W. Diesel and H.P. Lühr (1982), Lagerung und Transport wassergefährdender Stoffe, Erich Schmidt Verlag, Berlin (last updating 1990).
[34] EEC (1983,1987), COM(83)704/15.12.83 and discussions for this directive of Apirl 1987.
[35] OECD (1984), Emission standards for major air pollutants from energy facilities in OECD member countries.
[36] Commission of the European Communities (1990), Council Directive 82/501/EEC on the major accident hazards of certain industrial activities, EUR 12705.
[37] B. Kotsira-Douka, G. Drogaris, A. Skamagas, N. Skandalis and Th. Vasilopoulos (1986), "Upgraded refineries - new petroleum products and environment", report of the organization committee in the TEE conference, Athens, December 1986.
[38] A. Kumar and S. Mohan (1991), "Use of bulletin board system for air quality modelling", Environmental Progress, 10(1), F8.
[39] API (1983), Characterization of particulate emissions from refinery process heaters and boilers, API publication 4365.
[40] API (1977), Manual on disposal of refinery wastes. Volume on atmospheric emissions.

Chapter 9: sources and control for emissions of nitrogen oxides, API publication 931.

[41] D. Cobb, L. Glatch, J. Rund and S. Snyder (1991), "Application of selective catalytic reduction (SCR) technology for NO_x reduction from refinery combustion sources", Environm. Progr., 10(1), 49.

[42] API (1983), Particulate emissions from non-fired sources in petroleum refineries: a review of existing data, API publication 4363.

[43] J.D. Cunic, M.G. Bienstock and A.M. Edelman (1987), "Wet gas scrubbing: state of the art in FCCU emission control", Environm. Progr., 6(4), 267.

[44] API (1980), On evaporation loss from external floating roof tanks, API publication 2517.

[45] API (1962), On evaporation loss from fixed roof tanks, API publication 2518.

[46] API (1959), On evaporation loss from tank cars, tank trucks and marine vessels, API Bulletin 2514.

[47] API (1959), On hydrocarbon emissions from marine vessel loading of gasolines, API publication 2514A.

[48] API (1973), Hydrocarbon emissions from refineries, API publication 928.

[49] API (1976), Manual on disposal of refinery wastes. Volume on atmospheric emissions, Chapter 7: hydrocarbon emissions, API publication 931.

[50] U.S. EPA (1977), Control of refinery vacuum producing systems, wastewater separators and process unit turnarounds, EPA 450/2-77-025.

[51] U.S. EPA (1981), VOC fugitive emissions in petroleum refining industry, EPA 450/3-81-015a.

[52] U.S. EPA (1980), Assessment of atmospheric emissions from petroleum refining, EPA 600/2-80-075a-d.

[53] U.S. EPA (1983), Petroleum fugitive EIS. Emissions - background information for promulgated standards, EPA 450/3-81-015b.

[54] VDI (1983), Emission control oil refineries, VDI - Richtlinien 2440.

[55] U.S. EPA (1976), Revision of evaporative hydrocarbon emission factors, Radian Corporation, EPA 450/3-76-039.

[56] U.S. EPA (1983), Evaluation of the efficiency of industrial flares, EPA 600/2-83-070.

[57] ATV-VKS (1984), Grundsätze für den Bau und Betrieb von Abwasserreinigungsanlagen in Erdölraffinerien, A-701.

[58] L.J. Huber (1975), Proc. 9th World Petroleum Congress, 6, 257.

[59] Degremond (1979), Water treatment handbook, Halsted Press.

[60] CONCAWE (1990), Performance of oil industry cross-country pipelines in Western Europe, CONCAWE report No. 6/90, Brussels.

[61] W.L. Lyman, D.C. Noonau and P.J. Reidy (1990), "Cleanup of petroleum contaminated soils at underground storage tanks", Noyes Data Corporation, Pollution Technology Review, No. 195 (ISBN No. 0-8155-1258-9), Park Ridge, NJ.

[62] T.H. Pheiffer, T.J. Nunno and J.S. Walters (1990), "EPA's assessment of European contaminated soil treatment techniques", Environmental Progress, 9(2), 79.

[63] G.H. Shroff, A.F. Papa and J.M. Whalen (1985), "Emissions control at a coke-fired cogeneration plant", CEP, October 1985, 51.

[64] R. Hofmann (1991), Entschwefelung wie von selbst, VDI - Nachrichten, No. 26, 25 (28.06.91).

[65] R.J. Gleason and D.J. Helfritch (1985), "High-efficiency NO_x and SO_x removal by electron beam", CEP, October 1985, 33.

[66] J.T. Yeh, R.J. Demski, J.P. Strakey and J.I. Joubert (1985), "Combined SO_2/NO_x removal from flue gas", Environmental Progress, 4(4), 223.

[67] R.J. Walker and H.W. Pennline (1988), "Absorption, electrodialysis and additional regeneration in two flue gas SO_2/NO_x cleanup processes", Environmental Progress,

156

7(4), 215.

[68] J. Redman (1991), "Increasing sulphur recovery", The Chemical Engineer, 13 (April 1991).

[69] U.S. EPA (1983), Review of new source performance standards for petroleum refinery Claus sulphur recovery plants, EPA 450/3-83-014.

[70] API (1969), Manual on disposal of refinery wastes. Volume on liquid wastes.

[71] G. Drogaris (1986), "Environmental impact from HAR monternization", paper presented at the TEE conference on Upgraded Refineries - New Petroleum Products and Environment, Athens, December 1986.

[72] G. Drogaris (1988), "Environmental impact form HAR upgrading project", paper presented at Energex '88, Tripoli, November 1988.

[73] Kirk-Othmer (1983), Encyclopedia of chemical technology, 3rd ed., John Wiley and Sons, New York.

[74] U.S. EPA (1971), Background information for proposed new source performance standards, APTD-0711.

[75] VDI (1983), Emission control - nitric acid production, VDI - Richtlinien, No. 2295.

[76] A.F. Bais, C.S. Zerefos and I.C. Ziomas (1989), "Design of a system for real-time modelling of the dispersion of hazardous gas releases in industrial plants. Part I: emissions from industrial stacks", J. Loss Prev. Process Ind., 2, 155.

[77] M.H. Schlender and J. Rower (1989), "Soil gas survey techniques for the investigation of underground storage tanks", Environmental Progress, 8(4), 231.

[78] The Institution of Chemical Engineers and the Association of Cost Engineers (1988), A guide to capital cost estimating, 3rd ed., London.

[79] Der Bundesminister für Umwelt, Naturschutz und Reaktorsicherheit (1990), "Bekanntmachung der Neufassung des Bundes-Immissionsschutzgesetzes vom 14. Mai 1990", Bundesgesetzblatt, Teil I, No..23, 880.

[80] H.H. Shiel, J.H. Siegell and A.L. Jones (1991), "Modelling plant hydrocarbon emissions", The Chemical Engineer, No. 502 (15.08.91), 21.

[81] French Republic (1989), Installations registered for purposes on environmental protection, Act. No. 76-663 of July 19, 1976, Decree No. 77-1133 of Sept. 21, 1977.

[82] Minister for the Environment and the Prevention of Major Technological and Natural Risks (1989), Installations classified for the protection of the environment: storage of liquefied flammable gas, rubric 211, Ministerial Decree of 9 November 1989.

[83] H.J. Uth (1989), "Protection of areas in the vicinity of hazardous industrial plants in the FRG", in: Emergency Planning for Industrial Hazards, 3, edited by H.B.F. Gow and R.W. Kay, Elsevier Applied Science.

[84] (1978) Abstände zwischen Industrie- bzw. Gewerbegebieten und Wohngebieten im Rahmen der Bauleitung Runderlass v. MAQS, 25.07.1974/02.11.1977 MB1 NW, Seite 1688/SMB1.NW280.

[85] D. Lyddon (1984), "The role of EIA in physical land use planning", paper presented in the Seminar of the Greek Ministry of Land Use Planning, Housing and Environment, Athens, October 1984.

[86] H.S.E. (1989), Risk criteria for land-use planning in the vicinity of major industrial hazards.

[87] A.J. Muyselaar (1992), "The Dutch integral environmental zoning project", this book.

[88] F. Bro-Rasmussen (1992), "Environmental impact assessment and risk analysis in Denmark", this book.

[89] S.E.A.T.M. van der Zee, F.A.M. de Haan (1992), "Soil and ground water quality indicators", this book.

[90] M.J.F. van Pelt, A. Kuyvenhoven, P. Nijkamp (1992), "Sustainability, efficiency and equity: project appraisal in economic development strategies", this book.

[91] S. Cernuschi, M. Giugliano (1992), "Air quality assessment in environmental impact studies", this book.
[92] J. Jörissen, R. Coenen (1992), "The EEC Directive on EIA and its implementation in the EC Member States", this book.
[93] P. Stief-Tauch (1990), "Legislation in the European Communities", lecture presented to the Eurocourse on Sulphur Dioxide and Nitrogen Oxides in Industrial Waste Gases: Emission, Legislation and Abatement, Ispra, September 1990.
[94] A. Amendola (1992), "The EEC Directives on environmental hazards", this book.

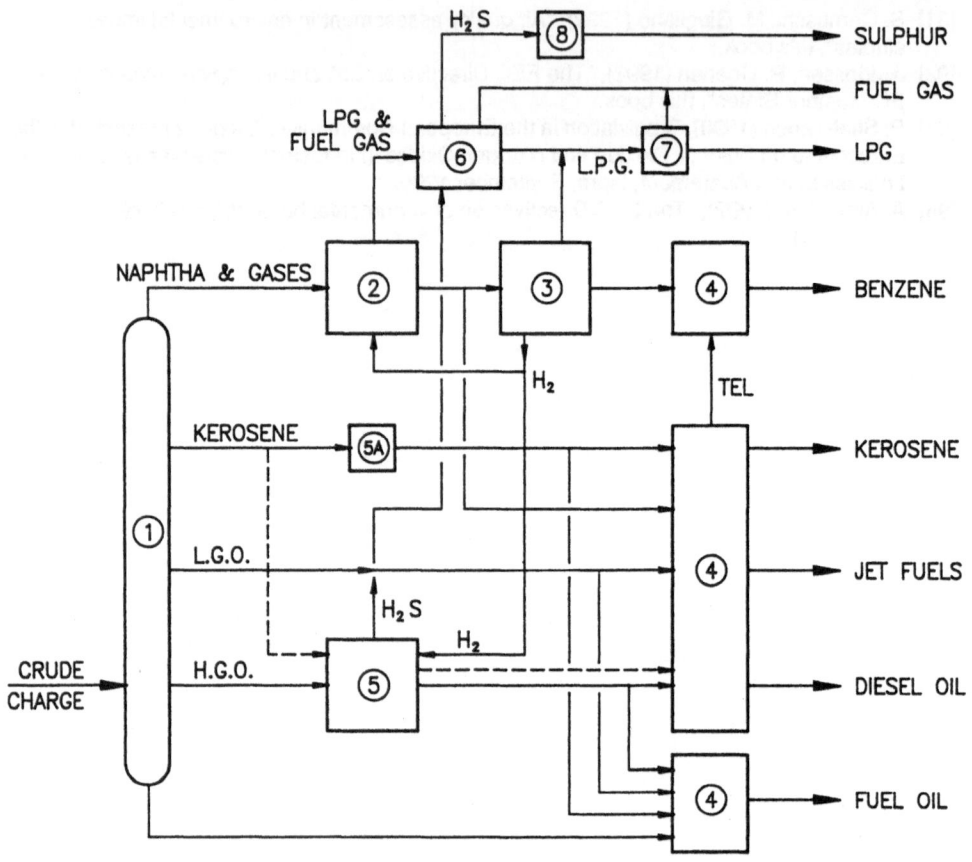

①	CRUDE DISTILLATION	⑤	HYDRODESULPHURIZATION
②	HYDROTREATING	⑥	H_2S SEPARATION
③	REFORMING	⑦	LPG RECOVERY
④	BLENDING	⑧	SULPHUR PRODUCTION
⑤ₐ	SWEETENING		

Figure 1. Typical block flow diagram of a hydroskimming refinery.

159

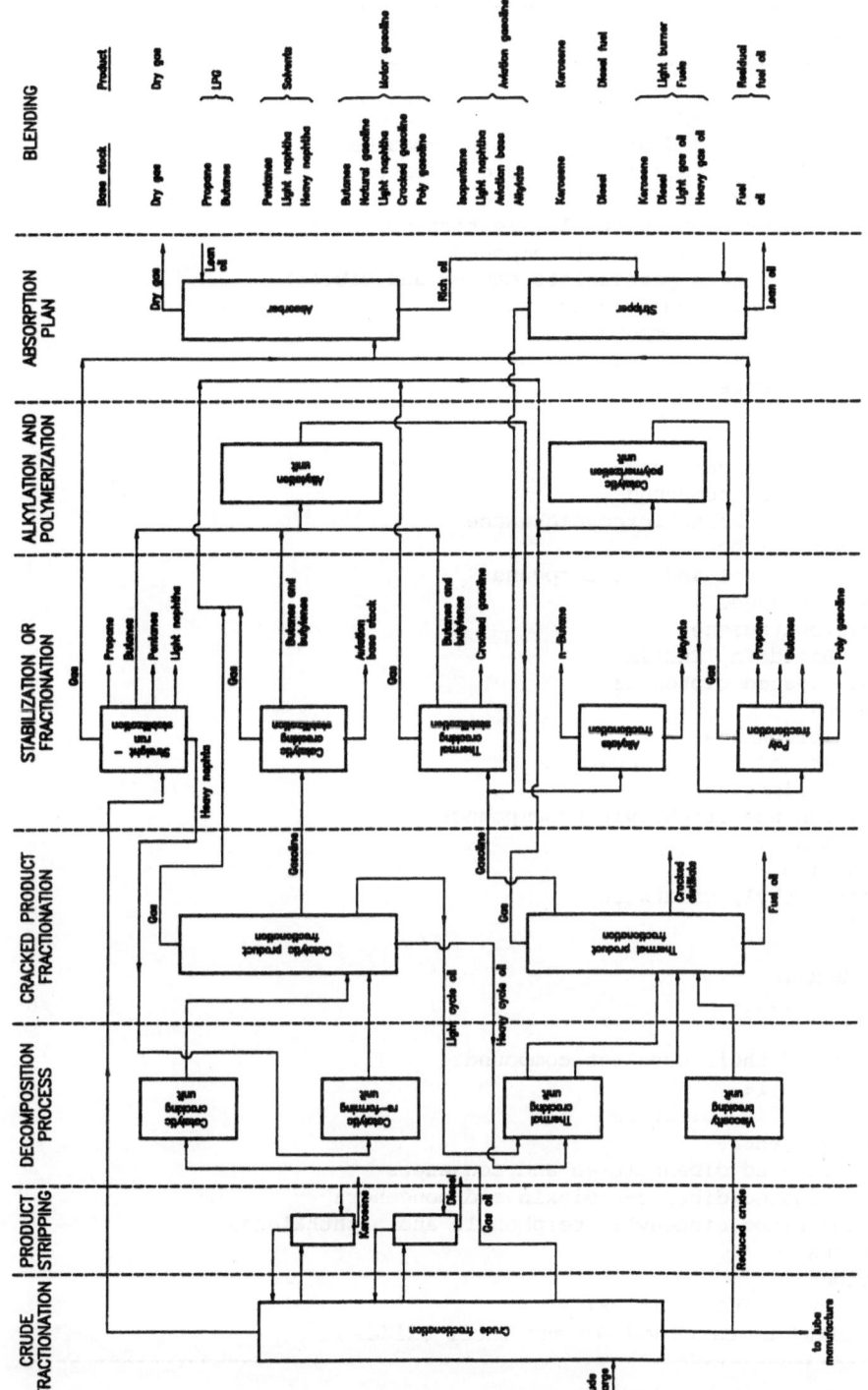

Figure 2. Typical block flow diagram of an integrated conversion refinery.

Table 1. Prescribed substances in the new Environmental Protection Act of the U.K. (Source [25]).

Release to air
Oxides of sulphur and other sulphur compounds
Oxides of nitrogen and other nitrogen compounds
Oxides of carbon
Organic compounds and partial oxidation products
Metals, metalloids and their compounds
Asbestos (suspended particulate matter and fibres),
    glass fibres and mineral fibres
Halogens and their compounds
Phosphorus and its compounds
Particulate matter

Release to water
Mercury and its compounds
Cadmium and its compounds
All isomers of hexachlorocyclohexane
All isomers of DDT
Pentachlorophenol and its compounds
Hexachlorobenzene
Hexachlorobutadiene
Aldrin, Dieldrin, Endrin
Polychlorinated biphenyls
Dichlorvos
1,2-dichloroethane
All isomers of trichlorobenzene
Atrazine, simazine
Tributyltin and triphenyltin compounds
Trifluralin
Fenitrothion
Azinphos-methyl, malathion
Endosulfan

Release to land
Organic solvents
Azides
Halogens and their covalent compounds
Metal carbonyls
Organo-metallic compounds
Oxidising agents
Polychlorinated dibenzofuran and congeners
Polychlorinated dibenzo-p-dioxin and congeners
Polyhalogenated biphenyls, terphenyls and naphthalenes
Phosphorus
Pesticides
Alkali metals and their oxides
    and alkaline earth metals and their oxides

Table 2. Summary of ambient air quality standards.

Pollutant	EEC	USA	California	WHO
SO_2	250 or 350 µg/m³ (*) (98% of samples not to exceed this value) 130 or 180 µg/m³ (**) (mean average of daily values over winter time: 1 Oct-31 Mar) 80 or 120 µg/m³ (***) (mean average of daily values over one year)	1,300 µg/m³ (yearly basis) 260-365 µg/m³ (daily basis) 60-80 µg/m³ (3h basis)	1,310 µg/m³ (hourly basis) 105 µg/m³ (daily basis)	100-150 µg/m³ (daily basis) 40-60 µg/m³ (yearly basis)
NO_2	200 µg/m³ (hourly basis not to be exceeded for 98% of time) 135 µg/m³ (hourly basis not to be exceeded for 95% of time) 50 µg/m³ (hourly basis not to be exceeded for 50% of time)	100 µg/m³ (yearly average)	470 µg/m³ (hourly basis)	400 µg/m³ (hourly basis) 150 µg/m³ (daily basis)
Particulates	250 µg/m³ (not to be exceeded for 98% of time)			
TSP	130 µg/m³ (mean average of daily values over winter time: 1 Oct-31 Mar) 80 µg/m³ (mean yearly average of daily values)	150-260 µg/m³ (daily basis) 60-75 µg/m³ (geometric mean, yearly average)	100 µg/m³ (daily basis) 60 µg/m³ (geometric mean, yearly average)	150-230 µg/m³ (daily basis) 60-90 µg/m³ (yearly average)

*The higher value applies if TSP < 150 g/m³, otherwise the lower value is applicable.
**The higher value applies if TSP < 60 g/m³, otherwise the lower value is applicable.
***The higher value applies if TSP < 40 g/m³, otherwise the lower value is applicable.

Table 3. Summary of major sources of air emissions within a refinery (Source mainly [26]).

Process units	Emission type
Crude unit including desalter*	None.
Vacuum unit*	Noncondensable vapours from steam condensers.
Catalytic cracking*	SO_2, NO_x, particulates, carbon monoxide, hydrocarbons, aldehydes and possibly ammonia.
Catalytic reforming*	Generally none, although some hydrocarbons may be released during regeneration. If regeneration is continuous, minor amounts of CO may be released.
Catalytic hydrocracking*	Emissions from catalyst regenerations which may release significant quantities of CO over relatively short time spans.
Alkylation*	None.
Light ends recovery*	Essentially none.
Catalytic hydrotreating*	Similar to catalytic hydrocracking.
Hydrogen manufacture*	None.
Solvent deasphalting	None.
Residual oil hydrodesulphurizing*	None, other than catalyst regeneration.
Asphalt blowing	None, assuming the vent gases, which are highly odorous, are vented before release.
Coking*	Wind-blown coke dust and emissions from the storage containers for the water used in cutting the coke.
Storage and blending	All storage tanks containing light hydrocarbons, including crude, have the potential to emit hydrocarbons to the atmosphere. Breathing losses can result from weathering or during emptying and filling. Also, the seals of floating roof tanks will leak to a certain extent.
Process heaters	Particulates, SO_2, CO, hydrocarbons, aldehydes, NO_2. The quantities will, of course, depend on the source of fuel gas, oil, residual oil, etc., the physical properties of each and the burner design (for NO_2 emissions, particularly).
Waste water treatment	API-separator (hydrocarbons).
Pressure relief and flare systems	Hydrocarbons.

*It is assumed that uncondensed gases from overhead of distillation columns are routed to the LPG (light ends) recovery or the fuel gas system, sourwater to the sourwater stripper unit and the overhead from the sourwater stripper is not vented directly to the atmosphere.

Table 4. Typical flue gas composition (% vol).

Substance	Fuel gas	Fuel oil
CO_2	6-9	10-13
H_2O	14-20	10-12
O_2	2.5-3.2	1.7-4.5
N_2	71-74	73-75

Table 5. Typical NO_x emission factor (Source [2]).

Fuel	Utility boiler ($>25x10^5$ kcal/h)	Industrial boiler
Natural gas	4400-8800 kg $NO_2/10^6$ Nm^3	2240 kg $NO_2/10^6$ Nm^3
Distillate oil	-	2.4 kg NO_2/m^3
Residual oil	5-12 kg NO_2/m^3	6.6 kg NO_2/m^3

Table 6. Typical emission factor from FCC regenerators (Source [2]).

Pollutant	Uncontrolled	El.Prec. + CO boiler
Particulates (kg/m^3 feed)	0.695 (0.267-0.976)	0.128 (0.020-0.428)
SO_2 (*) (kg/m^3 feed)	1.413 (0.286-1.505)	1.413 (0.286-1.505)
CO (kg/m^3 feed)	39.2	neg.
VOC (kg/m^3 feed)	0.63	neg.
NO_2 (kg/m^3 feed)	0.204 (0.107-0.416)	0.204 (0.107-0.416)
Aldehydes (kg/m^3 feed)	0.054	neg.
Ammonia (kg/m^3 feed)	0.155	neg.

*Depends mainly on the sulphur content of the FCC feed.
Values in parentheses indicate measured range.

Table 7. Particulate emission (kg/m3 feed) from FCC regenerators.

```
------------------------------------------------------------------
Emission control system      Source: API [42]      Source: EPA [2]
------------------------------------------------------------------

Internal cyclones                 0.870                 0.695
                               (0.048-4.136)         (0.267-0.976)

External cyclones                 0.03(*)                 -

CO boiler                         0.704                   -
                               (0.03-1.887)
Electrostatic preci-              0.135                   -
pitator (ESP)                  (0.01-0.278)

ESP and CO boiler                 0.085                 0.128
                               (0.026-0.431)         (0.020-0.428)

CO boiler and scrubber            0.026                   -
                               (0.024-0.028)
Complete regeneration
and SO₂ absorbing                 0.124(**)               -

catalyst
------------------------------------------------------------------
```

 *Based on one reference only.
**Based on two references only.
Values in parenthesis indicate measured range.

SILVIA: A DECISION SUPPORT SYSTEM FOR ENVIRONMENTAL IMPACT ASSESSMENT

A. Colorni and E. Laniado
Systems Theory Centre of the National Research Council
and Department of Electronics
Milan Polytechnic
Via Ponzio 34
I - 20131 Milan
Italy

ABSTRACT. A software package has been developed, containing a methodological framework and a decision support system for environmental impact assessment (EIA) problems. The package is named SILVIA (the acronym, in Italian, stands for Interactive Software for Environmental Impact Assessment) and includes three parts, concerning, respectively, the preliminary phase, the analysis phase and the decisional phase of an EIA procedure. The first part - which is named SERENA - is a data base containing information about Italian environmental legislation, check lists, existing data, available models and methods, previous case studies and bibliographic references. It is being developed at present and is not described here. The second part - which is named GAIA - is based upon spread sheet techniques, and makes it possible to create trees and matrices at different aggregations levels, to coordinate both the results of prediction models and the qualitative estimates and to represent the single steps of the analysis phase. The third part - which is named VISPA - is based upon multi-attribute analysis techniques, including: (a) elimination of dominated alternatives by the Pareto criterion, (b) preference analysis, (c) concordance and discordance analysis, (d) sensitivity analysis. The package has been developed on an IBM-PC and is user-friendly. The interactive procedure all over the steps makes it possible to really involve both the decision maker and the interested groups, in order to clarify the actual entity of conflicts and to provide information aiming and some more rationality and transparency in decision making.

1. INTRODUCTION

Environmental Impact Assessment (EIA) problems are generally characterized by several alternatives, by decision criteria that are not always easy to quantify and by the presence of interest groups with conflicting objectives. These problems thus require interdisciplinary work and great attention to the real development of the decision process [1].

It should be noted, however, that many of the environmental impact "assessments" performed in recent years consist in onerous collections of data, followed by the use of more or less sophisticated methods aiming at predicting some of the effects of a project on the natural, social and economic environment; they were, in fact, more "surveys" than assessments. Moreover, such "surveys" were performed according to different points of view, with no reference to a common standard: this makes comparison of different studies difficult and, even worse, means that it is often impossible for a public authority to really check the adequacy of the impact study. So, the final decision is often the product of intuition and the personal ability of

167

A. G. Colombo (ed.), Environmental Impact Assessment, 167–180.

the decision maker rather than the result of a decision process characterized by transparency and the participation of the groups involved.

In a complete environmental impact assessment one can distinguish, from a methodological point of view, three main phases.

The first is a preparatory phase, in which the problem must be correctly formulated: this implies, among other things, an analysis of the environmental standards, of the information available (data, cartography, but also the decision structure, the social groups involved, ...) and of similar previous cases.

The second is an analysis phase, which consists of the following steps: definition of the significant project alternatives and identification of the elementary activities associated with each of them; identification of the environmental sectors which could be affected by any of the alternatives and selection of a set of indicators for each of them; collection of data and, where possible, use of simulation models in order to obtain qualitative or quantitative estimates of the impact of elementary activities on the environmental indicators; aggregation of single estimates in order to forecast the impact of each alternative as a whole on each of the indicators.

Finally, there is an evaluation phase: its crucial aspects are the study of conflicts and the definition of the decision criteria.

The three phases, which do not necessarily follow in this order and can partially overlap, can be repeated several times, with successive levels of refinement.

The SILVIA project (SILVIA is the Italian acronym of Interactive Software for EIA) includes three software packages: SERENA, an expert system which is useful in the orientation and setting up phase of an environmental impact study; GAIA, the aim of which is to organize and represent the logic path and the set of operations necessary to coordinate the analysis and prediction phase; and VISPA, a decision support system based on multi-attribute analysis techniques. GAIA and VISPA can be used independently.

The programmes, which are user friendly, aim at making the technical phases of an EIA study transparent and repeatable. Transparency is obtained through continuous documentation and by saving all the elementary operations step by step; repeatability is guaranteed by the fact that it is possible to change some of the estimates and/or operations and to run the whole study rapidly. Transparency and repeatability are necessary conditions for real control by public authorities and for an effective participation of social groups in the decision process. From the point of view of both the decision maker and the designer, the proposed method is a tool for a logical organization of an EIA study, aiming at reducing time and cost and creating a common language. Here, we briefly present the characteristics of GAIA (Guide to Environmental Impact Analysis) and VISPA (Integrated Evaluation for the Choice between Alternative Projects).

2. GAIA (Guide to Environmental Impact Analysis)

2.1. The Basic Tools for the Analysis

In the analysis phase of an EIA one must in general study and represent sequences of cause-condition-effect relationships [2]: as an example, the ground level concentration of an atmospheric pollutant (effect) is the result of one or more emissions (cause) in a particular atmospheric situation (condition). For this phase, we propose an integrated use of three basic tools: check lists, trees and matrices.

2.1.1. Check lists. Check lists are normally used for the definition of the elements characterising each project alternative and/or the environment. For this purpose a series of disaggregations is performed: in order to describe a project alternative, it is necessary to identify a set of time steps and, for each of them, a set of elementary activities; in order to describe an environment, it is necessary to identify its different components and subcomponents (e.g., the component "atmosphere" can be further subdivided into "meteoclimatology" and "quality", "quality" into "micro" and "macropollutants", and so on). The lists so obtained have a hierarchical structure, i.e. an organization of the items at various levels. Check lists are not exhaustive and cannot be defined once and for all: as an example, the project activities depend on technology (and thus change with technological innovation) and the environmental components depend on the cultural and territorial characteristics of the area under examination.

2.1.2. Trees. Check lists can be represented by special graphs ("trees"), composed of nodes and arcs: the initial node (the only one without predecessors) is called the root and the terminal nodes (with no successors) are called the leaves; to describe the relationships existing between the nodes of a tree, we will use a genealogical terminology: if we are in a particular node, the node from which it originates is its father, the nodes with the same father are its brothers and the nodes which originate from it are its sons. A tree representing a check list can be enriched with further nodes in order to describe a temporal or spatial division or to take into account the possibility of different situations, such as atmospheric conditions or alternative scenarios.

2.1.3. Matrices. The use of matrices makes it possible to represent cause/effect relationships by crossing the nodes of two trees: a column tree, whose nodes indicate the "causes" and a row tree, whose nodes indicate the "effects". Each cell of a matrix thus relates a cause node with an effect node and may contain two pieces of information: a synthetic one, i.e. a number or a symbol, which represents the estimate (qualitative or quantitative) of the effect of the column node on the row node, and a descriptive one, i.e. a documentation of how the estimate was obtained and of its meaning.

2.1.4. The combined use of check lists, trees and matrices. The combined use of trees (enriched check lists taking into account various possible contexts and conditions) and matrices allows us to describe the typical cause/condition/effect relationships of the environmental impact analysis. As an example, the cause node "emission of SO_2" could be disaggregated so as to give rise to several sons associated to different atmospheric conditions (south wind, north wind, ...); similarly, for the effect node "ground level concentration of SO_2", one could have a geographical disaggregation (in district A, in district B, ...). In this example the matrix cells at the most disaggregated level could contain the estimates of the ground level concentration of SO_2 in each district and for each of the atmospheric conditions assumed.

A matrix relates all the nodes of one tree with all those of the other, independently of the hierarchical level in which they are found. Often, however, it is useful to choose, display (as rows or columns of the matrix) and make available for processing, only a subset of the nodes of a tree. The matrix on which one works is therefore generally a submatrix, representing the intersections between the selected subsets of the nodes of the row and the column trees: this allows working on the same matrix at various depth levels, i.e. selecting different submatrices.

2.2. Operations on Trees and Matrices

The main operations which can be performed on a pair of trees and the related matrix are the following.

2.2.1. Construction, modification and documentation of a tree. It is possible to insert, move and cancel both single nodes and subtrees; it is possible to save trees and subtrees so that they can be called again when necessary. One or more documentation screens can be associated with each node.

2.2.2. Selection of rows and columns to display in the matrix. It is possible to "switch on" or "switch off" both single nodes and subtrees of the row and column trees: only the cells concerning the pairs of nodes switched on are displayed in the matrix; the switched off nodes are ignored, but any information contained in non-displayed cells is kept in the memory and is available when required.

2.2.3. Insertion, modification and documentation of estimates. The matrix cells can contain both symbols or words (qualitative estimates) and numbers (quantitative estimates). It is possible to insert, modify or cancel the content of any cell at any time. One or more documentation screens can be associated with each cell containing an estimate.

2.2.4. Conversion of qualitative estimates to numerical values. The problem of performing (and documenting) a transformation of qualitative estimates into numbers is tackled by defining one or more "vocabularies". A vocabulary is a conversion table (from symbols and words to numbers) which can be created and assigned to one or more rows of the matrix. The conversion takes place by applying the related vocabulary to each row; it is a reversible operation, in the sense that one can restore the initial symbols or words at any time. Vocabularies (and relative assignments to rows) can be created, modified, cancelled, documented and saved so that they can be called again and applied when necessary.

2.2.5. Aggregations. Often the estimates obtained by the use of prediction models (qualitative or quantitative) are at a disaggregated level. Let us consider a submatrix whose columns and rows are the elementary activities of the construction phase of an alternative project (cause) and the environmental indicators (effects) respectively: in this case the matrix cells contain the estimates of the effects of each of the single activities on each of the environmental indicators. In order to estimate the overall effect of the set of the construction phase activities on each of the environmental indicators, it is necessary to perform an aggregation, i.e. to replace the columns representing single activities by a unique column representing the whole construction phase. The cells of this column contain, row by row, an estimate obtained by the application of a particular aggregation rule (minimum, sum, mean, weighted mean, ...). In the GAIA package it is possible to assign an aggregation rule to each row (in order to aggregate columns, as in the example cited) or to the columns (in order to aggregate rows); furthermore, it is possible to change, save and call again when necessary any set of aggregation rules (and the relative assignments).

2.3. The Architecture of GAIA

GAIA [3] is organized by projects: each project is defined by one or more alternatives, these in turn can be formed by a combination of constitutive elements.

2.3.1. Project alternatives. The presence of several feasible alternatives makes the decision process significant. As a minimum, there are two alternatives (e.g. to carry out a plan or not); in general, there can be strategic alternatives (e.g. a main road or a railway), which are investigated at the beginning of the decision process; intermediate alternatives (about location, technology, size), which are considered during the design process; minor alternatives (variants or mitigation actions) in the last phase of the decision process, concerning the executive project. Both the analysis phase, by means of the GAIA package, and the evaluation and choice phase, by means of VISPA, can be repeated at different depth levels and at different times.

2.3.2. Constitutive elements. A constitutive element is defined as a component of one or more alternatives, which can be studied, at least in part, autonomously. The subdivision of a main road into segments is a typical example of the use of constitutive elements: some of the impacts (those which do not depend on the project as a whole but only on whether a particular portion of territory is crossed or not) can be studied separately. If a segment is common to several route alternatives, the impacts estimated for it are valid for all the alternatives involved. A constitutive element is however not necessarily a "physical" component of an alternative: it is possible to separate, for instance, the operating phase actions from the construction ones, or the environmental indicators from the territorial, social or economic ones. In any case, the utility lies in dividing a large analysis into a number of homogeneous sections which are easier to handle.

2.3.3. Matrix sequences. An alternative or a constitutive element is represented in the GAIA package by a sequence of matrices, which follow, phase by phase, the logical path of the analysis. A matrix sequence may describe, for instance, a chain of cause/effect relationships: in this case the same tree can be a column tree (causes) or a row tree (effects) in successive matrices. As an example, a chain of cause/effect relationships is that defined by the following pairs of trees: project activity (causes) vs. environmental indicators; environmental indicators vs. environmental indicators (to take into account indirect effects); environmental indicators vs. human activities influenced.

Let us consider now a pair of trees and the related matrix representing a set of cause/effect relationships: it is possible to perform a series of operations on it (enrichment of the trees, selection of nodes, insertion of estimates, conversions, aggregations, ...). The results of each operation can in turn be memorised in a further sequence of matrices, whose rows and columns always have the same meaning. In this case each matrix represents a particular processing phase of the same set of cause/effect relationships.

2.3.4. Generation of alternatives. A matrix representing a complete alternative can be generated at any time by selecting a suitable set of constitutive elements. The programme compares the trees of the single constitutive element matrices and automatically composes the resulting complete alternative matrix, assigning a position in the general frame to each constitutive element with the following rule: if two constitutive elements have the same row tree, the programme places them side by side; if two constitutive elements have the same column tree, the programme forms them into a column. So, the row and column trees of the resulting matrix are the union of the rows and the columns present in at least one of the constitutive elements, respectively. Finally, the programme transfers the estimates contained in the cells of the single constitutive elements matrices into the corresponding cells of the resulting matrix.

2.3.5. The impact matrix. At the end of the analysis of an alternative it is possible to obtain a final vector, i.e. a single column representative of the complete alternative. In this case the column tree has just one node; if the nodes of the row tree are environmental indicators, the matrix cells contain an estimate of the effect of the complete alternative on them.

To compare the alternatives, the programme automatically constructs an "impact matrix", which contains a column for each final vector and a set of rows corresponding to the union of all the nodes present in at least one of the final vectors. The construction of the impact matrix poses a problem: in general the rows of different alternatives are not necessarily coincident. As an example, siting alternatives imply effects in distinct territorial areas; technological alternatives may have effects on different indicators. The cells of the impact matrix will thus be empty whenever an alternative (column) has no specific effect on an indicator (row): the correct content of any empty cell should be the estimate of the effect of the "do-nothing" alternative on the row indicator. For this reason the do-nothing (or "zero") alternative must always be present in the impact matrix and its "effects" must be estimated for all the rows. If the column representing the do-nothing alternative is full, the programme automatically assigns, row by row, its estimates to the empty cells of the impact matrix.

2.3.6. The connection with the VISPA package. An impact matrix created by means of the GAIA package can be used as an input to the VISPA package, which organizes the evaluation phase, in which the alternatives are to be compared. For this purpose the only condition is that the matrix must be a numerical one; qualitative estimates must be previously converted into numbers, possibly by means of vocabularies, as already described.

3. VISPA (Choice among Project Alternatives)

The VISPA package [4] is illustrated here by a description of its main logical steps.

3.1. Processing of the Initial Data

The impact matrix is the input data set on which the VISPA package works. It can be obtained from outside (e.g. by means of GAIA) or created autonomously inside VISPA and memorised. It is also possible to memorize and call again an impact matrix on which some operations have already been performed (such as normalization, aggregation and the application of value functions, which will be described below).

3.1.1. Normalization. In an impact matrix the indicators are generally measured in completely different units: the matrix may contain, as an example, social and economic indicators (such as the unemployment rate or the project cost) and environmental indicators (such as suspended particulate per cubic metre or the variety of animal species in an area) [5]. The first operation proposed is thus normalization, i.e. transforming the matrix elements into non-dimensional units, for instance numbers between 0 and 100.

Normalization is a linear transformation, which is carried out row by row. The most usual method is to divide all the cells of a row (i.e., the row indicator values corresponding to the different alternatives) by a suitable normalization value (usually the maximum value). This operation, although essentially technical, should be checked carefully. The numbers obtained depend in fact on the normalization value assumed, which in turn can depend on the alternatives considered. So, the numbers in the normalized matrix could be varied by deliberately introducing some alternative that is not really significant in the decision process, but which can influence the normalization values.

For this reason, in VISPA normalization is an optional operation, which can be performed in various ways, e.g. by introducing normalization values from outside or by defining a reference alternative.

3.2. Transformation of Indicators into Objectives

A crucial step in VISPA is the transformation of indicators into objectives, which can be performed by combining the following two operations: application of value functions and aggregation.

3.2.1. Value functions. The value function [6] of an indicator is defined in VISPA as the relationship between the values that the indicator itself can assume and a non-dimensional measure of the corresponding "satisfaction" or " benefit", expressed in numbers between 0 and 1. So, the application of a value function to a row of an impact matrix yields a set of numbers between 0 and 1, where 1 indicates a maximum "satisfaction" and 0 a minimum "satisfaction" for the effect of a column alternative on the row indicator considered.

In general, the value function may assume any form. In the VISPA package, the user can define a suitable value function for each indicator. As an example, Fig. 1 shows a "bell" function, whose parameters can be fixed by the user interactively with the programme.

3.2.2. Aggregation. It is not always possible to define a value function corresponding to a single indicator. Moreover, complex problems (in which several interests and groups are involved in the decision) are necessarily characterized by a large number of indicators, because, in general, each sector (atmospheric pollution, water pollution, social services, ...) is described by several indicators.

In these cases, a partial aggregation of indicators can be performed, in order to obtain a smaller number of rows which are representative of the various sectors (sector indices). As an example, concerning atmospheric pollution, SO_2, NO_x, suspended particulate and so on may all form a single global air quality index, which contains the information about the overall effect of the set of indicators considered. In general, the sector index is a non-linear function of the indicators, characterized by the presence of synergies.

Aggregation can also be used to reduce the number of objectives, obtaining an aggregated objective from a set of single ones.

3.3. Ranking the Alternatives: the Weighted Sum

In order to rank the decision alternatives, even if partially, it is necessary to estimate the relative importance of the objectives (rows) and thus indirectly of the sectors and interests that they represent. For this purpose, the following steps are made available in VISPA.

3.3.1. Elimination of dominated alternatives. The Pareto criterion makes it possible to eliminate inefficient alternatives, i.e. alternatives that are dominated by some other one: an alternative A is dominated by an alternative B if, for all the objectives considered, the performance of A is not better than that of B and for at least one of the objectives considered the performance A is worse than that of B.

As an example, Fig. 2 considers the performance of five alternatives in relation to two objectives to maximize. It can easily be seen that, as alternatives 1 and 4 are dominated by 2 and 5 respectively, the set of efficient alternatives comprises only alternatives 2, 3 and 5.

Once the programme has determined the dominated alternatives, the user can eliminate them: elimination is optional.

3.3.2. Attribution of weights. To suggest one (or more) vectors of relative weights of the objectives, the technical analyst, the decision maker, the sector experts and the social groups involved require an interactive support.

It has been noted that, even for a small number of objectives (less than 10), it is difficult for anybody to express a vector of relative weights directly. It is easier to perform a series of direct comparisons between pairs of objectives, answering a series of questions such as: how important is objective i compared with objective j ?

However, a technical problem arises: it is possible to obtain a vector of relative weights, equivalent to the set of pair comparisons, only if the answers obtained are "mathematically consistent" [7]. In particular, if the importance of objective i with respect to objective j is a_{ij}, the two properties of reciprocity ($a_{ij} = 1 / a_{ji}$) and consistency ($a_{ik} = a_{ij} \times a_{jk}$) must hold. In real cases it is difficult for the property of consistency to hold, especially if there are many objectives. It is however always possible, by using mathematical methods, to calculate a vector of relative weights, which is an interpretation of the inconsistent answers obtained.

In the VISPA package the user can express the weight vector either directly or by pairwise comparisons. Once a relative weight has been attributed to each row (i.e., to each objective), the VISPA package calculates for each column(alternative) its weighted sum, a number representing the overall behaviour of the alternative itself. So, it could be possible to evaluate all the alternatives under examination, to rank them and possibly to choose the best one. But the result is strongly dependent on the vector of assigned weights, i.e. on a subjective, uncertain and conflictual operation.

3.3.3. Sensitivity analysis. Expressing the weight vector is an uncertain and subjective operation, as it reflects somebody's preferences; it is a conflictual operation, as different decision makers or social groups may give different answers. Therefore, there is little sense in looking for "the optimum solution". Aiming at clarifying the real significance of existing conflicts, it may prove more useful to identify, for each objective, a critical weight interval, outside of which the ranking of alternatives and consequently the final choice would really change, as a function of the weight vector.

Keeping all the weights constant apart from one, VISPA computes the maximum variations (increase and decrease) of this weight that would not alter the ranking and thus not change the final choice. As an example, if the initial ranking among three alternatives were < A, B, C >, corresponding to the attribution of a weight $w_j = 0.15$ to objective j, a possible result of the sensitivity analysis might be:

$$0.06 < w_j < 0.30 \text{ , final choice: A,}$$

$$w_j = 0.06 \text{ , final choice: C,}$$

$$w_j = 0.30 \text{ , final choice: B.}$$

This is a real support to the decision maker, since the programme identifies the objectives such that a reasonable variation of their weight causes changes in the final choice, suggesting that further attention should be concentrated on them and on the really conflicting alternatives.

3.4. Other Ranking Methods and Elimination Phase

A controversial aspect of decision methods is the attempt to determine directly the final choice among a number of feasible but conflicting alternatives. It might prove more realistic to invert such a logic and, instead of looking for the best alternative directly, to eliminate gradually the worst ones, or at least those which are not satisfactory.

For this purpose, one idea is to calculate different rankings of the alternatives according to methods based on different logics (for instance, weighted sum and risk minimization): such a procedure makes it possible to identify the less significant alternatives (the ones placed in the last positions according to any possible ranking) and to eliminate them. Of course, after each elimination, the procedure (and thus the calculation of the various rankings) must be repeated to allow the necessary reconsideration of alternatives whose effects might have been "masked" by others now eliminated.

Let us now consider some of the ranking methods different from the weighted sum that are available in VISPA.

3.4.1. Concordance and discordance matrices. Some of the rankings created by VISPA are based on the calculation of the concordance and discordance matrices [8], whose generic elements c_{hk} and d_{hk} are a measure of "satisfaction" in choosing the alternative h instead of k and of "regret" in giving up alternative k due to h, respectively. Fig. 3 shows, as an example, an impact matrix composed of 3 alternatives and 4 objectives, together with a vector containing the weights attributed to the objectives. In order to compute the concordance and the discordance matrices, the programme compares the alternatives by pairs.

The concordance index c_{hk} of alternative h with respect to alternative k is obtained by summing the weights of the objectives for which h is preferred or indifferent to k. Fig. 4a shows the concordance matrix of the example of Fig. 3.

The discordance matrix can be constructed in a similar way. To calculate the discordance index d_{hk} of alternative h with respect to alternative k, the programme identifies the two rows (objectives) such that the product of the weight of the objective for the performance difference of the alternatives h and k is maximum, among the objectives for which k is preferred to h and among all the objectives respectively. The discordance index d_{hk} is the ratio between the two values thus obtained. The discordance matrix of the example of Fig. 3 is shown in Fig. 4b.

Using the information contained in the two matrices, the programme assigns to each alternative two numbers, the concordance and discordance absolute indices [9], which are computed according to the following definitions:

$$i_h^{(c)} = \sum_j c_{hj} - \sum_i c_{ih}$$

$$i_h^{(d)} = \sum_j d_{hj} - \sum_i d_{ih}.$$

The concordance absolute index of alternative h is a measure of how much it prevails over all the others: the higher its value, the more satisfactory alternative h is. The discordance index is a measure of overall regret in the case where the final choice is alternative h: the lower its value, the more satisfactory h is. Thus two further rankings of the alternatives, based on the absolute index of concordance and the absolute index of discordance respectively, are made available.

3.4.2. Weak dominance. Let us consider two thresholds, of concordance (S_c) and discordance (S_d), respectively: an alternative h can eliminate an alternative k if the two conditions

$$S_c \leq c_{hk}$$
$$d_{hk} \leq S_d$$

hold simultaneously.

Note that if $S_c = 1$ and $S_d = 0$ the two conditions above correspond to the classical Pareto dominance criterion; by decreasing S_c and increasing S_d this criterion is gradually relaxed: substantially a concept of absolute dominance (Pareto) is replaced by a concept of "weak" dominance. It is obvious that the result depends greatly on the threshold values S_c and S_d.

In VISPA the weak dominance concept is used in two ways. In the first, the user fixes the threshold values S_c and S_d and the programme shows the weakly dominated alternatives and, if required, eliminates them. In the second, the programme calculates, for each alternative h, the set of concordance and discordance threshold values such that h is weakly dominated by at least one other alternative. The results of this analysis are shown in Fig. 5. It is obvious that

176

the larger the area in the S_c-S_d plane in which alternative h (h = 1, 2, 3, 4) is weakly dominated, the less satisfactory the alternative is. So, a further ranking ("weak dominance ranking") is obtained.

3.4.3. Other ranking methods. Besides those illustrated up to now, the VISPA package creates a ranking based on a partial weighted sum: the user can select a subset of objectives, for which the programme calculates the weighted sums of the alternative performances and shows the corresponding ranking.

Furthermore, the programme computes two rankings based on the max-min logic: in these rankings, the alternatives such that the "best worst case" and the "best weighted worst case", respectively, occupy the highest positions.

Finally, the VISPA package shows the user all the rankings created in the various steps of the evaluation phase.

3.4.4. Elimination of alternatives. On the basis of information obtained during the various steps, the user can eliminate the unsatisfactory alternatives and/or the less conflicting objectives to concentrate further resources on a more thorough assessment of a reduced matrix (possibly by repeating the preceding steps with suitable variations).

4. CONCLUSIONS

An EIA study implies both technical aspects and subjective criteria. For this reason formalized models and software packages must be flexible, interactive and user friendly; moreover they must favour an integrated processing of qualitative and quantitative estimates.

The SILVIA package organizes an environmental impact study along a precise logical path. The analysis phase is made transparent, thanks to the possibility of documenting and memorizing all the steps, and repeatable, thanks to the possibility of calling the matrices again and rapidly repeating all the processing, modifying some estimates and operations if necessary. As for the evaluation phase, a set of different rankings is proposed, which are obtained according to the fundamental logics of maximizing an aggregate objective function or minimizing some measure of risk or inequality. The programme continuously interacts with the user, especially during the crucial steps of transforming the indicators into objectives and of attributing weights. Finally, the programme performs a sensitivity analysis and makes it possible to iterate the procedure by successive eliminations of the least satisfactory alternatives.

The methodology and the software packages which form the SILVIA project aim at introducing some rationalization and transparency into the decision process, by means of the experimentation of methods that imply a formalization of the procedures and an effective interdisciplinarity from the technical viewpoint and require an active participation of the groups involved from the social viewpoint. In this sense we hope that SILVIA can become a coordinating tool between different technical disciplines and an instrument of communication between the technical and the social worlds.

REFERENCES

[1] N. Lee, "Environmental Impact Assessment: a Review", Applied Geography, No. 3, 1983.
[2] L.W. Canter, Environmental Impact Assessment, McGraw-Hill, New York, 1977.
[3] A. Colorni, E. Laniado, GAIA, Clup, Milano, 1991.
[4] A. Colorni, E. Laniado, VISPA, Clup, Milano, 1988.

[5] H. Inhaber, Environmental Indices, John Wiley, New York, 1976.
[6] R. Keeney, H. Raiffa, Decisions with multiple objectives: Preferences and Value Tradeoffs, John Wiley & Sons, New York, 1976.
[7] T.L. Saaty, "Eigenvector and Logarithmic Least Squares", European Journal of Operational Research, 48, 1990.
[8] B. Roy, "Décisions avec Critères Multiples: Problèmes et Méthodes", Metra, 1, 1972.
[9] A. Goicoechea, D.R. Hansen, L. Duckstein, Multiobjective Decision Analysis with Engineering and Business Applications, John Wiley, New York, 1982.

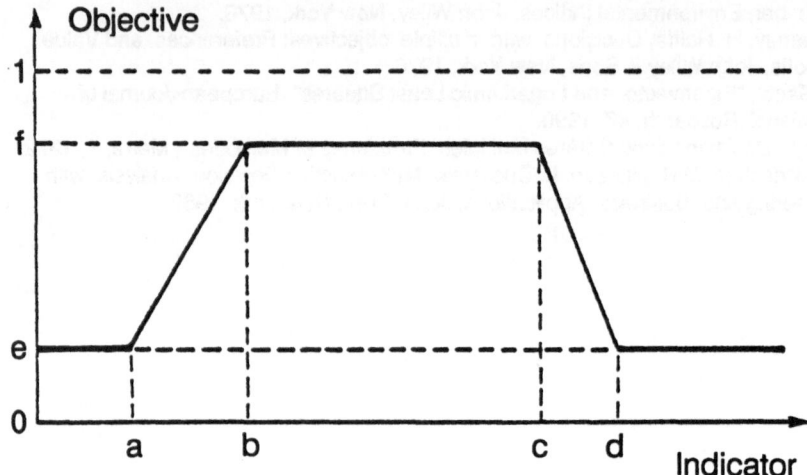

Figure 1. An example of a value function.

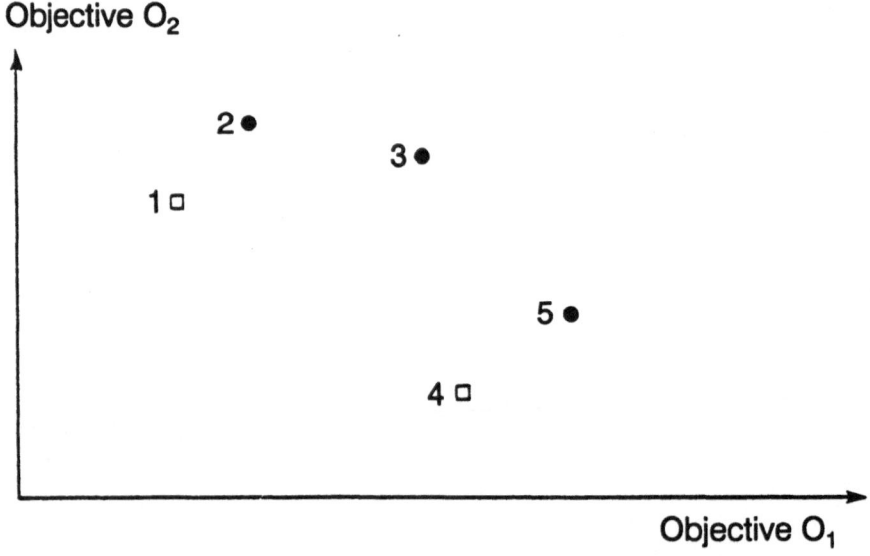

Figure 2. The Pareto criterion (see text for explanation).

	alt.A	alt.B	alt.C		weights
obj.1	1.0	0.6	0.4		0.3
obj.2	0.7	0.5	1.0		0.4
obj.3	0.8	1.0	0.8		0.2
obj.4	0.8	1.0	0.6		0.1

(a)　　　　　　　　　　　　　　　(b)

Figure 3. An impact matrix (a) and a weight vector (b).

	alt.A	alt.B	alt.C
alt.A	-	0.7	0.6
alt.B	0.3	-	0.6
alt.C	0.6	0.4	-

	alt.A	alt.B	alt.C
alt.A	-	0.33	0.66
alt.B	1	-	1
alt.C	1	0.30	-

(a)　　　　　　　　　　　　　　　(b)

Figure 4. Concordance (a) and discordance (b) matrices corresponding to the impact matrix of figure 3.

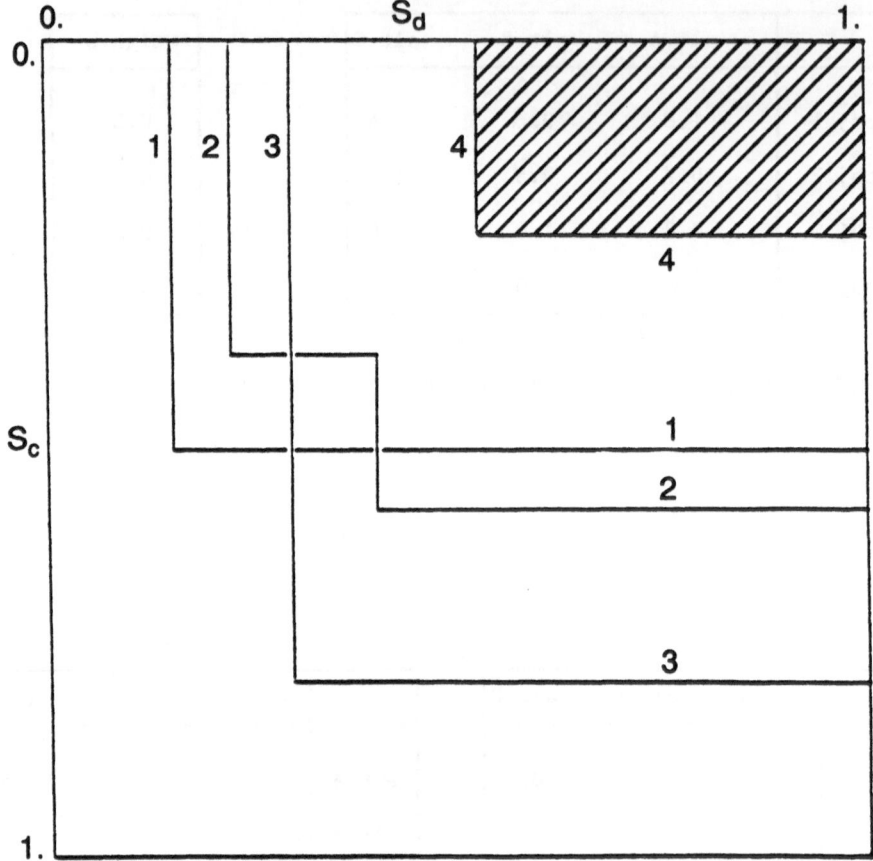

Figure 5. Weak dominance: the weak dominance area of an alternative is the upper right area. In the figure, the weak dominance of alternative 4 is sketched.

ENVIRONMENTAL INDICATORS AND MEASUREMENT SCALES

G. Volta
CEC, Joint Research Centre
Institute for Systems Engineering
and Informatics
I - 21020 Ispra (Va)
Italy

A. Servida
TECSA
Research and Development Division
Via Caravaggi
I - 24040 Levate (Bg)
Italy

ABSTRACT. The paper reviews the nature of indicators and indices as knowledge carriers in the Environmental Impact Assessment (EIA) process and revisits the theory of measurement in the perspective of its application to the knowledge relevant for EIA. In particular, the most critical issues related to the comparison and composition of indices are pointed out and discussed.

1. INTRODUCTION

EIA is a negotiated cognitive process which aims to help in reaching decisions related to a man-made intervention on the natural heritage. As such, it presents many facets and it can be modelled only by considering a large number of dimensions or variables: dimensions related to the actors (humans) and their values and dimensions related to the natural heritage. This multidimensionality is expressed by the proliferation of "indicators" and "indices" that identify objects, aspects, qualities, of particular relevance for the description of the human intervention, of the environment and of "changes" introduced in this environment. The set of indicators used when dealing with EIA, includes, as a subset, the physical quantities which are the object of the physical sciences. But, while the old family of physical indicators aimed at representing only the physical world, new families of indicators aim at capturing relevant aspects of the whole natural, social and psychological world. The proliferation of the indicators used in EIA is motivated, in the end, by the final objective: supporting a negotiation and assisting a collective decision. Therefore, the tendency to cover domains of knowledge which are as large as possible should be accompanied by precautions about the ways in which this knowledge is manipulated and formally structured. In particular, one must pay attention to the statistical analysis, composition and aggregation of indicators and indices, operations needed to make the negotiation and decision process manageable.

In our "scientific" world, measurement a privileged approach to formal manipulation and representation of knowledge. The practical aim of measurement is to map the knowledge onto the set of the real numbers i.e. to reduce all the qualities of the world, as it is perceived by our mind, to the particular qualities of a geometrical entity: the straight line. In order to be conscious of the value and of the pitfalls of this simplified representation, we must be aware that this representation arrives at the end of a long process, starting from the perception of the world through concepts, through their patterns built by operations of similarity and analogy then through linguistic expressions biased by the specific language used. In brief, we can say that measurement supposes a semantics, but does not replace a semantics.

A. G. Colombo (ed.), Environmental Impact Assessment, 181–188.
© 1992 ECSC, EEC, EAEC, Brussels and Luxembourg.

The foundations of measurement were a matter of investigation mainly by psychologists and mathematicians in the 1920s. The theory was well settled in the 1960s. But in that theory, based on rigourous mathematical concepts, the cognitive structure which is behind the formal operation of measurement is only superficially considered. In recent years, a vigorous cognitive science nourished by practical interests has emerged around the application of Artificial Intelligence, pushed to deepen that cognitive structure mainly under aspects such as imprecision, vagueness and uncertainty, which are characteristic of the socio-environmental disciplines. Moreover, computer science has developed techniques capable of formal representation and manipulation of knowledge directly in symbolic and logical terms without the need for a numerical transformation. These recent scientific and technological developments are expected to provide, in the future, a new framework, based on a better understanding of the cognition process, for measurement.

In this paper, the basic concepts and principles of measurement are summarized also taking account of this emerging perspective. Given these premises, the paper aims at reviewing the nature of indicators and indices, as knowledge carriers in the EIA process and looks against the well settled theory of measurement from the point of view of its application to attributes of the world which are relevant for EIA.

2. INDICATORS

The growth of attention to various attributes of the environment and the growth of monitoring activity has been accompanied by many efforts to categorize indicators. The most comprehensive effort in this direction, at the global environmental level, has been performed by the OECD [1] which is currently developing indicators in three specific areas: indicators for reporting on environmental conditions and trends, indicators for integrating environmental considerations into sectorial decision-making and indicators for incorporating natural resources into sectorial decision-making. Other categorizations have been proposed by USEPA [2] mainly in relation to its monitoring systems. These are numerous categorizations which are more oriented to describing specific environmental compartments: many examples are presented in this book. But, in our discussion, we would like to outline three possible categorization approaches that highlight semantic aspects.

1) The first categorization approach groups indicators according to the domain of specific knowledge used during the EIA process. EIA is a process carried out in various steps. Each step corresponds to the construction of a particular body of knowledge and therefore to a particular set of indicators.
- The first step is the description of the action that can have an impact on the environment. It is in general the construction and operation of a new technological system. The attributes of interest, in a systemic perspective, are the attributes that characterize its interfaces with other systems composing the environment. They are, in the first instance, attributes which are the object of physical and biological sciences.
- The second step is the identification of the systems composing the environment that can be affected. It is a problem, in some respects, symmetrical to the previous one: one changes the observer. The time at which the observation is made in general is different, however.
- The third step is the evaluation of the effects on these systems. It is still a value free evaluation, a simple search of cause-consequences relationships, a typical domain of the hard sciences.
- The fourth step is the evaluation of the impact of these effects (value connotation). This is a step where soft sciences, human sciences, i.e. social and economic indicators, inevitably enter.

- At the end we have the fifth step: a comparison of a modified (impacted) environment with a reference one (standard). It is the most complex step in which dimensions of a completely different nature enter the picture: individual and collective preferences, decision criteria, uncertainties in the representation of the "worlds" to be compared and synthetic indices.

These steps, in particular the third and the fourth ones, are not independent and they are in practice covered recursively. Values frequently emerge when seeking for the effects. The reference environment should include the hypothesis of some level of impact. The characterization of an action from the point of view of the impact supposes a preselection of the significant effects and values.

2) A second taxonomic approach is more linked to the knowledge structure and can be based on the relational distance of the indicator from the attribute of the world we consider relevant for our assessment. Indicators are signs that generically express a relation between two relata. The relation can be a short or a long chain, however.

- We can have indicators having a direct connotation of value. They are direct impact indicators, e.g. mortality indicators.
- Or, we can have indirect impact indicators. Their relevance is based on the assignment of a connotation value to an object that has no relevance of value per se. For instance, the concentration of a pollutant in the atmosphere has a connotation of value depending on the causal relationship between pollutant, concentration and damage to people.
- At the end, we can have "presumptive" indicators. The complexity of many real situations in the real world does not allow the investigation of the causal chain that connects the presence of an object, or an attribute, to an impact. In this case, the reasoning approach chosen is similar to the reasoning approach of a detective or of a judge. Analogy, similarity and pattern matching are the cognitive operations used for inference. It is the case of many biological indicators, investigated through global simulation and comparison. Time and economy constraints lead to the use of presumptive indicators. Authors have seen in the growth of the use of presumptive indicators the emergence of a new epistemological paradigm in social science [3]. This paradigm is also becoming more widely used in environmental science.

This classification is of particular relevance because it corresponds to a classical distinction between kinds of measurement: fundamental measurement, derived measurement and measurement by "fiat" [4].

3) A third approach to the classification of the indicators can be based on the consideration of their semantic level, nature or logic type. Semiologists distinguish various ways in which two systems of signs (e.g. linguistic expressions) can be linked. In some cases a system of signs can become the content of a second system of signs. When dealing with EIA indicators, we can disinguish indicators of the attributes of the world, that we can call indicators of first logic type, and indicators of some quality of the first ones, that we can call indicators of the second logic type. Uncertainty indicators and weights belong to this second category. The composition of indicators belonging to different logical levels poses particular logical difficulties as we will show below.

3. PRINCIPLES OF MEASUREMENT

While the indicator is defined (or designed as it would be) as a generic pattern embodying the state of knowledge (information) on an attribute relevant for the analysis being performed, the index is the outcome of the measurement process of the indicator. Thus, formally we may take the construction of an indicator and of its related index as the establishment of a correspondence between the empirical realities we are looking at and the set of numbers. The measure implies an empirical relational structure, defined as a generic set of real objects upon which one or more relations are defined and whose role is that of ordering paradigm (classification). Both the attribute relevant to the classification and the attribute-based procedure of comparing the elements in the set are then to be neatly defined.

With regard to environmental studies, an indicator could be conceived of as if it was an empirical relational structure. In this respect, measurement may be regarded as the construction of a isomorphism (i.e. a one-to-one relationship) from the empirical relational structure of interest into a numerical relational structure. That is, we simplify our conceptual model of the system while preserving the properties of interest. The isomorphism should preserve all the properties which hold in the empirical relational structure.

The important features of the real number series itself (as distinguished from the operations we can perform on the numbers) on which measurement relies are as follows:
- Numbers are ordered
- Differences between numbers are ordered
- The series has a unique origin indicated by the number "zero".

We shall call these characteristics: order, distance and origin. These characteristics form the basis to characterize the "scales" or measuring devices of practical interest [4]. Of the three characteristics listed, order is the only one that is invariably involved in measurement as it is usually conceived. In addition to order, a scale may possess either or both of the remaining characteristic of distance and origin. We then can distinguish between four types of scales, see Tab. 1.

Table 1. Types of scales

	No natural origin	Natural origin
No distance	Ordinal scale	Ordinal scale with natural origin
Distance	Interval scale	Ratio scale

A more rigourous mathematical presentation of basic measurement theory could show that, in principle, we can have an infinity of scales [5]. The scales introduced with the preceding qualitative considerations cover all practical cases, however.

In the ordinal scale the numbers are assigned to the various instances of an attribute (expressed by an indicator), so that the order of the numbers corresponds to the order of magnitude of the instances. With the interval scale, in addition to the order of the numbers corresponding to the order of magnitude of the various amount of the attribute, the size of the difference between pairs of numbers has meaning and corresponds to the distance (in a generic sense) between the pairs of amounts of the attribute. With the interval scale having natural origin, called the ratio scale as it is invariant under similarity transformation (see below), we have the additional restriction that the numbers assigned to instances correspond to the distances of these instances from the natural origin of the attribute. Ratio scale and interval scales correspond to what is commonly called quantitative or cardinal measurement. Ordinal scales correspond to qualitative or ordinal measurement. Ratio scale is favoured in physical sciences. Interval scales are also applicable in some cases, however: e.g. potential energy. Measurement by ratio scale is also called "extensive" measurement and is tacitly considered the ideal target. The numbers assigned in the isomorphism that allows a ratio scale are additive with respect to concatenation, so we can say that extensive measurement produces additive measures. Ordinal scales are abundant in the behavioural and social sciences and are becoming more and more important in environmental management.

4. INVARIANCE CHARACTERISTICS AND ADMISSIBLE STATISTICS

The assignment of numbers using a given scale has a practical meaning only if we know the degree to which this assignment is unique. We can cast this statement as the problem of finding the transformation of the scale that conserves the isomorphism between the empirical relational system represented and the formal numerical system representing it. In other words, we must find the transformation that does not cause a loss of empirical information.

The admissible transformations for the four scales are the following:

- For an ordinal scale, the admissible transformation is a monotonic increasing transformation.
- For an ordinal scale with natural origin, the admissible transformation is a monotonic increasing transformation that leaves the origin unchanged.
- For the interval scale, the admissible transformation is a linear transformation of the type $y = ax + b$, where "a" is any positive number and "b" is any finite number (linear or affine transformation).
- For a ratio scale, the admissible transformation is of the form $y = ax$, where "a" is any positive number (similarity transformation).

In general, the greater the isomorphism between the empirical relations and the formal model, the more restricted the range of invariant transformations. In this sense, the measurement by ratio scale, which accepts only a similarity transformation, is the most "representative". If the ratio scale is the most powerful one, why do we bother with other weaker forms of measurement? The reason is that the acceptable type of measurement depends on the structure of the empirical reality we intend to represent, and in many cases the structure cannot be represented by a ratio scale.

Admissible transformations are linked to the admissible descriptive statistics of a set of data obtained by some kind of measurement scale. The most known and used parametric statistics and parametric tests are not applicable to ordinal measurements. Only order statistics are applicable. For instance, to describe the central tendency, the median is applicable, while the mean is not applicable. Some common parametric statistics (mean, standard deviation, product-moment correlation, etc.), and statistical tests that do not assume the existence of a "zero", are applicable to data in an interval scale. The numbers associated with the ratio scale values are "true" numbers with the true "zero". Then, geometric mean, harmonic mean and any parametric statistic and statistical test are applicable. In the environmental field most of the indicators correspond to properties that can be measured only on an ordinal scale. Any statistical inference should be carefully chosen respecting the formal properties of the specific measurement type.

Stevens [6] was the first to recognize and emphasize the relation between the different scales, measures, and permissible operations on the measures (permissible statistics). Furthermore, he pointed out the fact that some aggregation operations are represented not by addition but weighted average. Precisely, we could say that for different scales the general pattern corresponding to the aggregation at empirical relational level are: the addition, for the ratio scale; the averages, for the interval scale and the ordinal operator max (min), for the ordinal scale. This could suggest an interpretation of the above correspondences in terms of semantic links between empirical aggregation and the represented set theoretical operation. In this respect, addition and ordinal operators represent just two particular cases of decomposable generalized union operators (given that the measures are defined in the interval (0,1)). The corresponding aggregations, at empirical structure level, can be recognized as figures of the above set compositions. On the other hand it is unclear, in the case of interval scales, which set aggregation could correspond to the average operators. I.e., for empirical relational structure the aggregation step can not be reduced to primitive set operations. In this respect, the interpretation of measurement scales in terms of set operations is still an open research programme.

5. COMPARISON AND COMPOSITION OF INDICES

The above brief introduction to the theory of measurement allows us to understand to what extent indicators could correspond to different relational structures, and why a simple numerical representation of them, the creation of indices, is not enough for their aggregation and use. Every indicator should be identified in terms of ordering relation and composition behaviour, and must be semantically evaluated. By semantic evaluation we mean the identification of the permissible transformations and operations which can be used to manipulate the numerical information.

Comparison and composition of indices can be considered as steps in the negotiation process in the sense that they are a problem of communication and negotiation between different sources or bodies of knowledge. I.e., any comparison and composition could be viewed as a "negotiation act" involving different bodies of knowledge in the broader context of the assessment. We may identify two classes of comparison and of composition: the symmetric and the asymmetric ones. Symmetric and asymmetric operations are qualified as "simple negotiation acts". In fact, they do grasp the simplest, or the most immediate, idea of negotiation between bodies of knowledge, as they are naturally associated to the idea of trading off and merging.

5.1. Symmetric Operations

Symmetric operations apply to indices corresponding to items of knowledge that we consider at the same level in the sense that they are independent and they cannot be linked in any way to some superior metaknowledge. Symmetric operations may present two cases: the indices belong to the same universe (they refer to the same attribute) or to different ones. The composition of indices belonging to the same universe of discourse is the consequence of the definition of the scale. All the considerations made earlier on admissible statistics, aggregations etc. are applicable. If the indices belong to different universes of discourse, we can consider two situations: the indices of the different universes are on the same scale or are on different scales. If they are on the same scale, and if this scale is a ratio scale, it is possible to define metric distances between attributes and therefore to define in a unique way the scale of the compound index. But, if the index scales are not of the same type, we cannot define a metric distance and we can compare only the weakest invariant between the various scales. For instance, if we want to compare and compose two indices, one on the ordinal scale and the other on the ratio scale, we can only keep the order preserving properties of the two scales and the statistics admissible for both (in this case: order statistics). This "reconduction approach" to the minimal invariant does not allow us to distinguish between "strong" and "weak" composition situations. A strong compositional situation is the case where all the indices to be composed, but one, belong to the strongest type of measurement scale. Conversely, in a weak compositional situation all indices but one belong to the weakest type of measurement scale. In both situations the reconduction approach will compare and combine the indices by means of the weakest invariant. This combination model forms part of a unfair negotiation process, in the sense that it disregards significant information.

5.2. Asymmetric Operations

The limitations of the symmetric composition can be overcome by asymmetric composition, which attributes different roles to the various indices. This is the case of fair negotiation, as all the available knowledge, including the objectives of the analysis, is exploited in driving the composition. Asymmetric composition can be assimilated to a multiobjective decision process.

The simplest case of asymmetric composition is the weighted average. It has been used, for a long time, as the prototype of possible composition patterns in all those cases where a common and equivalent level of composition does not exist. The compensation introduced by the weights should be referred to this idea.

However, weighted composition is not a solution for the problem of the composition of indices on different scales. To overcome the pitfall, the linear top-down procedure considered: from knowledge to indicator, to scale, to index, to compound index, could be replaced by a recursive representation process. The indices could be organized as a relational structure for which an ordering relation is defined, with direct reference to the knowledge. Along this line we no longer have, per se, a composition, but a revisitation of the cognitive process which generated the indices, in order to identify in it a relational structure that optimizes the use of the available information.

6. MEASUREMENT SCALES AND METAKNOWLEDGE

In the classification of indicators we mentioned the "second semantic level" indicators, i.e. indicators of some form of metaknowledge. These indicators raise particular problems of asymmetry in composition, which merit particular attention.

A metaknowledge widely used in EIA is expressed by "weights". Weights can be considered as indicators of the degree of importance given to indicators and indices. As such, weights have an analogy with utility. Weights are used in different circumstances for different purposes: ranking, averaging and normalizing. The epistemological statute of weights is still fairly undefined, so they are commonly used without paying attention to their intrinsic nature of results of measurement. In some cases the choice of a set of weights is implicitly considered a mapping from a variegated and not homogeneous set of indices to a ratio scale. In some other cases, weights are chosen to use just their ordinal properties. Only in the first case does the composition that we have mentioned as "weighted average" makes sense. Another metaknowledge is what we generically call "uncertainty". For a long time it has been assumed that the mapping of the empirical knowledge onto a formal numerical model is "crisp". This practical assumption gave no room for the consideration of imprecision and vagueness; aspects which are to some extent intrinsic to human knowledge. In recent years, the limitation of this assumption has been overcome assuming that we can consider as measurable attributes of the knowledge some of its qualities. In this way, measures of uncertainties could be systematically developed. The best known measure of uncertainty is subjective probability. B. De Finetti [7] showed how subjective probability can be measured on the ratio scale. But uncertainty does not always have the empirical properties (for instance, additivity) that justify a ratio scale. To deal with uncertainties in more general terms, new measures have been introduced under the umbrellas of so-called fuzzy logic, possibility logic, evidence theory, etc. [8]. This new scientific development has given a great incentive to the study of formal structures based on order relations. Indirectly, this development revealed the potential of measurement on the ordinal scale, if properly handled.

Uncertainty measures are currently used to create important compound indices: risk indices. Risk indices, for instance, are the result of a composition of a measure of damage and of a measure of uncertainty. This composition is a source of ambiguities whenever the characteristics of the indices combined are not clearly defined. In the framework of so-called expected utility theory, risk indices are the results of a symmetrical composition of indices on ratio scales. But the framework supposes a probability measure for the uncertainty and the independence of the two bodies of knowledge concerning damage and uncertainty. When probability measure is not justified, other composition approaches must be adopted, corresponding to the "negotiated" representation of the full body of the relevant knowledge.

7. CONCLUSIONS

The wide spectrum of knowledge implied in EIA makes the multiplication of indicators used inevitable but also imposes increased attention to the deep nature of knowledge and to the potential and pitfalls of its formalization. The main route to formalization, measurement, extensively investigated some 30-40 years ago, merits being revisited: it could help in avoiding an uncritical and ambiguous use of numbers.

Some parts of the classical measurement theory, like the part concerning weak types of measurement (ordinal scale) could be further developed, taking advantage of the advances in knowledge engineering.

The help that formalization by measurement of detailed properties can give in producing synoptic representations suitable for decision is limited, because it implies a decomposition-recomposition process with a great loss of information. The increased use of global indicators, i.e. indicators that represent a cognitive bypass to the limitations of any decomposition-recomposition approach, are, in this respect, fully understandable. The number attached to these global indicators must also be critically evaluated, however.

REFERENCES

[1] Organization for Economic Cooperation and Development (1991), Environmental Indicators, OECD, Paris.
[2] U.S. Environmental Protection Agency (1991), Ecological Indicators for the Environmental Monitoring and Assessment Program, USEPA, Research Triangle Park, NC.
[3] C. Ginzburg (1979), "Spie. Radici di un paradigma indiziario", in A. Gargani (ed.), Crisi della Ragione, Einaudi, Milano.
[4] W.S. Torgerson (1958), Theory and Methods of Scaling, Wiley and Sons, New York.
[5] D.H. Krantz, R.D. Luce, P. Suppes and A. Tversky (1971), Foundations of Measurement, Academic Press, New York.
[6] S.S. Stevens (1958), "Measurement and Man", Science, 127, 383-389.
[7] B. De Finetti (1974), Theory of probability, Vol. I and II, Wiley and Sons, New York.
[8] D. Dubois and H. Prade (1985), Théorie des Possibilités. Application à la Représentation des Connaissances en Informatique, Masson, Paris.

AIR QUALITY ASSESSMENT
IN ENVIRONMENTAL IMPACT STUDIES

S. Cernuschi and M. Giugliano
D.I.I.A.R. - Environmental Section
Milan Polytechnic
Via Fratelli Gorlini 1
I - 20151 Milan
Italy

ABSTRACT. The evaluation of the impact of atmospheric emission sources on air quality is normally conducted by comparison with air quality standards, most commonly defined in terms of statistical parameters of the concentration distribution. The present work reviews the general configuration of the standards and its main implications in the simulation of ground level pollutant concentrations from the source by atmospheric transport and diffusion models, with particular emphasis on the Gaussian plume equation and on the criteria utilized for its application in deriving concentration values suitable for a direct comparison with the standards.

1. INTRODUCTION

In environmental impact assessment studies, the alteration of air quality determined by a proposed or existing source is normally evaluated by comparison with air quality standards, defined in the environmental regulatory framework of most of the European countries, in terms of concentration values of the pollutants of interest that shall not be exceeded. The evaluation requires a proper description of the transport and diffusion of pollutants into the atmosphere in order to obtain, with a reasonable accuracy, the concentration levels determined by the emission source in any receptor of interest and in a form suitable for a direct comparison with the standards.

The problem to deal with is classical turbulent mass transport in a complex medium like the planetary boundary layer, further complicated by the possible variations in the emission regime (istantaneous, continuous constant or time-dependent) and in the spatial configuration (point, linear or areal) of the source. The phenomena are described in terms of three main components: the strength and the time regime of the emitting source, the atmospheric medium responsible for the transport and diffusion and the model for the numerical evaluation of the entire process. In the following chapters, each component is briefly reviewed, with particular emphasis on the gaussian model which represents one of the most commonly utilized models for dispersion calculations. The particular configuration of air quality standards, normally defined on a statistical basis, will also be illustrated, together with the consequent implications for the application of the dispersion model.

The notation used in the paper is summarised in Appendix A. Appendix B gives information on codes available for air quality assessment.

A. G. Colombo (ed.), Environmental Impact Assessment, 189–209.
© 1992 ECSC, EEC, EAEC, Brussels and Luxembourg.

2. THE EMISSION SOURCE

The strength of the emission and the effective release height are the most relevant data related to the source which influence the dispersion process.

The strength of the source is represented by the quantity of pollutant emitted per unit time, and is usually expressed in terms of mass flow rate for point sources, mass flow rate per unit length for line sources and mass flow rate per unit area for area sources. Emission data for the source to be evaluated and the pollutants of concern are mainly derived from compilations of emission factors available in the literature (U.S. EPA (1977), TNO (1980), Gaudioso et al. (1989)) or from extensive sampling measurements conducted on analogous existing sources. Emission standards imposed on the source could also, when they apply, be utilized, as well as the limits achievable with the control technology adopted for the source or normally required for particular source categories. The temporal variation of the emission strength, deriving either from normal fluctuations in the activity of the source (for example, residential space heating, semicontinuous industrial processes, traffic flow cycles) or from high abnormal releases of pollutants during equipment malfunctioning, should also be properly considered, especially in modelling short term concentrations of pollutants with acute effects.

The effective release height of the emission is given by the sum of the geometric height of the source and of the plume rise determined by momentum and buoyancy forces, if the emission is hot. The plume rise is a very important factor in determining maximum ground level concentrations, roughly proportional to the inverse square of the effective stack height, since it could increase the geometric height by a factor of 2 to 10 times.

Plume rise H is generally described by mathematical models with the following structure:

$$\Delta H = E \ x^b / u^a \tag{1}$$

where E is dependent on the thermal and kinetic energy of the emission, x is the downwind distance from the source and u is the mean wind velocity at the height of emission release. Some of the models most commonly utilized are reported in Tab. 1 (Seinfeld (1986)). The formulations of Briggs, with the well-known "2/3" dependence of plume rise with downwind distance, have been shown to agree with a great bulk of field and laboratory data, and consequently are the most frequently applied.

3. THE GAUSSIAN MODEL OF ATMOSPHERIC TRANSPORT AND DIFFUSION

The mathematical description of the spatial and temporal variation of a pollutant released into the atmosphere can be obtained through different approaches. With the statistical approach, the path of individual particles transported and diffused by the atmosphere is described by statistical functions, based on the fact that diffusive motions have a certain random or stochastic nature. With the similarity approach, models are formulated by dimensional analysis techniques through the assumption of a "similar" behaviour of the atmosphere from one place or time to another. In the gradient transport or K-models, the general equation of turbulent diffusion, derived from a material balance on an infinitesimal volume of atmosphere, is resolved. The exact solution cannot be obtained analytically, so the models are further subdivided into numerical models, derived from numerically approximated solutions of the complete equation, and analytical formulations obtained from the exact solution of approximate forms of the general equation.

The Gaussian model belongs to the latter category. It is derived from the analytical solution of the general equation, assuming constant values of the turbulent diffusion coefficients K and

of the mean wind velocity, and is widely utilized in the description of atmospheric transport mainly because:
- it gives results that agree with experimental data as well as any model;
- it is conceptually fairly simple and has an easily intelligible mathematical structure;
- it is consistent with the random nature of turbulence, which is the dominant factor of the dispersion process;
- other so-called theoretical formulations contain large amount of empiricism in their final stages.

The basic gaussian equation for the variation in the downwind concentration C of the pollutant in any receptor with coordinates x, y, z with respect to the base of the source is:

$$C(x, y, z) = \frac{Q}{2 \pi u \sigma_y \sigma_z} \exp\left[\left(-\frac{y^2}{2\sigma_y^2}\right)\right] \left\{ \exp\left[-\frac{(z-H)^2}{2\sigma_z^2}\right] + \right.$$

$$\left. + \exp\left[-\frac{(z+H)^2}{2\sigma_z^2}\right] \right\}$$

(2)

where $C(x, y, z)$ is the estimated concentration (mg m^{-3}), Q is the emission strength (mg s^{-1}), u is the mean wind velocity (m s^{-1}) at the effective release height H (m) and σ_y (m) and σ_z (m) are the diffusion coefficients in the horizontal and vertical direction, respectively. The formulation assumes a total reflection of the plume by the ground, accounted for with the last term in square brackets in (2) by assuming an "image" source at distance H beneath the surface.

The dispersion parameters σ_y and σ_z are the standard deviations of the gaussian concentration profiles of the pollutant in the horizontal (x - y) and vertical (z - x) planes of the plume cross section (Fig. 1, Stern (1976)). In general, they are dependent on the standard deviation of the wind velocity fluctuations in the y and z directions, perpendicular to the downwind direction x, on the travel time of the pollutant and on universal functions of a certain number of parameters, generally difficult to obtain, which define the main characteristics of the planetary boundary layer in which the dispersion takes place. Several studies are available which, through theoretical and empirical approaches, derive correlations between σ_y and σ_z and the different types of atmosphere categorized, according to its potential capacity of dispersion, into stability classes defined in terms of easily obtainable meteorological measurements. The most widely used stability classification scheme was developed by Pasquill, and is reported in Tab. 2: the atmosphere is divided into 6 different classes based on five wind velocity ranges, three classes of global radiation for daytime conditions and two classes of net radiation for nigh-time conditions.

On the basis of Pasquill's stability classification, the dependence of σ_y and σ_z on downwind distance is generally fairly well described by simple power-law relationships:

$$\sigma_y = a_y x^{b_y}$$

(3)

$$\sigma_z = a_z x^{b_z}$$

(4)

where the parameters a_y, a_z, b_y and b_z are dependent on the stability class (Hanna et al.

(1982)); more complex expressions, proposed by some authors (Seinfeld et al. (1986), Hanna et al. (1982)) are still derived from modifications of the previous equations. Tab. 3 reports more commonly utilized values of the numerical parameters in (3) and (4) together with the averaging times utilized for their estimation: the latter are also those to which the concentration values obtained with the model are referred.

One of the main positive features of the Gaussian model lies in its capability of being easily adapted to the description of situations which are different from the schematization utilized for the derivation of the basic equation (2). The most important modifications refer to the evaluation of mean concentrations over long time periods, to the description of plume dispersion in a non-homogeneous atmosphere and to the application to sources with geometrical configurations other than a point (linear and areal sources).

The calculation of concentrations averaged over long time intervals (typically the annual mean), and hence with climatological significance, must consider the variations in wind velocity and direction and in the atmospheric stability during the time period of concern. Meteorological data normally available are organized in terms of joint frequency tables of wind speed and direction and Pasquill stability categories, with each value representing the time fraction of the year during which the combination wind speed - wind direction - stability occurs. Typical sets of data include 5 wind velocity classes, 6 Pasquill categories and 16 wind directions, represented by the standard 22,5° degree sectors of the wind rose, for a total of 480 different combinations. The basic Gaussian equation is modified by assuming, for every sector of the wind rose, a random distribution of the wind direction: this leads to a uniform horizontal distribution of the pollutant in each sector, with a concentration C_L given by (Turner (1969)):

$$C_L(x, u, \theta, S) = \frac{2Q}{(2\pi)^{.5} u \sigma_z(u, S)(2\pi x / n)} \exp\left[-\frac{H^2(u, S)}{2\sigma_z^2(u, S)}\right] \tag{5}$$

where u is the mean wind speed representative of each wind velocity range, θ is the wind direction sector, S is the Pasquill stability category and n is the number of wind direction sectors (generally 16). The mean overall concentration $C_{LT}(x)$ for any receptor at a distance x from the source is then calculated by the summation of all values of $C_L(x, u, \theta, S)$ weighted with the frequency of occurrence $f(u, \theta, S)$ of every corresponding combination of wind speed u and direction θ and stability class S, derived from the joint frequency table:

$$C_{LT}(x) = \sum_{u, \theta, S} f(u, \theta, S) \, C_L(x, u, \theta, S) \tag{6}$$

Equation (6) gives thus the climatological mean value of the concentration, and is the basic expression of most long term Gaussian models.

In some situations, the dispersion of the plume in a non-homogeneous atmosphere is of particular significance: maximum ground level concentrations in this condition could be even twice as high than the values calculated for homogeneous conditions. To simulate the restriction on vertical diffusion of the emission by the stable atmosphere at the top of the mixing layer, the basic Gaussian equation is modified by taking into account the reflections of the plume from the ground surface and at the top of the mixing layer (Turner (1969)):

$$C(x,z) = \frac{Q}{2\pi u \sigma_y \sigma_z} \sum_{-N}^{+N} [\exp -\frac{1}{2}(\frac{z-H+2NL}{\sigma_z})^2 +$$

$$+ \exp \frac{1}{2}(\frac{z+H+2NL}{\sigma_z})^2]$$

(7)

where L (m) is the mixing height and N is the number of reflections. The method is known as the multiple plume image method, and the summation over N should be extended from $-\infty$ to $+\infty$: in practice, the series term converges rapidly and only the first reflections can be considered (N from -3 to +3).

Dispersion from line and area sources can still be simulated by slight modifications of equation (2). Typical line sources are represented by traffic on roads and highways or by a series of point sources located on a straight line and near to each other. In such situations, the absence of horizontal dispersion is taken into account by considering a uniform concentration profile in the crosswind direction: equation (2) is then modified, for a source of infinite length perpendicular to the wind direction, with the same criteria as those utilized in deriving the mean concentration C_L for the long term model (equation (6)):

$$C(x) = \frac{2Q}{(2\pi)^{.5}\sigma_z u} \exp(-\frac{H^2}{2\sigma_z^2})$$

(8)

where Q is the emission rate per unit length of the source (mg(m s)$^{-1}$). Equation (8) can still be applied, with proper modifications, to finite crosswind sources or to wind directions not perpendicular to the source.

Area sources typically encountered in dispersion calculations include storage piles, waste dumps and liquid storage or treatment ponds. An area source representation also constitutes the basis of urban diffusion models (Hanna et al. (1982)), in which the urban area is subdivided in a square grid pattern. Most simple approaches in modelling this type of emissions (Cernuschi and Giugliano (1989)) consider the area as a virtual point source, located upwind of the area boundary at a distance that gives the plume of the virtual source the initial lateral and vertical dispersion of the real emission at the effective source location. The concentration is then still evaluated by the basic equation (2), and the virtual distance, which depends on the source lateral and vertical dimensions, is simply added to the downwind distance x in the calculation of the dispersion coefficients σ_y and σ_z. The initial dispersion of plumes from area sources can also be taken into account, in more complex formulations, by considering the source as a finite crosswind line source (U.S. EPA (1987)): modifications of equation (8) are then utilized with the same addition of a vertical virtual distance in the σ_z evaluation, as previously outlined.

The basic Gaussian equation has also been utilized, with proper modifications, for the evaluation of the deposition of pollutants on the ground surface. Deposition phenomena are very important for pollutants with significant effects on the deposition surface, on vegetation and on soil, and for compounds which have toxic and cumulative properties that lead, through their transport and accumulation in the environment, to indirect pathways of human exposure.

The basic processes of the atmospheric deposition of pollutants include dry and wet removal. The former is active through mechanisms of turbulent diffusion, sedimentation and impact which, in the absence of precipitation, transport the contaminant to the deposition surface; the latter involves atmospheric precipitation (rain, snow, hail) as the transport medium to the surface through scavenging mechanisms. The most common schematization of the process, valid for particulate matter and highly soluble gases in the aqueous phase, describes the deposition with a first order irreversible mechanism: the ground deposition D is thus considered linearly dependent with the atmospheric concentration C of the pollutants through a parameter v, which is the deposition velocity of the pollutant:

$$D = C \, v \qquad\qquad (9)$$

If C is in mg m^{-3} and v is in m s^{-1}, the deposition D is in mg m^{-2} s^{-1}, i.e. mass deposited per unit time and surface area.

The deposition velocity depends on the type of pollutant considered, on the meteorological conditions at the local microscale and on the characteristics of the deposition surface. Despite extensive field and laboratory experiments and detailed theoretical calculations (Hanna et al. (1982), Sehmel (1980), MacMahon and Denison (1979), Seinfeld (1988)), there is still much uncertainty connected with this fundamental parameter.

The dry deposition velocity is basically dependent on the size of the deposited material. For particulate matter with diameter greater than about 10 μm, gravitational effects prevail and v can be well approximated with the Stokes law terminal settling speed of the particles. For particles with diameter less than about 10 μm and for gaseous pollutants, the deposition process is also influenced by Brownian motion, with an increasing relative importance with the decrease in particle dimensions and with consequent deposition velocities much higher than those predicted by Stokes law. When the pollutant has been transported to the surface it is there retained by physical (adsorption, impact) and chemical (reactive adsorption or absorption, photosynthesis, biological processes) mechanisms. Although various models have been proposed for the theoretical evaluation of the deposition velocity (Seinfeld (1988)), the complex relationships existing between the different parameters of concern have been established mainly through empirical observations (McMahon and Denison (1979), Sehmel (1980), Nicholson (1988)), which point out the importance of the density and the dimensions of the particles, of the terrain roughness, of the local micrometeorological conditions (stability, wind speed, humidity) and, especially for gaseous compounds, of the pollutant reactivity and the physico-chemical nature of the deposition surface.

Dry deposition is calculated through equation (9), by the deposition velocity and the ground level concentration C(x, y, 0) of the pollutant:

$$D = v \, C(x, y, 0) \qquad\qquad (10)$$

C(x, y, 0) can still be evaluated by a Gaussian formulation, corrected to account for the progressive depletion of pollutants from the plume due to deposition. One of the most commonly applied techniques involves the subtraction, from the effective release height, of a term representing the tilt in the plume axis deriving from large particle settling and the application, to the image source term, of a reflection coefficient which incorporates the effect of deposition by allowing only a partial reflection of the plume from the ground surface:

$$C(x,y,0) = \frac{Q}{2\pi u \sigma_y \sigma_z} \exp\left[\left(-\frac{y^2}{2\sigma_y^2}\right)\right] \left\{ \exp\left[-\frac{(H-xv/u)^2}{2\sigma_z^2}\right] + \right.$$

$$\left. + \alpha \exp\left[-\frac{(H-xv/u)^2}{2\sigma_z^2}\right] \right\} \qquad\qquad (11)$$

The reflection coefficient α is a function of the same parameters which influence the deposition process, and is generally evaluated through empirical correlations with the deposition velocity v (U.S. EPA (1987)).

The wet deposition velocity is usually evaluated through a washout coefficient W, defined as the ratio between the concentration of pollutant in the aqueous precipitation phase and in the atmosphere at some reference height (Hanna et al. (1982)):

$$W = C_W/C_0 \tag{12}$$

Wet deposition D_W can thus be evaluated by multiplying the concentration in the precipitation and the rainfall rate P:

$$D_W = C_W P \tag{13}$$

where P should be expressed in proper units. The washout coefficient can thus be utilized to define the wet deposition velocity v_W, by analogy with the dry process, combining equations (12) and (13):

$$D_W = W C_0 P = v_W C_0 \tag{14}$$

This can then be utilized like v to develop models for the evaluation of the wet deposition, substituting for C_0 expressions derived from the modification of the basic Gaussian equation as in (11).

Wet deposition is sometimes also described by assuming a first order decrease in the pollutant concentration with time during the precipitation event:

$$C_t = C_0 \exp(-\Lambda t) \tag{15}$$

where Λ (t^{-1}) is the scavenging coefficient, i.e. the specific removal rate of the pollutant from the atmosphere, and t is the time since the precipitation began. The deposition D_W is then given by

$$D_W = \int_0^Z \Lambda C \, d_z \tag{16}$$

where Z is the depth of atmosphere in which the scavenging takes place. If rain is assumed to fall through a Gaussian plume, C can be substituted by expressions similar to those utilized in evaluating dry deposition (Hanna et al. (1982)). Both the washout ratio W and the scavenging coefficient Λ are theoretically complex functions of droplet size distribution, physical and chemical characteristics of the particle or gas and precipitation rate, and are normally evaluated by reference to values calculated from field experimental data (McMahon and Denison (1979)).

4. AIR QUALITY STANDARDS

The configuration of air quality standards is, in general, a matter of compromise between the requirement of simple references for the evaluation of data monitored by control networks, the need to consider the different mechanism of the pollutant effects, mainly on human health

(WHO (1987)), the time and strength characteristics of the emission sources and the possible interactions in the atmosphere which eventually result in the production of secondary contaminants. The consequence is a more or less complex structure of limits, which are normally oriented to the control either of acute episodes of contamination, with limits established on the extreme values of the concentration averaged over short time intervals (1 h - 4 h), or of base pollution levels, with limits established over longer time periods (typically the annual mean). Air quality standards for Italy are reported, as a typical example, in Tab. 4.

The evaluation of the impact on air quality of atmospheric emissions is generally conducted through three different steps:

1. identification of the base level of air quality characteristics in the area where the source will be located;
2. evaluation of the expected effect on air quality resulting from the activity of the new source;
3. comparison of the simulated scenario with existing air quality standards or with other reference situations.

The direct approach for the evaluation of the first step makes use of data derived from air monitoring networks. Dispersion models for the simulation of the pollutant distribution can be utilized alternatively or to integrate the monitored data for areas outside the range of the network. The quality of the emission source inventory, which generally must also include sources which are difficult to evaluate like vehicular traffic, represents a critical requirement for the simulation exercise (Giugliano and Cernuschi (1991a)). For the purpose of the evaluation of base air quality levels, long term mean concentration values obtained with climatological models previously illustrated can be generally considered adequate; anyhow, for a significant presence of particular sources with intense activity over restricted time periods (residential space heating, vehicular traffic), the concentrations of the pollutants of concern averaged over shorter times should also be properly estimated with short term dispersion models.

The utilization of an atmospheric transport model which should be applicable to the configuration of the emission source of concern, in deriving concentration values consistent with the structure of air quality standards, is the basic premise for working out the second step. For the comparison with standards configured as those reported in Tab. 4, the evaluation should be conducted by:

a. a short-term model with results expressed in terms of hourly concentration values;
b. the time matched hourly sequence of meteorological parameters which, for the area of concern, should represent the characteristics of the atmosphere over an entire year and with climatological significance;
c. the hourly sequence of the emission source strength over the year, if its variation is considered to be significant.

Bearing in mind that every yearly simulation will result, for each receptor point, in 8760 hourly concentration values, the evaluation would also normally require an in-line software for the estimation of the main parameters of the descriptive statistics for the entire set of simulated concentration values. The appropriate parameters required by the air quality standard (maximum value, extreme percentiles, mean and median values) can thus be calculated for each receptor to carry out the direct comparison. The distribution of the same parameters in the area of concern in terms, for example, of concentration contours, are hence generally utilized as the representation of the air quality arising from the simulated scenarios (Fig. 2, 3 and 4).

The application of dispersion models in simulation exercises should obviously take into proper consideration the uncertainties contained in the evaluation, either as inherent uncertainty related to the stochastic nature of atmospheric turbulence, thus impossible to eliminate, or as reducible uncertainty deriving from the approximations included in the model itself (incomplete physical description of all the phenomena involved, space and time not properly resolved, etc.) or from the quality of input meteorological and emission related data (Benarie (1987),

Venkatram (1981)). The uncertainties acquire a significant effect on the estimation of extreme values (highest and second highest, extreme percentiles), most commonly included in air quality standards: a difference of a factor of 2 between observed and estimated values is generally considered to be a good indication of the acceptable behaviour of the model, as a consequence of the inherent simulation limits contained in it.

The simulation of hourly concentration values implies, as its major drawback, the availability of the time correlated sequence of meteorological data and the statistical manipulation of very large sets of concentration values. The same precision level of the resulting information can also be obtained with an alternative simpler approach, which utilizes climatological long term dispersion models. Meteorological parameters, as previously illustrated, are in this case normally required in terms of synthetic data, represented by the joint frequencies of occurrence of selected combinations of hourly wind speed and direction, atmospheric stability and the associated mixing layer depth. Input data are thus drastically reduced and the concentration values obtained, for every receptor point, with their corresponding frequencies of occurrence during the time period but without any reference to the temporal sequence of the event. From this limited set of data grouped by frequency classes, it is then fairly simple to derive the statistical parameters required for the comparison with the standards and, eventually, also indications on the type of distribution which best fits the data itself. However, the effective time series of the resulting hourly values is not already available, and the statistics for different time averages, for example 24 hour mean and extreme percentiles as required in Italian SO_2 and suspended particulate matter standards (Tab. 4), cannot longer be calculated. The evaluation can still be conducted with the aid of the well-known relationships empirically observed between maximum concentrations at different averaging times, and expressed as:

$$C_M(t_1) = C_M(t_0) (t_1/t_0)^{-b}$$
(17)

where $C_M(t_1)$ and $C_M(t_0)$ are the maximum concentrations at averaging times t_1 and t_0, respectively, and b is a parameter depending on the pollutant and the sampling location, with values generally in the range 0.15 - 0.30 (Larsen (1969), McGuire and Noll (1971), Drufuca and Giugliano (1978)).

The methodological approach illustrated up to now is well suited for the evaluation of the alteration of air quality due to the most common criteria pollutants. For particular sources, great significance can also be acquired by the emission of non-criteria micropollutants which, because of their high chemical stability and toxicity characteristics, are able to reach human subjects and adversely affect their health even a long time after they have been removed from the atmosphere. The exposure to these compounds, mainly toxic metals (As, Cd, Cr, Hg, Pb, Se) and organic polynuclear (PAHs) or halogenated compounds (PCBs, dioxins and furans), takes place through multiple pathways, with the direct exposure from contaminated air inhalation that frequently proves the least relevant (Travis and Hattemer-Frey (1987), Stevens et al. (1989), Levin et al. (1991)). This results, both in the definition of air quality standards and in the evaluation of the environmental impact, in the necessity for a novel approach which should also consider indirect exposure pathways properly.

A brief schematization of the method to be adopted, for which extensive details can be found in the literature (Levin et al. (1991), Giugliano and Cernuschi (1991b), Hattemer-Frey and Travis (1991)), is reported in Fig. 5. The evaluation of ground level concentrations and depositions of the toxic contaminant are the fundamental basis for the subsequent estimation of direct and indirect human exposure, respectively. The simulation can be performed by climatological models: the toxic action of the compounds considered is essentially related to chronic effects so the long term values, generally averaged on an annual basis, can conveniently be utilized. The distribution of the contaminant in the different environmental compartments through which the main routes of exposure originate (soil, water, food) is subsequently evaluated by further environmental transport models. The efficiency of different direct (air inhalation, dermal contact) and indirect (soil and water ingestion, dietary intake) pathways in determining the total exposure of human subjects is estimated through biochemical and metabolic information, and the value obtained utilized to quantify the health related risk. For toxic compounds without known or suspected carcinogenic effects (for example, Pb and Hg), risk is evaluated by comparison with reference dosages with no adverse effect (ADI, Admissible Daily Intake or NOAEL, No Observed Adverse Effect Level), whereas for carcinogenic substances, even if only suspected (for example, As, Cd and dioxins), risk is expressed in terms of the probability that an individual will develop cancer after exposure to the calculated dosage, over a 70 year mean assumed lifetime, derived from toxicological models of dose-response experimental data (Hattemer-Frey and Travis (1991), Ricci et al. (1985), Colombi et al. (1990)). Final results can be conveniently represented as risk maps (Fig. 6) in the area of concern. The values obtained, even with all the uncertainties contained in the evaluation, can be utilized as a valuable index of the health impact of the emission, and represent an important contribution for the decision procedure on the acceptability of the source.

5 . CONCLUSIONS

The evaluation of air quality impact of atmospheric emissions is normally conducted, for the most common criteria pollutants, through a comparison with air quality standards. Ground level concentrations of the pollutants emitted by the source can be conveniently simulated with a Gaussian approach, widely utilized in atmospheric dispersion modelling for its conceptual simplicity, easy of application and good agreement with experimental measured data. Gaussian models are formulated for sources with different geometric configurations (point, linear and areal) and are easily adapted to the description of situations most commonly encountered in practice (non-homogeneous atmosphere, deposition calculations, concentrations averaged over long time periods). In order to allow a direct comparison with air quality standards, usually defined on a statistical basis, Gaussian models should strictly be applied in short term versions, together with statistical in-line software for the evaluation of the main parameters of the concentration distribution (annual mean, maximum values, extreme percentiles) for each receptor of interest and for any time average of concern. Long term climatological versions of the model can alternatively be applied, with a consistent reduction in the data set to be manipulated and the same precision level of the results obtained: however, for the evaluation of concentrations at averaging times other than hourly, this simpler approach should be supported by empirical relationships derived for the area of concern between extreme and mean concentrations at different averaging times.

For toxic non-criteria micropollutants, the evaluation of air quality impact should also consider pathways of exposure, as well as the direct human exposure pathway represented by the inhalation of contaminated air. This results in a novel and more complex approach whose fundamental basis is still the evaluation of ground level concentrations and depositions of the contaminants, which can be performed with climatological Gaussian models, but that also requires the evaluation of the subsequent distribution of the toxic compound in the different environmental compartments and of the human exposure, and the resulting dosage, through which the health related risk is quantified.

REFERENCES

Benarie, M.M. (1987), "The limits of air pollution modeling", Atmospheric Environment 21, 1-5.

Cernuschi, S. and Giugliano, M. (1989), "Assessment techniques for gas emission and dispersion from waste landfills", in T.H.Christensen, R. Cossu and R.Stegmann (eds.), Sanitary landfilling: process, technology and environmental impact, Academic Press, London (U.K.), pp. 437-451.

Colombi, A. et al. (1990), "Valutazione dell'impatto ambientale - l'analisi della componente salute", Difesa Ambientale 5, 54-59.

Drufuca, G. and Giugliano, M. (1978), "Relationship between maximum SO2 concentration, averaging time and average concentration in an urban area", Atmospheric Environment 12, 1901-1905.

Gaudioso, D. et al. (1989), Guida ai fattori di emissione degli inquinanti atmosferici, Report ENEA RT/STUDI/89/07, Rome (Italy).

Giugliano,M. and Cernuschi, S. (1991a), "La stima dei dati di emissione di inquinanti atmosferici", Ingegneria Ambientale XX, 356-361.

Giugliano, M. and Cernuschi, S. (1991b), "La valutazione quantitativa del rischio di inquinanti atmosferici tossici e persistenti. Il caso dell'incenerimento di rifiuti solidi", Ingegneria Ambientale, accepted for publication.

Hanna, S.R. et al. (1982), Handbook on atmospheric diffusion, U.S. Dept. of Energy, Technical Information Service, Oak Ridge, Ten. (U.S.A.).

Hattemer-Frey, H.A. and Travis, C.C. (1991), Health effects of municipal solid waste incineration, CRC Press, Boca Raton, Fla. (U.S.A.).

Larsen, K. (1969), "A new mathematical model of air pollution concentration averaging time and frequency", J. Air Pollution Control Assoc. 19, 24-30.

Levin, A. et al. (1991), "Comparative analysis of health risk assessment for municipal solid waste combustors", J. Air Waste Manageme. Assoc. 41, 20-31.

McGuire, T. and Noll, K. (1971), "Relationship between concentrations of atmospheric pollutants and averaging time", Atmospheric Environment 5, 291-298.

McMahon, T.A. and Denison, P.J. (1979), "Empirical atmospheric deposition parameters: a survey", Atmospheric Environment 13, 571-585.

Nicholson, K.W. (1988), "The dry deposition of small particles: a review of experimental measurements", Atmospheric Environment 22, 2653-2666.

Ricci, P. et al. (1985), "Regulating cancer risks", Environmental Sci. and Technology 19, 473-479.

Sehmel, G.A. (1980), "Particle and gas dry deposition: a review", Atmospheric Environment 14, 983-1011.

Seinfeld, J.H. (1986), Atmospheric physics and chemistry of air pollution, J. Wiley, New York, N.Y. (U.S.A.).

Stern, A.C. (ed.) (1976), Air pollution - vol. 1, Academic Press, New York, N.Y. (U.S.A.).

Stevens, J.R. et al. (1989), "Environmental pollution - a multimedia approach to modeling human exposure", Environmental Sci. and Technology 23, 1180-1186.

TNO (1980), Handbook of emission factors, Ministry of Public Health and Environmental Protection, The Hague (Netherlands).

Travis, C.C. and Hattemer-Frey, H.A. (1987), "Human exposure to dioxin from municipal solid waste incineration", Waste Management and Research 9, 151-156.

Turner, B. (1969), Workbook of atmospheric dispersion estimates, U.S. Department of Health, Education and Welfare, Public Health Service, Consumer Protection and Environmental Health Service, National Air Pollution Control Administration, Cincinnati, Oh. (U.S.A.).

U.S. EPA (1977), Compilation of air pollutant emission factors, Publ. AP- 42, Ann Arbor, Mich. (U.S.A.).

U.S. EPA (1987), Industrial source complex (ISC) dispersion model user's guide, 2nd ed. (revised), U.S.EPA 450/4-88-002a, Research Triangle Park, N.C. (U.S.A.).

Venkatram, A. (1981), "Inherent uncertainty in air quality modeling", Atmospheric Environment 21, 1-5.

WHO (1987), Air quality guidelines for Europe, WHO Regional publications, European series No. 23, Copenaghen (Denmark).

APPENDIX A. NOTATION

C   pollutant concentration (mg m^{-3})

C_M maximum ground level concentration (mg m^{-3})

C_W pollutant concentration in aqueous precipitation (mg m^{-3})

C_0 pollutant concentration in the atmosphere at a reference height (mg m^{-3})

d   stack diameter (m)

D   pollutant deposition (mg s^{-1} m^{-2})

D_W wet deposition (mg s^{-1} m^{-2})

E   proportionality constant for plume rise evaluation (units dependent on plume temperature and atmospheric stability, see Tab. 1)

F   buoyancy flux (m^4 s^{-3})

g   acceleration due to gravity (m s^{-2})

H   effective release height (m)

L   height of the mixing layer (m)

n   wind direction sectors

N   number of reflections of a Gaussian plume by the ground surface and the top of the mixing layer

p   atmospheric pressure (kPa)

P   rainfall intensity (mm h^{-1})

Q   emission strength (mg s^{-1} for point sources, mg s^{-1} m^{-1} for line sources, mg s^{-1} m^{-2} for area sources)

T_a ambient temperature (° K)

T_s emission temperature (° K)

u   mean wind velocity (m s^{-1})

v   deposition velocity (m s^{-1})

v_W wet deposition velocity (m s^{-1})

V_s stack gas velocity (m s^{-1})

W   washout coefficient

x   downwind distance from the emission source (m)

y   crosswind distance from the emission source (m)

z   vertical distance from plume centreline (m)

α   reflection coefficient

ΔH   plume rise (m)

Λ   scavenging coefficient (h^{-1})

σ_y horizontal diffusion coefficient (m)

σ_z vertical diffusion coefficient (m) .

APPENDIX B. CODES FOR AIR QUALITY ASSESSMENT

Numerous air quality models are now available for a wide variety of specific applications. The US-EPA recommend the models of the UNAMAP series (version 6) available as an ASCII magnetic tape containing FORTRAN codes and test data for 31 Air Quality Simulation Models as well as associated documentation. These packages are also available in personal computer and "user friendly" forms from software houses. Some characteristics of the most popular UNAMAP models are listed here.

SCREEN	Short-term Gaussian steady-state algorithm estimates concentrations of stable pollutants that have been historically applied in urban areas.
CRSTER	Estimates ground-level concentrations resulting from up to 19 collocated elevated stack emissions.
CDM - 2.0	This climatological dispersion model determines long-term quasi-stable pollutant concentration.
ISCLT	(Industrial Source Complex Long-Term) - is a steady-state Gaussian plume model which can be used to calculate long-term pollutant concentrations from an industrial source complex.
VALLEY	A steady-state, univariate Gaussian plume dispersion model useful for calculating concentrations on any type of terrain.
COMPLEX 1	A multiple point source code with terrain adjustment using sequential meteorological data to calculate concentrations using the VALLEY algorithm.
RTDM	Third level screen model to augment VALLEY (first level screening) and COMPLEX 1 (second level) for emission sources in terrain above source height.

Two models designed in Italy are also available: the gaussian DIMULA in short term and climatological version by ENEA of Rome and the KAPPA G model by FISBAT of Bologna which uses the Gaussian and K theory approach.

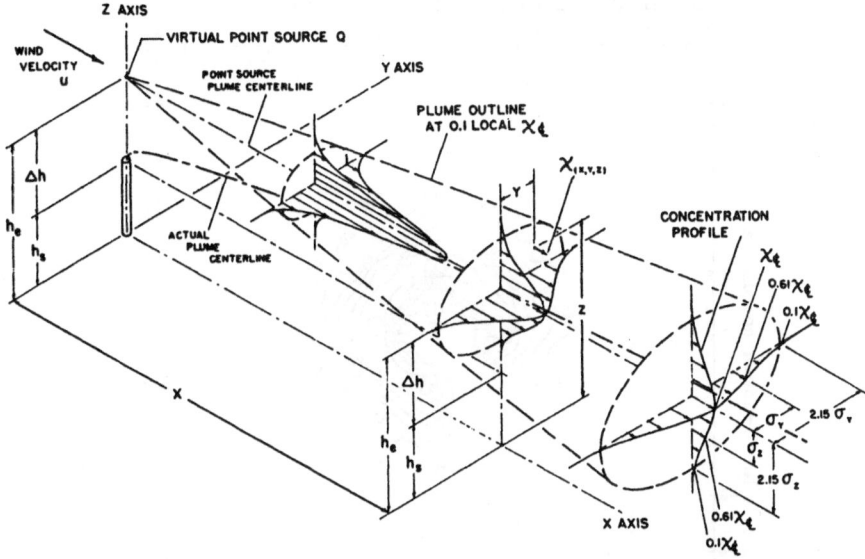

Figure 1. Spatial representation of the concentration profiles in a Gaussian plume formulation.

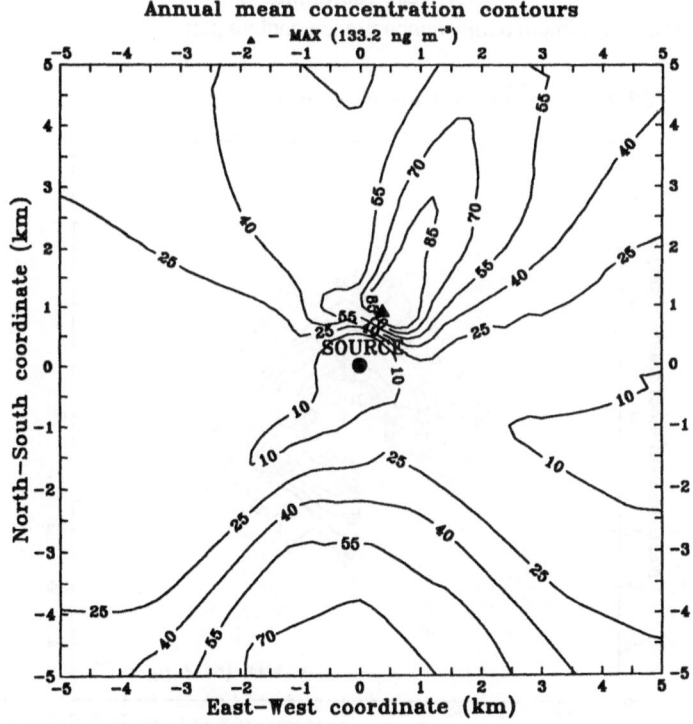

Figure 2. Example of annual mean concentration contour plot.

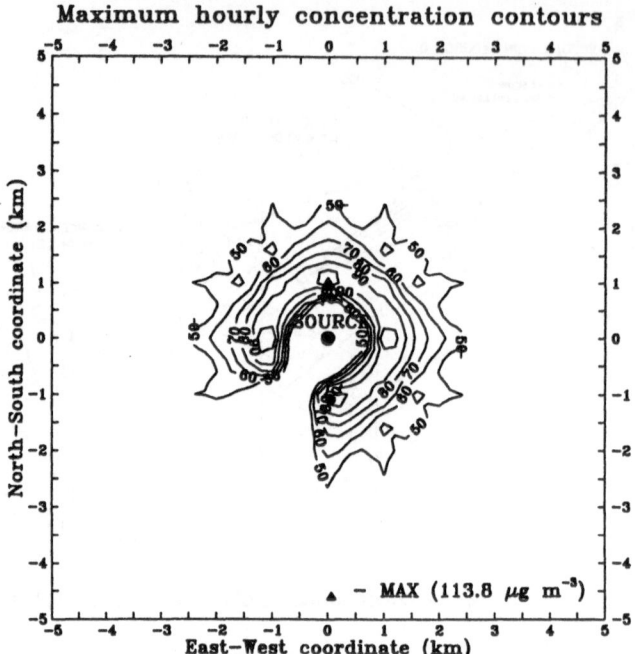

Figure 3. Example of maximum hourly concentration contour plot.

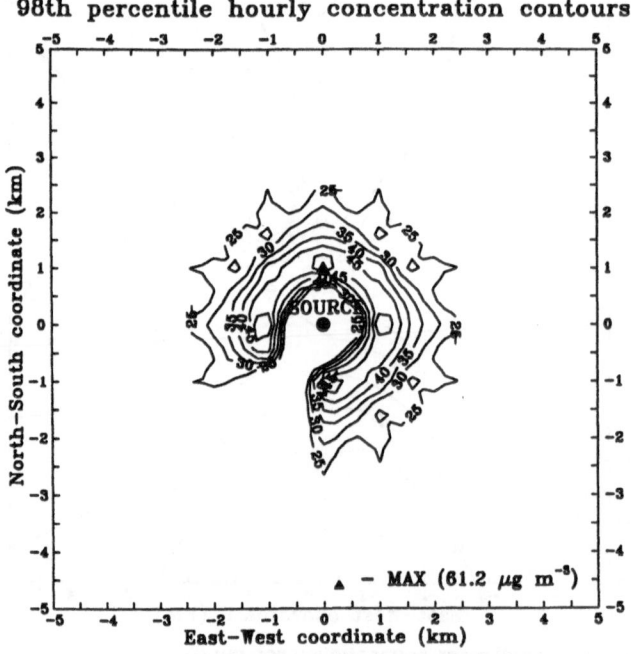

Figure 4. Example of 98th percentile hourly concentration contour plot.

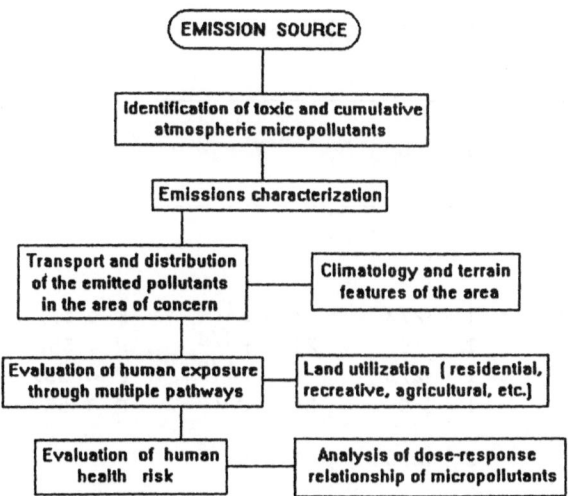

Figure 5. Schematization of the general methodology for the evaluation of health impact from toxic and cumulative micropollutants.

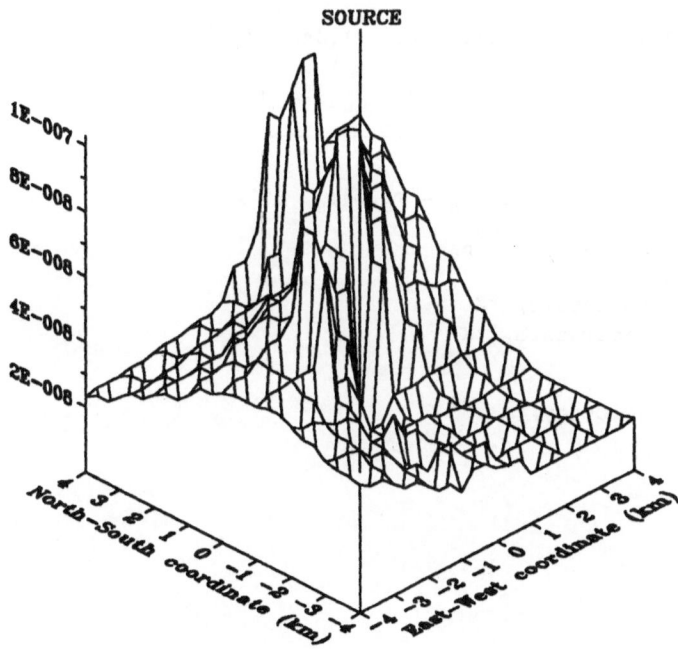

Figure 6. Example of a cancer risk map.

206

Table 1. Summary of plume rise mathematical models with structure
$\Delta H = E\, x^b/u^a$.

AUTHOR	ATMOSPHERIC STABILITY	a	b	E	CONDITIONS
Hot plumes					
Briggs	Neutral and	1	2/3	$1.6\, F^{1/3}$	$F < 55,\ x < 49\, F^{5/8}$
	unstable	1	0	$21.4\, F^{3/4}$	$F < 55,\ x \geq 49\, F^{5/8}$
		1	2/3	$1.6\, F^{1/3}$	$F \geq 55,\ x < 119\, F^{2/5}$
		1	0	$38.7\, F^{3/5}$	$F \geq 55,\ x \geq 119\, F^{2/5}$
	Stable	1	2/3	$1.6\, F^{1/3}$	
Cold plumes					
Asme	All categories	1.4	0	$d V_s^{1.4}$	$V_s > 10$ m s^{-1}
					$V_s > u$
					$\Delta T < 50\ ^\circ$K
Briggs	Neutral	2/3	1/3	$1.44\, (d V_s)^{2/3}$	$V_s/u \geq 4$
		1	0	$3\, d V_s$	$V_s/u \geq 4$

d = stack diameter, m
F = buoyancy flux, $g d^2 V_s (T_s - T_a)/4 T_s$, m^4 s^{-3}
g = acceleration of gravity
p = atmospheric pressure, kPa
P_0 = 101.3 kPa
T_a = ambient temperature, $^\circ$K
T_s = emission temperature, $^\circ$K

$\Delta T = T_s - T_a$
V_s = stack gas velocity, m s^{-1}

Table 2. Meteorological conditions defining Pasquill-Gifford stability classes.

Wind velocity u at 10 m (m s⁻¹)	Night-time stability classes N = net radiation (mW cm⁻²)			Daytime stability classes R = solar global radiation (mW cm⁻²)			
	$-14<N\leq-4.2$	$-4.2<N\leq-2.1$	$-2.1<N\leq0$	$0\leq R<15$	$15\leq R<29$	$29\leq R<58$	$58\leq R<\infty$
$0\leq u<2$	F	F	D	D	B	A-B	A
$2\leq u<3$	F	E	D	D	C	B	A-B
$3\leq u<4$	E	D	D	D	C	B-C	B
$4\leq u<6$	D	D	D	D	D	C-D	C
$6\leq u<\infty$	D	D	D	D	D	D	C

Table 3. Coefficients in Gaussian plume dispersion parameter correlations(1), (Seinfeld, 1988).

Source	Averaging time (min)	Coefficient	Stability class					
			A	B	C	D	E	F
Pasquill-Gifford	10	a_y	0,443	0,324	0,216	0,141	0,105	0,071
		b_y	0,894	0,894	0,894	0,894	0,894	0,894
ASME	60	a_y	0,41	0,36		0,32		0,31
		b_y	0,91	0,86		0,78		0,71
		a_z	0,40	0,33		0,22		0,06
		b_z	0,91	0,86		0,78		0,71
Klug	10	a_y	0,469	0,306	0,230	0,219	0,237	0,273
		b_y	0,903	0,885	0,855	0,764	0,691	0,594
		a_z	0,017	0,072	0,076	0,140	0,217	0,262
		b_z	1,380	1,021	0,879	0,727	0,610	0,500
Pasquill-Gifford	10	I_y	-1,104	-1,634	-2,054	-2,555	-2,754	-3,143
		J_y	0,9878	1,0350	1,0231	1,0423	1,0106	1,0148
		K_y	-0,076	-0,0096	-0,0076	-0,0087	-0,0064	-0,0070
		I_z	4,679	-1,999	-2,341	-3,186	-3,783	-4,490
		J_z	-1,7172	0,8752	0,9477	1,1737	1,3010	1,4024
		K_z	0,2770	0,0136	-0,0020	-0,0316	-0,0450	-0,0540

$$\sigma_y(x) = a_y\, x^{b_y}$$

$$\sigma_y(x) = \exp\left[I_y + J_y \ln x + K_y (\ln x)^2\right]$$

$$\sigma_z(x) = a_z\, x^{b_z}$$

$$\sigma_z(x) = \exp\left[I_z + J_z \ln x + K_z (\ln x)^2\right]$$

(1) Application restricted to downwind distances not exceeding 10 km.

Table 4. Air quality standards for Italy.

Pollutant	Standard Sampling period	Standard Averaging time	Value
SO_2	year	24 h	Median = 80 µg m^{-3}
			98th percentile = 250 µg m^{-3}
	winter		
	(Oct-Mar)	24 h	Median = 130 µg m^{-3}
NO_2	year	1 h	98th percentile = 200 µg m^{-3}
Ozone (O_3)	month	1 h	Not to be exceeded[*] more
			than one time = 200 µg m^{-3}
CO	--	8 h	Mean = 10 mg m^{-3}
	--	1 h	Mean = 40 mg m^{-3}
Pb	year	24 h	Mean = 2 µg m^{-3}
F_2	--	24 h	Mean = 20 µg m^{-3}
	month	24 h	Mean = 10 µg m^{-3}
Suspended particulate matter	year	24 h	Mean = 150 µg m^{-3}
			95th percentile = 300 µg m^{-3}

[*] If this value is exceeded, total non-methane hydrocarbons should not exceed mean concentration averaged over 3 consecutive hours of 200 µg m^{-3}.

SURFACE WATER QUALITY INDICATORS

P. J. Newman
Water Research Centre Medmenham
Henley Road, Marlow, Bucks
UK - Buckinghamshire, SL7 2HD
United Kingdom.

ABSTRACT. Recent political pressures, often by environmental groups, have forced many of the Governments of the individual Member States to take a more positive approach to the management of the quality of surface waters within their country. This has increased markedly the need for monitoring the quality of surface water throughout the Community over the last few years. In addition the Community has introduced many water and environmental directives designed to protect the quality of the surface waters of the Community which will also lead to increased monitoring. Such monitoring programmes can be very complex and can generate so much data that presentation and interpretation becomes quite difficult. Accordingly, classification schemes, based on water quality indicators, have been devised by most Member States to convert the mass of data produced into water quality classes. These, in turn, can be used to demonstrate compliance with the requirements of EC Water Directives and to reveal temporal and spatial trends in surface water quality. Most schemes currently in use in the Member States for the monitoring and management of surface water quality are based on physico-chemical parameters, for example, suspended solids, biochemical oxygen demand, dissolved oxygen and ammoniacal nitrogen levels. It is most likely though that these will be complemented in the near future by biological/ecological techniques since these tend to provide a more overall assessment of the impact of water pollution and other environmental degradation. The paper examines the various approaches to the assessment of surface water quality and reviews a range of such schemes currently in use in the Community. Details of methods used to represent the data obtained from the use of these schemes on maps of water quality are also given. Finally recommendations are made for possible future surface water classification schemes.

1. INTRODUCTION

Man has always had a need to gauge the quality of surface waters and this he has done by the use of a wide range of quality indicators. Perhaps the first of such indicators utilised by early man would be the examination of the suitability of a sample of surface water for drinking by assessing the colour, taste and odour and the presence (or more importantly, the absence) of suspended or floating matter.

Since then many other indicators of surface water quality have been introduced to assess the 'fitness for use' of given bodies of surface water, whether it is for drinking or for other uses such as agricultural, industrial or recreational purposes. For example, even as far back as the middle of the last century attempts were made to develop a scheme for the classification of surface water quality which was based on biological indicators. Thus, the workers, Kolenti (1842) and Cohn (1853) (both quoted in Lieberman, 1962) observed that organisms present in polluted waters are different from those that occur in clean water and accordingly were able to

A. G. Colombo (ed.), Environmental Impact Assessment, 211–233.
© 1992 ECSC, EEC, EAEC, Brussels and Luxembourg.

construct an assessment scheme of water quality. During the intervening period a wide range of biological and chemical classification schemes have been derived and introduced in the individual countries of the European Community. Now, for example, it is usual to find more than one such scheme in operation at the same time in a given Member State. Over the last few years, all European Community Governments have been obliged to adopt a more positive approach to the management of the quality of the surface waters within their respective country. There are many reasons for this change in approach but probably the most important are:

- increasing public environmental concern;
- impact of the implementation of EC Water Directives;
- recent drought conditions experienced in many parts of the Community.

As a consequence most Member States have introduced programmes consisting of water quality indicators to monitor the quality of their surface waters. These are used to evaluate, for given water bodies, the quality (overall or 'fitness for use'), compliance with relevant EC Water Directives and whether the quality is increasing or decreasing.

The European Commission has plans to introduce many additional Directives designed to protect and improve further the surface waters of the Community. These will specify not only surface water quality criteria for specific pollutants, for example, as Daughter Directives of the EC Framework Directive 76/464/EEC dealing with the discharge of certain dangerous substances into the aquatic environment, but also to establish high ecological quality for all surface waters throughout the Community. The results of these initiatives, and particularly that concerning the proposed ecological directive, will increase the requirement for monitoring still further.

There is probably no ideal or universal indicator of surface water quality and accordingly many tend to be used. Whilst a single indicator will provide some assessment of the quality, seeking to improve that assessment by introducing additional indicators greatly increases the complexity of the procedure. Furthermore, whilst it is relatively easy to devise schemes which show whether the quality of a given stretch of surface water either passes or fails a particular standard, introducing intermediate classes to indicate whether the quality is improving or decreasing with time, also adds to the complexity.

In spite of these difficulties most of the monitoring programmes introduced by the Member States for the assessment of surface water quality tend to be quite complex and to involve many indicators of water quality (parameters). These generate so much data that they are difficult to handle and to interpret. To overcome these difficulties, Member States have attempted to devise schemes by which the data obtained from applying a particular monitoring procedure to a given stretch of water are compared with standardised quality criteria for recognised classes of water quality. In that way, the overall quality of the waters being examined can be described by a simple classification term. Schemes of this type also enable changes in quality of the water in question to be identified more easily and to allow the intercomparison of the quality of two or more stretches of water to be carried out.

A recent survey of surface water classification schemes carried out by the author (Newman 1988) showed that Member States seem to dislike simple schemes even when they are really adequate for the purpose required and there does seem to be an inclination to develop ever increasingly complex schemes. This point is amply illustrated by the French authorities who have modified, on several occasions, an already comprehensive scheme involving some 30 physical, chemical and biological parameters (Multipurpose Scale Scheme, see Section 3.1.1.) so that either good quality or conversely, poor quality waters can be classified more accurately. This had the effect of turning a scheme consisting of only four quality classes into one involving six and later seven quality classes. Even though this development may lead to an improvement in the definition of the quality of certain waters, it is questionable whether it is worth the additional assessment involved. Furthermore, the complexity of the modified schemes would probably preclude their general use.

In the development of any classification scheme there is clearly a need to strike a balance between simplicity, comprehensiveness and general applicability. It is important that any scheme introduced should be sufficiently comprehensive to demonstrate adequately the quality of the surface waters in question. However, the introduction of classification schemes containing a large number of classes can be somewhat artificial. For, in reality, unless the water quality apparently represented by a particular class can be, at least in theory, reproduced in an aquarium in the laboratory, and be shown to be different to adjacent classes, it probably has no real significance in practice for water quality. Furthermore, the perceived quality can equally appear to increase (or to decrease) during subsequent monitoring even if the real quality does not change, because the higher (or lower) class is not significantly different from the originally determined class.

2. BASIS OF SURFACE WATER QUALITY CLASSIFICATION SCHEMES

There are two main methods for the assessment of surface water quality; these are, physico-chemical and biological.

Physico-chemical assessment of water quality involves the measurement of parameters such as colour, suspended solids, biochemical oxygen demand, dissolved oxygen and ammoniacal nitrogen levels. Often the concentrations of a wide range of potential pollutants are also determined for many classification schemes.

Biological assessment on the other hand relies on the fact that pollution of a water body will cause changes in the physical and chemical environment of that water and that those changes will disrupt the ecological balance of the system. Thus, by measuring the extent of the ecological upset, the severity of the pollution can be estimated. The species most commonly used for the assessment of river water quality are the larger and more easily seen and sampled invertebrates which colonise the substrata of rivers. Such animals are collectively referred to as macro-invertebrates, of which the main constituents are often the young aquatic stages of certain insects. Within this bottom dwelling community the sensitivity and tolerance to pollution varies considerably from species to species. Thus, for example, some species are very sensitive to reductions in dissolved oxygen level e.g. stonefly larvae and will not be found in areas where the oxygen levels are not consistently high. Other species, on the other hand, such as some freshwater oligochaetes and the larvae of some chironomid midges, are very tolerant in this respect.

Likewise some species require particular physical conditions in order to survive and thrive, e.g. clean surfaces or clean gravel interstices. Any change in conditions such as silting, will significantly reduce or even eliminate that species. This is a characteristic feature of polluted environments where reductions in overall community diversity and an increase in the density of tolerant species will be observed. Thus, the quality of water can be assessed by means of a field inspection of the macro-invertebrate communities inhabiting the sub-stratum noting the relative proportions of sensitive, less sensitive, tolerant and most tolerant species and comparing the results found with the expected ratios from unpolluted habitats of a similar type. Some species of fresh water macro-invertebrate arranged in order of their increasing tolerance to organic pollution is given in Figure 1. It is also interesting to note that the macro-invertebrate (c), Gammarus pulex, changes colour when it dies and on the basis of this colour change can be used to detect sources of pollution.

Each method for the assessment of surface water quality has its own particular application, advantages and disadvantages. These are summarised in Table 1 (taken from An Foras Forbartha, 1984) from which the authors conclude that a combination of both techniques is preferable to either alone.

Physico-chemical methods involving the analysis of single samples may be considered to provide a 'snap-shot' assessment of water quality, but provide in the long term an 'average' indication of water quality. However, they are unlikely to detect pollutant discharges that are either irregular or surreptitious. Routine physico-chemical analyses can be relatively expensive in terms of the equipment needed and the number of analyses required to achieve results with a comparable reliability to those achieved by biological surveillance. Their particular advantage is that the analyses involved are precise, discriminating and quantitative and therefore suitable for water management since they provide the basic information required by licensing authorities for the assessment of compliance with prescribed standards.

Biological techniques, on the other hand, seek to assess the accumulative impact of pollutants on the ecosystem. These effects are largely independent of the time(s) during which the pollutants were discharged and therefore this approach provides an 'integrating' assessment of water quality. This is an important advantage of biological techniques. They do, however, suffer from the limitation of being poor in terms of their precision and discrimination of the pollutants involved. They are also said to be open to operator bias.

Most Member States at present rely heavily on physico-chemical methods of assessing surface water quality. However, there is a large resurgence of interest in biological techniques and a number of Member States are currently developing them. Of particular note is a computer based system entitled RIVPACS (River Invertebrate Prediction and Classification System) for the prediction of the invertebrate fauna to be expected at unpolluted sites from a limited number of environmental features. The predicted value can then be compared against the observed value to give a relative assessment of water quality at any given site. The system (developed by the UK Institute of Freshwater Ecology, Wright et al., 1989) is currently being tested in the field.

The interest in biological schemes will increase greatly in the near future with the adoption (expected Spring 1992) of the proposed EC directive on the ecological quality of surface water throughout the Community. This seeks to maintain and, if found necessary, to improve the quality of all surface waters so that they are of high ecological quality and, as a consequence, will entail extensive monitoring.

3. SURFACE WATER QUALITY CLASSIFICATION SCHEMES USED IN EC MEMBER STATES

Examples of surface water quality classification schemes used in EC Member States are discussed in this section. It should be noted that these have to be chosen to illustrate particular features of such classification schemes and do not represent a comprehensive listing.

3.1. Rivers

3.1.1. Physical-chemical quality indicators. Most Member States operate physico-chemical classification schemes for the assessment of the quality of their river waters. For example, in:

a) **Benelux countries.** In these countries (Belgium, Luxembourg and The Netherlands) the quality of river waters is assessed by means of the "water oxygen balance". For this, the following three key indicators (parameters) of water quality are considered:
- percentage oxygen saturation
- biochemical oxygen demand
- ammoniacal nitrogen content

Assessment of the water oxygen balance at a given location on a water course is obtained in the following way:

First, the water samples taken are analysed for the individual parameter listed above. Then, for each sample, points for each parameter are assigned by referring the values found with the point-score system shown in Table 2.

These individual point-scores are then summed to give an overall assessment of water oxygen balance for the particular water sample (which will vary from 3 to 15 points with the lower values indicating water of higher quality).

Finally, an overall assessment of the water oxygen balance at the sampling point is determined by averaging the total point-scores found for individual samples analysed. The average point-score, so derived, can then be used to determine the water quality class at the sampling location by reference to the values given in Table 3. It will be seen that, in this case, the classification scheme based on the water oxygen balance consists of five water quality classes.

b) **England and Wales**. A similar scheme to the Water Oxygen Balance is used for rivers and canals in England and Wales. Here, the National Water Council River Classification Scheme (UKNWC 1981) is used. This is based on definitions of quality criteria required for certain potential uses of water and consists of four main quality classes (with the highest quality class being subdivided). Details of these quality criteria expressed in terms of dissolved oxygen and ammonia levels, biochemical oxygen demand (as for the Water Oxygen Balance) as well as toxicity to fish (Alabaster and Lloyd 1980) are given in Table 4. The table also outlines the relationship between the quality classes and the NWC River Class descriptions established by the scheme. Explanatory notes associated with the table are also given. Reference is also made in the "Quality criteria" column of the table to the correlation between water quality as defined by this scheme and that required by the EC Directive 80/778/EEC concerning the quality of surface water intended for the abstraction of drinking water.

c) **France**. The main classification scheme used in France for rivers and canals is the Multipurpose Scale (Duport and Margat, 1983). This has been in use in various forms since the early 1970s and although not an official scheme, is widely used by the relevant organisations such as the "Agences de Bassin" and the Secretariat for the Environment. It is based on quality objectives for specific uses of the water and consists of four quality classes (plus a "non-classification" grading, used in those cases where the water quality is worse than Class 3, i.e. excessively polluted and of little value for any use). These classes are described in Table 5.

The quality of water for each class is described by reference to given ranges of values of a large number of physical, physico-chemical, chemical and biological parameters. These are shown in Table 6. It will be seen that the boundaries of the values of the parameters for the various classes are most often based on those given in the EC Directive dealing with water quality, for example, Directive 75/440/EEC concerning the quality required of surface water for abstraction of drinking water.

It was mentioned in Section 2 that, in spite of the comprehensive nature of the Multipurpose Scale, the French authorities have made several attempts to describe more precisely the quality of river water by devising additional classification schemes. One such scheme, the Scale II Scheme, was developed by Bureau and Duport (1975) and is claimed to more accurately describe the good quality waters of the Multipurpose Scale, Classes 1A and 1B. It also meets an important criteria for classification schemes, that is the number of classes should be approximately the same for each parameter and that the number of measurements to describe each class should be comparable. Scale II consists of six quality classes and these are shown in Table 7.

Not content with these schemes, the French have also developed, i) the **Scale III Scheme** (Duport and Margat, 1983) consisting of 7 classes and claimed to be able to more accurately describe not only the good quality waters of the Multipurpose Scale, but also the poorer waters and also ii) a **Scheme for higher quality waters** (Duport and Margat, 1983) consisting only of

5 classes but which considers new parameters and creates a further division of Class 1A of the Multipurpose Scale. Furthermore, there are plans to introduce into the schemes refinements to allow for the river stage (i.e. ranging from the upper to the lower reaches) to be taken into account. For brevity, details of these schemes have not been included in the paper. Further information can be obtained from Newman (1988).

3.1.2. Chemical Index for monitoring river water quality. An attempt to assess river water quality on the basis of chemical criteria has been made by Bach (1980) in Germany. This was based on earlier work carried out in the USA (Brown et al., 1970) and in Scotland (Scottish Development Department 1976 and Bolton et al., 1978). It involves the measurement of a number of chemical parameters (indicators of water quality) for the water samples (see Table 8) followed by the combination of the values found into one number - the Chemical Index - which represents the overall quality of the sample. The Chemical Index is a multiplicative index which has the following form:

$$CI = \prod_{i=1}^{n} q_i{}^{w_i} = q_1{}^{w_1} q_2{}^{w_2} \quad \ldots \quad q_n{}^{w_n} \tag{1}$$

in which

CI   is the Chemical Index. It is a dimensionless number on a continuous scale from 0 to 100, with 0 for the worst and 100 for the best quality waters,

n   is the number of parameters,

q_i is the sub-index for the ith parameter, a dimensionless number between 0 and 100 and a function of the ith parameter,

w_i is the weighting for the ith parameter, a number between 0 and 1 and such that the sum of the respective weightings is equal to 1,

i.e. $\sum_{i=1}^{n} w_i = 1.$ (2)

The Chemical Index for a given sample point is calculated in the following manner:

(i)   Firstly, each parameter is assigned a weighting (w) (Table 8). These weightings may be considered to be a measure of the priority/importance of each parameter.

(ii)   Then, each parameter is assigned a sub-index value (q) derived from the values found for that parameter for a given water sample by reference to pre-determined calibration curves (one for each parameter). These curves are shown overleaf (Figures 2-9).

(iii)   Finally, the value of CI is calculated by substituting the respective values of w and q into Equation 1 given above.

An example of a calculation for the Chemical Index is shown in Table 9. It will be seen from the values of the various parameters determined for this water sample that it can be described as moderately polluted and this is reflected in the relatively low value for the Chemical Index. It is interesting to note that several workers (Landwehr, 1974 and LAWA, 1976) report relatively close correlation between the Chemical Index and various biological indices.

3.1.3. Biological indicators of water quality. A number of biological quality assessment schemes have been introduced in the Community. However, and as mentioned in Section 2, the interest in this approach is likely to increase markedly over the next few years. Most of the schemes introduced so far are very complex and would cause difficulties in routine use. It is this factor that will need to be addressed during the development of further schemes. Perhaps, the simplest scheme in use and which is described below to illustrate the features of the method is the Belgian Biotic Index Method (De Pauw and Vanhooren, 1983). This method relies on the consideration of two pollution effects, firstly the reduction in community diversity and secondly the progressive loss of certain groups of clean water fauna as a result of increasing pollution. Thus, assessment of water quality by this method requires a survey of the species of macro-invertebrates present at the sampling point, noting their relative abundance. Then, by reference to Table 10, prepared by Tuffery and Verneaux (De Pauw and Vanhooren, 1983) on the basis of susceptibility of certain taxonomic groups to pollution, the biotic index can be determined. The biotic index obtained in this way ranges from 0 (major pollution preventing survival of invertebrates) to 10 (natural ecological conditions). For convenience, however, the results are classified according to a six point scheme which shows the relationship between biotic index and water quality (Table 11).

The main advantage claimed for the biological method is that assessment can be deduced from a single set of results, whereas for the water oxygen balance (see Section 3.1.1) a minimum of 20 measurements, spread over several months, is required.

3.2. Lakes

Most Member States use a version of the "Fixed Boundary" eutrophication scheme, first proposed by OECD (1982) for the classification of lake water quality. Details of the background to the scheme and of the version used in Ireland are given below.

All lakes tend to undergo a process of natural eutrophication, i.e. increase in nutrient supply and productive capacity, as their volumes decrease - due to the accumulation of organic and inorganic sediments. The process of natural eutrophication is relatively slow perhaps lasting several thousands of years. However, when the nutrient supply is significantly increased by discharges of pollutants (usually domestic sewage or fertilisers - which are much richer in plant nutrients, especially phosphorus, than natural leachates), the process is greatly speeded up and excessively high levels of productivity can occur (referred to as artificial or accelerated eutrophication). It can be seen, therefore, that the state of eutrophication of a lake as compared with that for an unpolluted lake in a similar location is a measure of the pollution present and can be used as a means of classifying the water quality of the lake.

The original distinction between oligotrophic and eutrophic was made in a quantitative manner and referred to a number of biological and physico-chemical features which were easily observable. The comparison is summarised in Table 12. OECD in its study, however, attempted to devise a quantitative classification which was more suitable in the context of water quality classification and management. It considered the question of lake classification and the resulting distinction of different trophic categories. These are shown in Table 13. It will be seen that, in addition to the three basic categories already discussed above, the OECD study group distinguished two further trophic states: ultra-oligotrophic lakes (lakes of extremely low productivity, usually located in remote mountainous areas) and hyper-eutrophic (seriously polluted lakes subject to an excessive degree of artificial enrichment and producing extremely dense growths of algae).

3.3. Estuaries and Coastal Waters

In England and Wales, as well as in Ireland, the quality of estuaries and coastal waters is assessed on the basis of a classification scheme established in 1981 (UKNWC). This scheme takes into account the uses to which the estuary may be put, amenities and industry. In the scheme points are first allocated for biological, aesthetic and chemical quality to zones within the estuary or coastal water. Details of the allocation of these individual points scores are given in Table 14. These individual scores are then summed against the following classification to arrive at an overall description of the water quality of the estuary:

Classification (Class)	Description	Number of points
A	Good quality	30 - 24
B	Fair quality	23 - 16
C	Poor quality	15 - 9
D	Bad quality	8 - 0

4. METHODS OF REPRESENTATION OF DATA OBTAINED

There are a number of methods used throughout the Community for the representation of the data obtained on maps of water quality. It should be noted that these apply equally to show overall quality as well as the level of individual indicator (parameters) of water quality.

a) **Colour coding**. For this, a range of colours are used to show waters of different quality classes. The colours usually chosen are in the same sequence as that found in a rainbow (see Tables 3 and 5) and range from blue for waters of very good quality to red for waters of very poor quality. The system works well in practice and waters of most concern, i.e. those grossly polluted, are easily identified.

b) **Use of dots of varying size**. For this the quality of water along a water course is indicated by a dot whose size is varied according to the extent of pollution - the greater the extent of the pollution, the larger the size of the dot used. Again, the system is shown to be effective in practice since waters that are heavily polluted are easily identified. It also avoids the not inconsiderable additional costs involved with colour printing necessary for method a) above.

c) **Use of different shaped symbols**. This is as for method b) above, but instead of varying the size of the dot with the water quality, different shaped symbols are used to denote waters of particular water quality classes.

5. RECOMMENDATIONS FOR FUTURE SURFACE WATER CLASSIFICATION SCHEMES

From an analysis of the various surface water quality classification schemes used throughout the Community, it would appear that the following points should be taken into account in the development of future water quality classification schemes:

- The new schemes should be complementary to the objectives of EC Water Directives and not based solely on nationally established standards;
- They must have general applicability whilst being sufficiently sensitive to be capable of assessing local waters in order to identify areas at risk and those requiring attention. Accordingly, they should be constructed on a hierarchical approach rather than a simpler "compliance - non-compliance" basis. It would also be most convenient if a given scheme could be used for all surface waters of a given type, for example, a

classification scheme for rivers should, if possible, be suitable for all rivers regardless of stage and flow (see comment on this topic below);
- The new schemes should be sufficiently simple for easy use whilst at the same time adequately comprehensive for the purposes for which they were introduced;
- They should be based on i) the three basic parameters of pollution, i.e. percentage oxygen saturation, biochemical oxygen demand and ammoniacal nitrogen content, and ii) the ecological quality;
- Allowance should be made in river classification schemes for the relationship between water quality, river stage and flow. This may mean individual schemes being developed for different stages and/or flow of a river (see comment above).

REFERENCES

Alabaster, J.S. and Lloyd, R. (editors) (1980), Water quality criteria for freshwater fish, Butterworths, London.

An Foras Forbartha (National Institute for Physical Planning and Construction Research) (1984), The National Survey of Irish Rivers 1982-83. Biological investigations of quality in selected rivers and streams.

Bach, E. (1980), "A chemical index for the surveillance of river water quality", Deutsche Gewässerkundliche Mitteilungen 24, 102-106.

Bolton, P.W., Currie, J.C., Tervet, D.J. and Welsh, W.T. (1978), "An index to improve water quality classification", Water Pollution Control 271-284.

Brown, R.M., McLelland, N.I., Deininger, R.A. and Tozer, R.Z. (1970), "A water quality index - do we dare?", Water and Sewage Works , 117, 335-343.

Bureau, M. and Duport, L. (1975), Investigation of the results of national inventory of water pollution using equivalence factor analysis method. French Ministry of Agriculture.

De Pauw, N. and Vanhooren, G. (1983), "Method for biological assessment of watercourses in Belgium", Hydrobiologia 100, 153-168.

Duport, L. and Margat, J. (1983), Measurement of the quality of waters - methods used to show the quality, Report No DPP/SE.B/LD/SB, French Ministry of the Environment.

Landwehr, J.M. (1974), "Water quality indices - construction and analysis", PhD dissertation, The University of Michigan.

LAWA (1976), River water quality maps of the German Federal Republic.

Liebmann, H. (1962), Handbuch der Frisch wasserund Abwasserbiologie, Vol I, 2nd edition, R. Oldenburg, München.

Newman, P.J. (1988), Classification of surface water quality, Heinemann Professional Publishing, Oxford.

OECD (1982), Eutrophication of waters, monitoring, assessment and control, OECD, Paris.

Scottish Development Department (1976), Development of a water quality index, Report No. ARD3.

UK National Water Council (UKNWC) (1981), River quality - the 1980 survey and future outlook, HMSO, London.

Wright, J.F., Armitage, P.D., Furse, M.T. and Moss, D. (1989), "Prediction of invertebrate communities using stream measurements", Regulated Rivers: Research and Management 4, 147-155.

Figure 1. Some species of fresh water invertebrates each magnified 5 times in order of increasing tolerance to organic pollution.

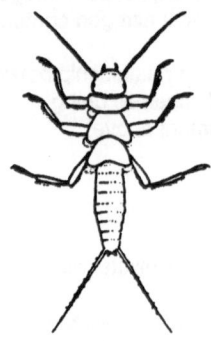

(a) *Dinocras cephalotes* (Plecoptera)

(b) *Ecdyonurus venosus* (Ephemeroptera)

(c) *Gammarus pulex* (Amphipoda)

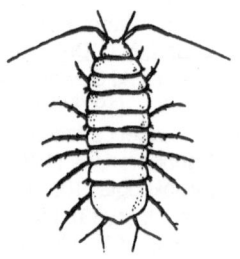

(d) *Asellus aquaticus* (Isopoda)

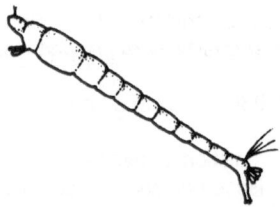

(e) *Chironomus riparius* (Diptera)

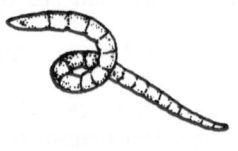

(f) *Tubifix tubifex* (Oligochaeta)

Figures 2 - 9. Chemical index - calibration curves.

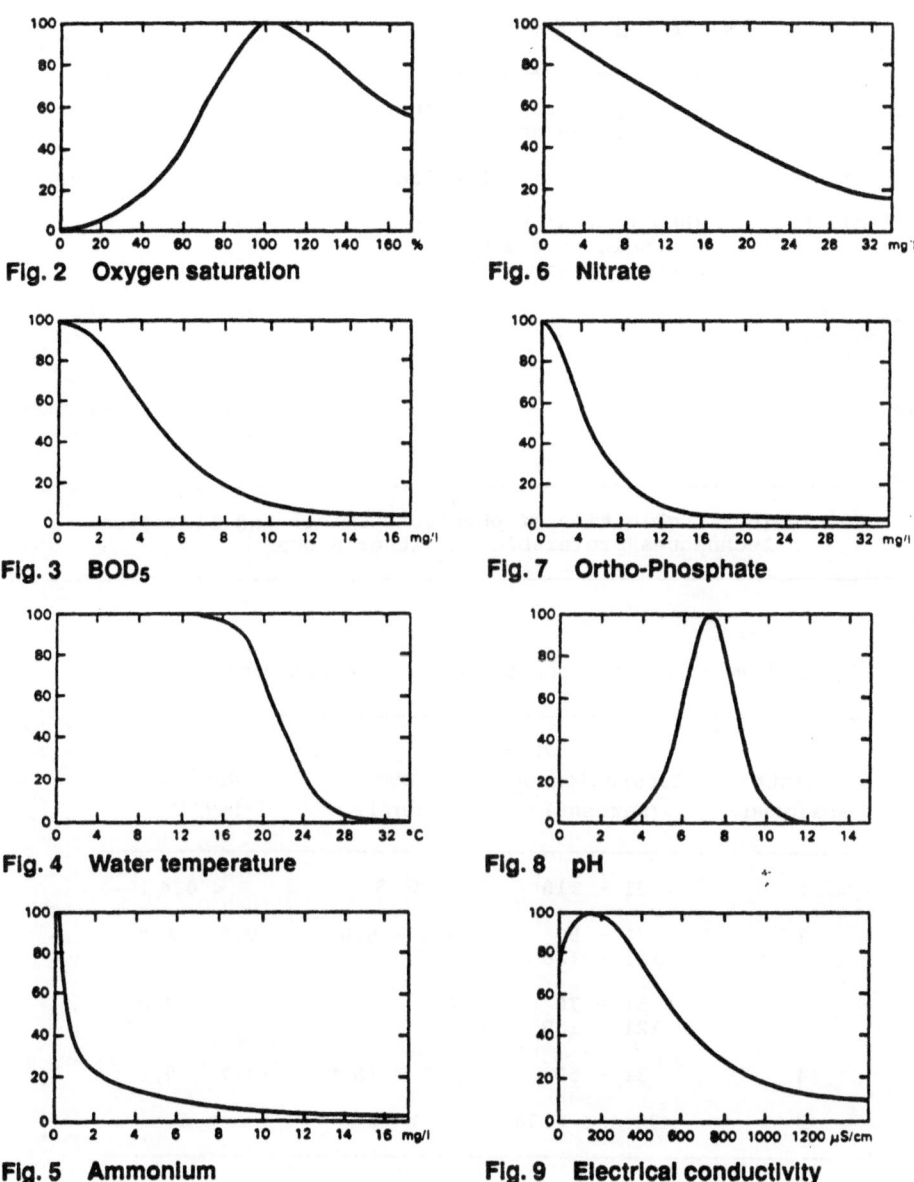

Fig. 2   Oxygen saturation

Fig. 6   Nitrate

Fig. 3   BOD₅

Fig. 7   Ortho-Phosphate

Fig. 4   Water temperature

Fig. 8   pH

Fig. 5   Ammonium

Fig. 9   Electrical conductivity

222

Table 1. Comparison of water quality assessment by biological and physico-chemical techniques (taken from An Foras Forbartha, 1984).

Realm		Performance Chemical	Biological
Precision	(i.e. pollutant concentration assessment)	Good	Poor
Discrimination	(what kind of pollution)	Good	Poor
Reliability	(how representative is a single or a limited number of samples)	Poor	Good
Measure of ecological effects		No	Yes
Cost		Relatively high	Relatively low

Conclusion: Combination of physico-chemical and biological techniques preferable to either alone

Table 2. Point-score system for assessment of water oxygen balance.

Points assigned	Saturation O_2 (percent)	BOD (mg/l)	$NH4^+-N$ (mg/l)
1	91 - 110	≤ 3	< 0.4
2	71 - 90 111 - 120	3.1 - 6.0	0.5 - 1.0
3	51 - 70 121 - 130	6.1 - 9.0	1.1 - 2.0
4	31 - 50	9.1 - 15.0	2.1 - 5.0
5	<30 and >130	> 15	> 5

Table 3. Relationship between water quality and total points score for water oxygen balance.

Total points score for water oxygen balance	Quality class	Colour code
3 - 4.5	1 - very good	blue
4.6 - 7.5	2 - good	green
7.6 - 10.5	3 - average	yellow
10.6 - 13.5	4 - poor	orange
13.6 - 15	5 - very poor	red

Table 4. River quality classification in England and Wales.

River Class	Quality criteria	Remarks	Current potential uses
1A Good Quality	Class limiting criteria (95 percentile) (i) Dissolved oxygen saturation greater than 80% (ii) Biochemical oxygen demand not greater than 3 mg/l (iii) Ammonia not greater than 0.4 mg/l (iv) Where the water is abstracted for drinking water, it complies with requiurements for A2* water (v) Non-toxic to fish in EIFAC terms (or best estimates if EIFAC figures not available)	(i) Average BOD probably not greater than 1.5 mg/l (ii) Visible evidence of pollution should be absent	(i) Water of high quality suitable for potable supply abstractions and for all other abstractions (ii) Game or other high class fisheries (iii) High amenity value
1B Good Quality	(i) DO greater than 60% saturation (ii) BOD not greater than 5 mg/l (iii) Ammonia not greater than 0.9 mg/l (iv) Where water is abstracted for drinking water, it complies with the requirements for A2* water (v) Non-toxic to fish in EIFAC terms (or best estimates if EIFAC figures not available)	(i) Average BOD probably not greater than 2 mg/l (ii) Average ammonia probably not greater than 0.5 mg/l (iii) Visible evidence of pollution should be absent (iv) Waters of high quality which cannot be placed in Class 1A because of the high proportion of high quality effluent present or because of the effect of physical factors such as canalisation, low gradient or eutrophication (v) Class 1A and Class 1B together are essentially the Class 1 of the River Pollution Survey (RPS)	Water of less high quality than Class 1A but usable for substantially the same purposes
2 Fair Quality	(i) DO greater than 40% saturation (ii) BOD not greater than 9 mg/l (iii) Where water is abstracted for drinking water it complies with the requirements for A3* water (iv) Non-toxic to fish in EIFAC terms (or best estimates if EIFAC figures not available)	(i) Average BOD probably not greater than 5 mg/l (ii) Similar to Class 2 of RPS (iii) Water not showing physical signs of pollution other than humic colouration and a little foaming below weirs	(i) Waters suitable for potable supply after advanced treatment (ii) Supporting reasonably good coarse fisheries (iii) Moderate amenity value
3 Poor Quality	(i) DO greater than 10% saturation (ii) Not likely to be anaerobic (iii) BOD not greater than 17 mg/l. This may not apply if there is a high degree of re-aeration	Similar to Class 3 of RPS	Waters which are polluted to an extent that fish are absent or only sporadically present. May be used for low grade industrial abstraction purposes. Considerable potential for further use if cleaned up
4 Bad Quality	Waters which are inferior to Class 3 in terms of dissolved oxygen and likely to be anaerobic at times	Similar to Class 4 of RPS	Waters which are grossly polluted and are likely to cause nuisance
X	DO greater than 10% saturation		Insignificant watercourses and ditches not usable, where the objective is simply to prevent nuisance developing

Notes
(a) Under extreme weather conditions (eg flood, drought, freeze-up), or when dominated by plant growth, or by aquatic plant decay, rivers usually in Class 1, 2 and 3 may have BODs and dissolved oxygen levels, or ammonia content outside the stated levels for those Classes. When this occurs the cause should be stated along with analytical results.
(b) The BOD determinations refer to 5 day carbonaceous BOD (ATU). Ammonia figures are expressed as NH_4.
(c) In most instances the chemical classification given above will be suitable. However, the basis of the classification is restricted to a finite number of chemical determinands and there may be a few cases where the presence of a chemical substance other than those used in the classification markedly reduces the quality of the water. In such cases, the quality classification of the water should be down-graded on the basis of biota actually present, and the reasons stated.
(d) EIFAC (European Inland Fisheries Advisory Commission) limits should be expressed as 95 percentile limits.

* EEC category A2 and A3 requirements are those specified in the EEC Council Directive of 16 June 1975 concerning the Quality of Surface Water Intended for Abstraction of Drinking Water in the Member State.

Table 5. French multipurpose scale.

Class	Quality objectives for specific uses of water	Quality description	Colour coding
Class 1A	Includes waters considered free from pollution and suitable for purposes with the most demanding quality requirements.	Excellent	Blue
Class 1B	These waters are of slightly lower quality, but are nevertheless suitable for all purposes.	Good	Green
Class 2	Fair quality, adequate for irrigation, industrial purposes and the production of drinking water after thorough treatment. It may generally be used for watering animals. Fish can live in it normally, but their reproduction may be random. Leisure pursuits connected with water are possible when no extended contact is involved.	Passable	Yellow
Class 3	Mediocre quality, just suitable for irrigation, cooling waters and navigation. Fish may exist in these waters but will be random during periods, for example, of low flow and high temperature.	Mediocre	Orange
No classification	Waters where the maximum admissible Class 3 for one or more parameters is exceeded. They are considered as unsuitable for the majority of purposes and may constitute a threat to public health and the environment.	Bad	Red

Table 6. Values of parameters associated with quality classes of the "Multipurpose Scale".

Parameter	Quality class 1A	1B	2	3
Conductivity µS/cm at 20 °C	<400	400 to 750	750 to 1500	1500 to 3000
Temperature °C	<20	20 to 22	22 to 25	25 to 30
pH	6.5 to 8.5	6.5 to 8.5	6 to 9	5.5 to 9.5
SS (mg/l)	<30	<30	<30	30 to 70
O_2 dissolved (mg/l)	> 7	5 to 7	3 to 5	) aerobic medium
				) must be
O2 dissolved as %				) maintained
saturation	>90%	70 to 90	50 to 70	) continuously
BOD_5 (mg/l)	<3	3 to 5	5 to 10	10 to 25
Oxidisability (mg/l)	<3	3 to 5	5 to 8	
COD (mg/l)	<20	20 to 25	25 to 40	40 to 80
NH_4 (mg/l)	<0.1	0.1 to 0.5	0.5 to 2	2 to 8
NO_3 (mg/l)			<44	44 to 100
Total N (Kjeldahl)	<1	1 to 2	2 to 3	>3
Fe (mg/l)	<0.5	0.5 to 1	1 to 1.5	
Mn (mg/l)	<0.1	0.1 to 0.25		0.25 to 0.50
F (mg/l)	<0.7	0.7 to 1.7	0.7 to 1.7	>1.7
Cu (mg/l)	<0.02	0.02 to 0.05	0.05 to 1	>1
Zn (mg/l)	<0.5	0.5 to 1	1 to 5	>5
As (mg/l)	<0.01	<0.01	0.01 to 0.05	>0.05
Cd (mg/l)	<0.001	<0.001	<0.001	>0.001
Cr (mg/l)	<0.05	<0.05	<0.05	>0.05
CN (mg/l)	<0.05	<0.05	<0.05	>0.05
Pb (mg/l)	<0.05	<0.05	<0.05	>0.05
Se (mg/l)	<0.01	<0.01	<0.01	>0.01
Hg (mg/l)	<0.0005	<0.0005	<0.0005	>0.0005
Phenols (mg/l)	-	<0.001	0.001 to 0.05	0.05 to 0.5
Detergents (mg/l)	<0.2	<0.2	0.2 to 0.5	>0.5
Chloroform-extractable material (mg/l)	<0.2	0.2 to 0.5	0.5 to 1	>1
Coliforms (per 100 ml)	<50	50 to 5000	5000 to 50000	
E.coli (per 100 ml)	<20	20 to 2000	2000 to 20000	
Faecal strep. (per 100 ml)	<20	20 to 1000	1000 to 10000	
Difference between biotic and normal index	1	2 or 3	4 or 5	6 or 7

Note. Use of scale:
- Each parameter assigns a given quality to the water if the most unfavourable
  measurement reaches the corresponding lower limit shown in the table for at
  least 10% of the measurements.
- The worst of the various qualities provided by the parameters is selected to
  give an overall assessment of the water quality.

Table 7. Scale II quality of surface waters.

Classes	1	2	3	4	5	6	Units
Water temperature	<5	5-10	10-15	15-20	>20		°C
pH	<7	7-7.5	7.5-7.75	7.75-8	>8		
Conductivity	<200	200-300	300-400	400-500	>500		µS/cm
Decantable material	<0.001	0.001-0.05	0.05-0.15	>0.15			cm^3/l
Dissolved oxygen	<6	6-8	8-10	10-12	>12		mg/l
% saturation	<70	70-80	80-90	90-100	>100		%
BOD2	0-1	1-2	2-3	>3			mg/l
BOD5	0-2	2-3	3-4	4-6	>6		mg/l
Oxidisability	0-1	1-2	2-3	3-5	>5		mg/l
Flow	0-5	5-15	15-40	40-100	>100		m^3/s
Suspended solids	0-6	6-12	12-18	18-35	>35		mg/l
COD	0-10	10-15	15-25	25-50	>50		mg/l O_2
Cl^-	0-10	10-15	15-20	20-30	>30		mg/l
$SO_4^=$	0-10	10-15	15-25	25-50	>50		mg/l
NH_4^-	<0.01	0.01-0.3	0.3-0.5	0.5-1	>1		mg/l
NO_3^-	<3	3-5	5-10	>10			mg/l
NO_2^-	<0.02	0.02-0.08	0.08-0.15	0.15-0.3	>0.3		mg/l
Na	<6	6-10	10-15	15-25	>25		mg/l
K	<1.5	1.5-2.5	2.5-3.5	3.5-5.5	>5.5		mg/l
Ca	<20	20-40	40-60	60-80	80-100	>100	mg/l
Hg	<3	3-5	5-7	7-10	10-15	>15	mg/l
Bicarbonate	<100	100-150	150-200	200-250	250-300	>300	mg/l
Phenol	(presence/absence)						mg/l
Phosphate	<0.1	0.1-0.3	0.3-0.5	0.5-1	>1		mg/l
Detergents	<0.01	0.01-0.1	0.1-0.25	>0.25			mg/l

Table 8. The parameters used for the Chemical Index and their respective weightings.

	Parameter	Unit	Weighting
1.	Saturation with oxygen	%	0.20
2.	BOD_5	mg/l	0.20
3.	Water temperature	°C	0.08
4.	Ammonium ion NH_4^+	mg/l	0.15
5.	Nitrate ion NO_3^-	mg/l	0.10
6.	Ortho-phosphate o-PO_4^{3-}	mg/l	0.10
7.	pH		0.10
8.	Electrical conductivity	µS/cm	0.07
	n = 8		$\Sigma = 1.00$

Table 9. Computation of the Chemical Index.
River: Nahe, Sampling point: mouth, Date: 21.9.88.
River: Nahe, Sampling point: mouth, Date: 21.9.88.

Parameter	Unit	Value	q_i	w_i	$q_i^{w_i}$
O_2 saturation	%	57	40	0.20	2.091
BOD_5	mg/l	5.3	45	0.20	2.141
Water temp	°C	16.5	96	0.08	1.441
NH_4^+	mg/l	3.6	16	0.15	1.516
NO_3^-	mg/l	27.8	24	0.10	1.374
o-PO_4^{3-}	mg/l	8.4	1	0.10	1.000
pH		5.9	53	0.10	1.487
Conductivity	µS/cm	550	56	0.07	1.325
Flow	m^3/s	6.3		$\Sigma=1.00$	CI=26.5

Table 10. Standard table to determine the biotic index.

Faunistic group			total number of systematic units present				
			0-1	2-5	6-10	11-15	≥16
			B i o t i c			i n d e x	
1 Plecoptera or Ecdyo- nuridae (+ Hepta- geniidae)	1 2	several SU(*) only 1 SU	– 5	7 6	8 7	9 8	10 9
2 Cased Trichoptera	1 2	several SU only 1 SU	– 5	6 5	7 6	8 7	9 8
3 Ancylidae or Ephemeroptera exept Ecdyonuridae	1 2	more than 2 SU 2 or < 2	– 3	5 4	6 5	7 6	8 7
4 Aphelocheirus or Odonata or Gammaridae or Mollusca (except Sphaeridae)	0	all SU mentioned above are absent	3	4	5	6	7
5 Asellus or Hirudinea or Sphaeridae or Hemiptera (except Aphelocheirus)	0	all SU mentioned above are absent	2	3	4	5	–
6 Tubificidae or chironomidae of the thummi-plumosus group	0	all SU mentioned above are absent	1	2	3	–	–
7 Eristalinae (= Syrphidae)	0	all SU mentioned above are absent	0	1	1	–	–

(*) SU = Number of systematic units observed of this faunistic group.

Table 11. Relationship between biotic index and water quality.

Class	Biotic index	Colour	Water quality
I	10-9	blue	lightly or unpolluted
II	8-7	green	slightly polluted
III	6-5	yellow	moderately polluted - critical situation
IV	4-3	orange	heavily polluted
V	2-0	red	very heavily polluted
-	0	black	study impossible complete lack of bio-indicators.

Table 12. Comparison of the main biological and physico-chemical features of oligotrophic (unproductive) and eutrophic (productive) lakes.

Feature	Oligotrophic lake	Eutrophic lake
Weed and algal growth	poor-moderate	good-abundant
Dominant types of phytoplankton in summer	chlorophyceae (green algae) bacillariophyceae (diatoms)	myxophyceae (blue-green algae)
Reed growth in bays	sparse	may be abundant
Invertebrate fauna	sparse, diverse dominated by insects	usually abundant and diverse dominated by non-insect types
Fish	dominated by game fish (trout, salmon, char); growth poor	usually dominated by coarse fish (roach, bream, rudd, perch) and game fish may be absent; growth good
Water transparency	high in the absence of colour	low to moderate particularly in summer
Nutrient concentrations	low	relatively high
Deoxygenation of lower layers	never extensive even when the lake is thermally stratified	may be extensive especially when lake is thermally stratified

Table 13. OECD study group on lake eutrophication - lake classification scheme.
Suggested values for total phosphorus [P], mean concentrations of chlorophyll-a [chl],
maximum concentrations of chlorophyll-a, [chl max], and for mean, [$\overline{sec}$], and minimum,
[sec min], transparency to distinguish between lakes of different trophic status (OECD 1982).

Trophic category	[P] mg/m^3	[chl] mg/m^3	[chl max] mg/m^3	[$\overline{sec}$] m	[sec min] m [1]
Ultra-oligotrophic	< 4	< 1	< 2.5	> 12	> 6
Oligotrophic	< 10	< 2.5	< 8	> 6	> 3
Mesotrophic	10 - 35	2.5 - 8	8 - 25	6 - 3	3 - 1.5
Eutrophic	35 - 100	8 - 25	25 - 75	3 - 1.5	1.5 - 0.7
Hyper-eutrophic	> 100	> 25	> 75	< 1.5	< 0.7

[1] These values will only apply in lakes which are free of significant water colour.

233

Table 14. Allocation of points for Estuary quality.

Biological Quality

Points awarded if the estuary meets this description

(Scores under a, b, c and d are summed)

(a) Allows the passage to and from freshwater of all relevant species of migratory fish, when this is not prevented by physical barriers. (Relevant species include salmonids, eels, flounders and cucumber smelts etc). — 2

(b) Supports a residential fish population which is broadly consistent with the physical and hydrographical conditions. — 2

(c) Supports a benthic community which is broadly consistent with the physical and hydrographical conditions. — 2

(d) Absence of substantially elevated levels in the biota of persistent toxic or tainting substances from whatever source. — 4

Maximum number of points — 10

Aesthetic Quality

(One description only is chosen)

(a) Estuaries or zones of estuaries that either do not receive a significant polluting input or which receive inputs that do not cause significant aesthetic pollution. — 10

(b) Estuaries or zones of estuaries which receive inputs which cause a certain amount of aesthetic pollution but do not seriously interfere with estuary usage. — 6

(c) Estuaries or zones of estuaries which receive inputs which result in aesthetic pollution sufficiently serious to affect estuary usage. — 3

(d) Estuaries or zones of estuaries which receive inputs which cause widespread public nuisance. — 0

Chemical Quality

(One value only is chosen)

Dissolved oxygen exceeds a saturation value of:

60%	10
40%	6
30%	5
20%	4
10%	3
below 10%	0

SOIL AND GROUND WATER QUALITY INDICATORS

S.E.A.T.M. van der Zee and F.A.M. de Haan
Department of Soil Science and Plant Nutrition
Wageningen Agricultural University
6700 EC Wageningen
The Netherlands

ABSTRACT. Quality standards are used in The Netherlands to evaluate whether soil and ground water have been contaminated and whether this is serious. These standards are based on either total content (soil) or solution concentrations (ground water). We show that neither choice is adequate for assessing the hazard of contamination. Alternatives and the associated complications are mentioned.

1. INTRODUCTION

During the past decades, awareness has grown that soil and ground water quality are threatened by contamination with a variety of components. These components include salts (in arid regions or coastal plains) and nutrients due to excessive application of manure, pesticides, heavy metals, chlorinated and other hydrocarbons, etc. This awareness has resulted in the development of quality standards to assess whether the soil/ground water system has been contaminated, and to what extent. However, the development of such standard criteria is far from trivial because of several complicating factors.

One of the complicating factors is that the soil/ground water system is so complex. Thus, because of the presence of a solid phase that interacts with the compounds present in solution, the system usually exhibits a buffering capacity. This buffering capacity is mostly larger than for the surface water or atmosphere compartments, although it differs for the various compounds of interest. Hence, when we define the buffering capacity as the potential to counter perturbations imposed on the system at hand, the vulnerability for e.g. adverse effects of contamination depends on the compound.

An additional complication is that the system itself is spatially variable. These geographical differences may be profound. Consequently, the buffering capacity varies from place to place. Furthermore, the objective of setting quality standards is usually to prevent or recognize threats of adverse contamination effects. This implies that such standards should also consider adverse effects in future. In view of the dynamics of the soil/ground water system this time dependence may be hard to evaluate. Recognizing this dynamic aspect, agreement was reached during the C.E.C. Conference on "Scientific basis for soil protection in the European Community" [Barth and L'Hermite, 1987] that the evaluation should preferably be based on effects that can be expected from the presence and behaviour of pollutants/contaminants. Clearly, this requires a quantitative risk assessment approach taking the various functions into account.

In this paper first the different functions and the potential threats due to contamination are discussed. Then the present standards for soil and ground water quality are reviewed. A major

235

A. G. Colombo (ed.), Environmental Impact Assessment, 235–259.
© 1992 ECSC, EEC, EAEC, Brussels and Luxembourg.

part is devoted to considering the standards critically from the point of view of their aims. Although most examples are given for the Dutch situation this does not imply that the present discussion is not applicable for other countries. However, in other countries a larger variety may be expected with regard to e.g. subsoil geology. Hence, a larger diversity of parent material and reference values as discussed later in this contribution may be found. The main message of this contribution concerns the identification of appropriate quality indicators (for soil and ground water). Such quality indicators may be either abiotic or biotic factors that may be of use in assessing whether soil or ground water quality is acceptable or not. Because of the generality of the discussion presented, the limitation of most examples to the Dutch case does not limit the relevance for other countries.

2. THREATENED FUNCTIONS

When we consider the concepts of contamination and pollution we mean in the present context that adverse effects are expected. Thus, we do not consider an action but a situation. This distinction is important because many compounds become contaminants only because the dose (load) to the system is large. That we are dealing with a relative scale becomes apparent when we realize than many compounds are naturally present in soil and ground water, even though this may be at trace levels. Whereas this is not the case for xenobiotic substances like dioxins or natural compounds like cadmium, often compounds are essential in the biosphere.

For the heavy metals zinc and copper we may therefore distinguish three ranges of their presence in e.g. soil (Figure 1). At too low levels, where bio-availability is suboptimal, these compounds are in the deficiency range. The yield or crop quality will then be limited by the bio-availability of the (micro)nutrient. At larger contents, the contents in soil may be adequate for e.g. crop growth and we speak of the optimal range. At still larger levels adverse effects may occur and we consider the system to be contaminated.

With regard to crop growth the adverse effect may represent phytotoxicity, which leads to reduced yields, as well as too large concentrations in the plant tissue. In the latter case the range is suboptimal from an economic (product quality) point of view rather than from a plant-nutrition point of view. However, both situations may be realistic. For instance, the compounds Cu and Zn may lead to toxicity effects in plants. For grassland, however, Cu-contents may still be acceptable for the crop itself, but may lead to toxicity effects for sheep, when the grass is destined for consumption by sheep. For other compounds, crop quality constraints are based on tissue concentrations in view of potential hazardous effects for human and animal health, rather than on crop yields. These compounds (e.g. Cd, Pb) may accumulate in large concentrations before possible yield reductions are expected [Ferdinandus et al., 1989, Wiersma, 1985]. This implies that quality constraints should reflect both direct and indirect effects of contamination. In the previous example, indirect effects were associated with bio-accumulation and toxicity in food chains, for soil used in agriculture.

In principle soil and ground water have different functions that may be threatened by soil contamination. Examples of threatened soil functions are:
- living platform. This applies for example to risks associated with mouthing behaviour (pica) of children (play grounds, gardens, parks) [Brunekreef, 1985], and economic/ psychosocial effects when residential areas are contaminated and inhabitants are faced with loss of property or property value. Also real human toxicity effects may necessitate urgent measures that require relocation of residents;
- growing medium. The crops grown in agricultural areas that were contaminated may show yield or quality reduction. Also natural vegetation may be treatened (acid deposition, eutrophication). More indirect effects can be expected due to (in)direct uptake by grazing animals [Bremner, 1981], and release of contaminants

in the food chain;
- ecological function. Partly this was considered above, as vegetation may react to even slight changes in the soil nutrient status. Adverse effects on e.g. earthworms [Ma, 1983] may also result in decreased soil fauna or flora activity and thereby affect the reproductive capacity of soil.

Important functions of ground water may be:
- resource for drinking water. Contamination may make ground water useless for direct distribution as drinking water, necessitating costly purification measures, if purification is possible at all;
- surface water recharge. In view of the various functions of surface water, its quality also needs to be protected. Recharge by ground water rich in e.g. nutrients may lead to surface water eutrophication. This may result in a poor recreational value of surface water, fish mortality, toxicity (by toxin production) caused by excessive algae growth;
- construction material. This applies equally to soil and ground water, as the solid phase of both compartments may be useful for the construction of buildings, roads, etc. Contamination may reduce their suitability for this purpose. It is important to realize that contamination of (ground) water implies contamination of the solid phase also.

Of course other functions may be identified. An example is the water filtering function of soil, to protect ground water quality. In some cases this filtering (purifying) capacity is negligible. An example is the release of nitrate which originates from excessive manure applications to soil. Because nitrate hardly sorbs onto the solid matrix it is readily leached from soils that have a limited denitrification potential. Consequently, nitrate fluxes into ground water may be large. Indeed, in some predominantly sandy areas in The Netherlands, the nitrate quality standard laid down by the EC for the safeguarding of human health (drinking water) is exceeded. Sometimes, this standard (50 mg nitrate per litre) is exceeded down to 70m below the soil surface. For other compounds the filtering capacity of soil is (extremely) large. An example is phosphate which will not be readily leached but accumulates in the solid phase instead. The onset of leaching may take a long time. However, for such compounds the problem resembles a time bomb [Stigliano et al., 1991]. Once the situation has deteriorated such that (sudden) adverse effects can be expected, it appears to be very difficult, if at all possible, to cure this situation [Van der Zee, 1988, van der Zee et al., 1990b].

The complications that may arise due to pollution of the soil and ground water compartments are huge. These compartments are vital for the global element cycles, hence for the functioning of the biosphere. Moreover, they are often buffered against perturbations (i.e., both contamination and reclamation) by liquid/solid interactions. Such buffering varies spatially (or geographically) as well as for each compound. In the sequel, we will evaluate current Dutch standards for soil and ground water quality for some substances and give an assessment of some shortcomings of the current regulations.

3. QUALITY STANDARDS

In view of the urgency of the situation, such as polluted soil and ground water underlying urban areas that sometimes resulted in evacuating the houses for sanitation operations, measures were developed in the Dutch Soil Cleanup Act. To present guidelines for assessing the severity of the situation the A-B-C values were defined. These signal values represented (A < B < C) :
- A-value: background value; no problems expected,
- B-value: a level of contamination requiring further investigation,
- C-value: a level of contamination that requires further action in view of possible serious effects.

Because these values were the only available reference, they were maintained rigidly despite their preliminary nature. Although the values were not soundly based on a quantitative risk assessment, they became legally enforced, which lead to much commotion, damage, economic losses, etc. For instance, newly developed residences were sometimes demolished on the basis of contamination exceeding the C-value, without further (quantitative) risk analyses. Furthermore, contaminated soil became chemical waste once it was excavated for construction purposes and thereby required costly sanitation or disposal. Sanitation was enforced until contaminants were present at contents lower than or equal to the A-value, which was never intended to be used as such. Because of the excessive costs of such operations, more serious pollution cases may remain unaddressed. In view of the discussion in the sequel, we mention that the A-B-C values are based on total element contents in the soil (i.e., mg / kg).

4. NEW REFERENCE VALUES FOR SOIL

During the last decade much research was done to appreciate the relevance of particular heavy metal levels in the context of soil quality. A systematic study by Edelman [1984] analysed heavy metal contents in long term nature reserves. Because, for instance, atmospheric deposition is a cause of diffuse contamination over most of Europe, the heavy metal contents found are not to be considered natural background values. Instead, such contents represent almost uncontaminated or low level contaminated soils.

Whereas the natural background depends on the parent material, i.e., the mineralogical composition of primary and secondary minerals, which is relatively well known, human activities resulted in a spatially variable addition of e.g. heavy metals. For illustration the diffuse atmospheric deposition rates for some heavy metals in The Netherlands are shown in Figure 2. Clear spatial differences that are related with industrial activities are observed. Therefore, the data obtained by Edelman [1984] reveal the "current background values".

The current background values may not be reference values for good soil quality, because this would require the assurance that no adverse effects are due to such heavy metal levels. This assurance is still lacking. One of the main conclusions that may be drawn for the data set is the dependency of heavy metal levels on clay content and organic matter content [Lexmond and Edelman, 1986, 1987], in topsoils (0-10cm). Their model developed to quantify this dependency (see also De Haan et al. [1990]) accounts for the effects of organic matter content (humus, H) on the bulk density. An illustration of the features for some heavy metals is provided in Figure 3, where the metal content (mg kg^{-1}) is related with the clay content (g kg^{-1}). Soil samples with H < 0.25 kg kg^{-1} (.) were distinguished from those with H > 0.25 kg kg^{-1} (o). Full lines denote the relationships and dashed lines an upper bound range for nature reserve topsoils. In addition a horizontal dashed line represents the original A-values specified in the Dutch Soil Clean-up Act, which appear to overestimate the current backgrounds.

For easy reference, the new reference values are presented in relation with L and M in Table 1. Note that in the equation of Table 1, both L and M are in % by mass.

An inventory of heavy metal contents in Dutch arable soils was made by Van Driel and Smilde [1982] and Wiersma et al. [1985], using topsoil samples (0-20cm). An illustration is shown in Figure 4. The indicated lines represent the dashed upper bound range of Figure 3. Whereas for Cr and Ni the situation is comparable to nature reserve areas, for Cd this is not the case. High values are, among other factors, due to the high Cd-contents of phosphate fertilizer.

5. STANDARDS AND EFFECTS

For the current standards for soil and ground water quality we deal with several problems. The first problem is that they inadequately reflect effects. Although the C-values may have been partly motivated taking effect or no-effect levels into account, it is useful to realize that they are generic as well as based on total amounts. By generic we mean that the C-values are not geographically diversified, as are the A-values of e.g. heavy metals. The latter were related to organic matter and clay content of the soils, which vary geographically. We observe that the C-values are therefore the same for soils with a large and with a small buffering capacity alike.

Provided the total contents of a contaminant are the same for two different soils, it is therefore possible for the effects to be different. This is not always the case, as direct ingestion or inhalation of soil or wind-blown dust may lead to the same effects in both cases. Then, the total content may be what matters.

When, however, the effect of contamination is controlled by bio-availability or mobility, the crucial parameter is not the total content. Instead, the parameter of interest is related closely to the concentration (c) in solution (or gas phase) [De Haan and Van Riemsdijk, 1986]. The relationship between this key parameter (c) and the total content (T) is controlled by soil chemical processes. Mathematically, we may express the relationship by

$$T = \rho q + \theta c \tag{1}$$

where q is the sorbed amount, ρ is the dry bulk density, θ is the volumetric water fraction, and other symbols are as defined earlier. Whereas standards are often formulated in terms of T, effects are related with c. To obtain an impression of the discrepancy between these quantities, we analyze (1) in more detail, for the contaminant cadmium.

Soil sanitation is required in The Netherlands in case of "serious risks" for human health or for the environment. Examples of pathways of contaminants present in soil or ground water to the biosphere are e.g. transport to roots of plants where uptake occurs, or transport to surface water. The equation describing this process mathematically is

$$\frac{\partial T}{\partial t} = \nabla \cdot [\theta D \nabla c - \theta v c] + \Phi \tag{2}$$

where D is the dispersion tensor, v is the interstitial flow velocity and Φ is a production/consumption term. The first term on the right hand side is the divergence of the flux. Hence, the flux towards roots etc. depends on concentration instead of on T. Only when accumulation is of interest we quantify T, which comprises e.g. adsorption, and precipitation. Considering cadmium, adsorption is most important. It was shown by Van der Zee and Van Riemsdijk [1987] and Boekhold et al. [1990] that Cd adsorbs according to the Freundlich equation, just as Zn, Cu. This equation reads

$$q = k c^n \tag{3}$$

Based on earlier work by Chardon [1984], see also De Haan et al. [1987], it was known that the parameter k depends on a variety of properties (see Figure 5). Most important for sandy soils appeared to be the organic matter content and the soil-pH. These two factors can be given explicitly in (3), namely

$$q = k^* oc (H^+)^{-0.5} c^n \tag{4}$$

Now k^* accounts for all factors except oc and pH (or proton activity in solution, (H+)).

In a recent study Boekhold et al. [1991] showed that (4) accurately describes the effects of the explicitly given factors oc, pH, and Cd-concentration. Determining k^* and n for two soil samples by batch experiments gave the information needed for predicting transport experiments (Figure 6). The agreement between the independent predictions (i.e., not fitted) and the experimental data appears to be good, and this gave confidence in the model (eq. 4). Subsequently, along a transect in a Cd-contaminated field, measurements were made of T, c (denoted Cd_S), oc, and pH. The results are shown in Figure 7 and reveal quite different patterns for all parameters. The main message is that using only the spatial pattern of T, the pattern of c, which is of prime interest, cannot be predicted. This means that current assessment of T-values may suggest a large risk (if T is large) at some place. However, the real risk may be relatively small, because at that spot c may be small.

However, using (4) it should be possible to relate T and c when the data for oc and pH are also taken into account. Indeed, calculating for each location the T-value, using the oc, pH, and Cd_S-value (=c) at that particular location, we obtain an excellent prediction of the T-pattern, as is shown in Figure 8. This exercise suggests that, for this atmospherically contaminated field, negative effects will be about the same for all parts of the field, because c does not reveal a clear pattern. Because of spatial variations of oc and pH, the total content (T) may be quite variable, because the buffering capacity varies spatially. An illustration of oc-variability for a field is shown in Figure 9.

In general, the assessment of risks may not be based on measurements of total content of contaminants only. Among others because a total content for a soil with a low buffering capacity is more serious than for a soil with a large buffering capacity, this capacity or the contaminant concentration in solution itself may have to be assessed. A similar treatment may be given for ground water. A difference is that for ground water it is usually simpler to measure the concentration than the total content. However, as elementary texts on solute transport show, both the velocity with which contaminants move through soil and ground water, and the time needed for sanitation actions, depend on the buffering capacity. Hence, also for ground water we should not make the limitation to set standards for either T or c. For both content and concentration, related by the buffering capacity, standards are needed. In this respect it is useful to observe that soil standards based on T, and ground water standards usually based on c, may be incompatible in practice. This is the subject of the next section.

6. STANDARDS AND SUSTAINABILITY

Standards for soil and ground water quality should be based on sustainable use of these resources. This involves the anticipation of adverse effects for the future. Consequently, practices that are currently not recognized as harmful, may nevertheless constitute a problem in the future. When the future risks are not considered acceptable, their cause(s) should be eliminated now (even if current risks are acceptable). A method for identifying such long term problems was developed by Ferdinandus et al. [1989] and Van der Zee et al. [1990a]. Illustrating this method for heavy metals, we consider a ploughed soil layer. For this layer, a balance equation may be formulated given by

$$\frac{dT}{dt} = \frac{dI}{dt} - \frac{dJ}{dt} - \frac{dP}{dt} \tag{5}$$

where the accumulation rate (dT/dt) is balanced by the rates of input (I), leaching (J), and plant uptake (P). Input is due to e.g. atmospheric deposition, application due to the presence in fertilizer, manure, pesticides, etc. For instance, it is well known that manure and fertilizer contain heavy metals [De Boo, 1988, Ferdinandus, 1989]. Pig manure may have large Cu-concentrations, and phosphate fertilizer may contain much cadmium. The same is the case for natural precipitation as is shown in Table 2. In this table the profound input rate due to the use of sewage sludge can also be seen. Sewage sludge contains large amounts of all heavy metals. We now consider different crops, as indicated in general terms in Table 2. For realistic Dutch crop rotation schemes, fertilization was assumed to be exactly in agreement with the NPK-requirements of the crops. However, the source of these elements (N, P, K) was varied. In one scheme the requirements were met by fertilizer. In another scheme, manure or sludge was applied till one of N, P, or K requirements (e.g. N) were met. The remaining demand of the two other elements (e.g., P and K) was fulfilled with commercial fertilizer. Because the heavy metal contents of manure, sludge, and fertilizer are reasonably well known, the input of Table 2 could be calculated. The input by precipitation are average values for The Netherlands.

To quantify the leaching rate (dJ/dt) and the uptake rate (dP/dt) is a problem, as it requires the quantification of a sorption equation e.g. as was done by Boekhold et al. [1990] and a relationship between plant uptake rate and the solution concentration. However, we can easily assess the acceptable removal rates, when we assume that (1) ground water recharge quality remains equal to the current mean ground water quality, and (2) removal by crops is exactly in agreement with crop quality standards. Whereas (1) defines the "stand-still" principle for ground water and quantifies (dJ/dt), (2) is related with (dP/dt) for the case crop quality is just acceptable. For the Dutch situation these removal rates are also shown in Table 2, and may be considered the maximum permissible removal rate.

We observe that immission rates are larger than removal rates in many cases. This implies that on average (for The Netherlands), heavy metal accumulation will occur until a steady state is reached (Figure 10). At this steady state, input equals removal and T is constant. Then we may expect, for the cases of Table 2 where input is larger than removal, that either crop quality or ground water recharge quality (or both) become sub-standard. It may take a long time to reach a steady state [Boekhold and Van der Zee, 1991; Ferdinandus et al., 1989], but quality standards may also be exceeded relatively fast, depending on the soil and environmental conditions. Anyway, this example shows that current standards should be obeyed also on the long term. This implies that the quality standards for air (represented by atmospheric deposition in the example), soil, ground water and surface water, should in principle be consistent.

To make the different standards consistent with each other necessitates the formulation of criteria regarding the acceptable fluxes between different compartments. These fluxes are difficult to quantify because they require, among other information, good knowledge of the contaminant concentrations in solution or gas phase. These depend on the sorption behaviour of the contaminant which is often poorly known and which differs for each soil type. The work by Boekhold et al. [1991] for heavy metals, and by Van der Zee et al. [1990b] for phosphate gives an impression how a consistent set of standards may perhaps be obtained, but their approach involves further research. Consequently, it may be necessary to aim at reducing input by formulating standards for the input rates of contaminants at the soil surface, at the ground water level, etc. The quality restriction at the ground water level was implicitly made by the EC in regulations for pesticide concentrations in ground water. After all, most pesticide screening approaches are based on the concentrations that are expected to leach the ground water level.

If these expected concentrations exceed the EC-quality standard of 0.1 mg m^{-3} [Council European Communities, 1980], the pesticide in question is likely to be prohibited. Examples of pesticides screening models were given by Jury et al. [1983], Rao et al. [1985] and Van der Zee and Boesten [1991]. They illustrate how quality standards may be translated mechanistically in soil as well as pesticide properties. These properties are evaluated in terms of expected effect (i.e., the probability that concentrations in ground water will exceed EC-standards). Obviously, similar approaches are also needed for other contaminants. For heavy metals one might consider screening models as developed by Ferdinandus et al. [1989] and Boekhold et al. [1990] to assess the hazard of exceeding quality standards of (i) soil, (ii) ground water, (iii) crop etc. in the steady state regime.

7. BIOLOGICAL QUALITY ASSESSMENT

A disadvantage of quality assessment based on contaminant mobility (concentration in solution) is that, although it is related to e.g. current water standards, it does not specify effects. A necessary step therefore is to specify the effect in terms of mortality, tissue concentration etc. for biota. At the moment, a large number of studies describe such relationships with respect to the total contaminant content in soil. Examples for plant uptake were given by e.g. Lexmond [1980] for copper, Kuboi et al. [1986], Bjerre and Schierup [1985], and Busch [1985] for cadmium.

In the past decade, much research was dedicated to identifying biological indicators for soil quality. At the same time, it may be emphasized that the ecological functioning of soil depends on the activity of soil biota. Micro-organisms are particularly important for element cycling in general, and nutrient cycling in particular. Micro-organism activity depends, among other things, on the moisture content of soil. Other soil biota, of which earthworms, such as Lumbricus rubellus, form the main part of biomass in soil, are likewise important for mineralizing organic matter. Furthermore, both their mechanical activity and their excretion products contribute significantly to soil structure stabilization. Such stabilization is of major importance for the maintenance of a good aeration and drainage. Because of the intensive contact with soil (e.g. through the intestines), earthworms are in principle useful as a bio-indicator [Van Hook, 1974]. In addition, earthworms serve as food for e.g. birds and moles. Elevated contaminant contents of earthworms are therefore important in relation to food chain contamination.

In view of the aspects mentioned above, the effects of soil contamination on soil biota have received considerable attention. For micro-organisms, the effects may be visually apparent in the case of inorganic matter accumulation. This is indicative of a reduced ability to decompose organic matter. Effects that are visually less easy to observe may be the inhibition of respiration, nitrogen mineralization and nitrification. Of these, respiration in particular is a sound general indicator of microbiological activity. For the processes mentioned above, the ranges of various heavy metals that produce adverse effects are given in Figure 11. These ranges are obviously rather qualitative. For instance, soil type may affect the effect level profoundly. This is illustrated for Pb-contamination effects on respiration in Figure 12, for different soils.

The degree of inhibition of respiration by heavy metal contents depends on biotic as well as abiotic factors. The main important abiotic factors may differ for the various heavy metals. Doelman [1990] summarized the primary (1) and secondary (2) controlling abiotic factors for different heavy metals. He found significantly different control for Cd (1. clay, 2. organic matter), Cu and Pb (1. Fe, 2. Mn), Ni (pH), and Zn (1. Fe, 2. clay). To illustrate the complications that may arise in effect assessment, even for one soil, the effect of Cu on nitrogen transformation may be considered. At our laboratory, the adverse effects of increasing Cu-activity and decreasing pH were investigated (personal communication Lexmond, unpubl. data). The decrease of NH_4 due to NO_3-production depended sensitively on both Cu-activity and pH (Figure 13). However, this appeared to be the case only when Cu was added almost simultaneously as NH_4. When Cu was applied to the soil weeks to months earlier, only significant pH-effects were observed and no significant effects of Cu-contents were established. Possibly, this reveals that adaption of the micro-organisms took place during the incubation with Cu. It is equally feasible, though, that Cu was immobilized by reactions with the solid phase. This would reduce Cu bio-availability and allow the microbial population to restore itself after short term Cu-toxicity. Which possibility is most likely is hard to state, and may anyway be different for other soils, pollutants, etc. It is of interest that when nitrogen was applied in the form of manure neither pH nor Cu-activity affected nitrogen conversions.

In general, contamination may lead to the elimination of the most sensitive microbial species as well as to increased resistance of the population. These phenomena should not be neglected in experimental work. The addition of easily available contaminants may give the wrong impression that effects are profound. After adaptation has occurred, microbial activity may be affected less than short term experiments suggest. Whether contaminants are bio-available may depend on time. Thus contaminants may in due time be effectively immobilized, as was suggested by the Cu/N-conversion example. Recently, Rijnaerts et al. [1990] showed that immobilized HCH (hexachlorocyclohexane) may be degraded by micro-organisms, but the rate at which this occurs may depend on diffusion from aggregates towards the micro-organisms.

It is not our intention to say that lethal effects are of importance only for the assessment of soil quality. It was shown for earthworms by Ma [1982] that adverse effects on cocoon production, food consumption, and growth (by weight) may occur at sublethal copper contents. Earthworms may accumulate e.g. Cd and Zn, whereas no accumulation of Cr, Mn, Ni, or Cu was established [Van Hook, 1974]. However, Ten Napel [1988] found a concentration factor (Cu-worm/Cu-soil) of 7.8 $(Cu-soil)^{-0.6}$, with Cu-soil in mg/kg. Again bio-availability as related to concentrations in solution may affect toxicity, because Ma [1982] observed that toxicity of Cd, Zn, and Pb increased when pH was decreased. This can be explained by an increase in solubility and mobility of these metals, as it would be in agreement with the generally found decrease of sorption at higher acidity. A very sensitive indicator may be cocoon production. For a sandy calcareous soil, low in organic matter and with a mean Zn-content of 47 mg/kg, effects of Zn additions as given in Figure 14 were found at our laboratory [personal communication Lexmond, unpubl. data]. For reference, the dune soil had a pH 7.1, 1.5% organic matter, 2.5% free calcium carbonates, and was predominantly sandy (99% of the mineral fraction).

Dose-effect relationships have also been developed for other organisms. The biomonitoring of earthworms has already been mentioned by Ma [1983]. Soil nematodes appear to be promising for biological soil quality assessment as they have been adequately classified, can be easily isolated from soil and can be counted automatically [Bongers, 1988, 1990]. With a recently developed index to evaluate the perturbation of the soil nematode communities, whether an ecological soil typology can be developed for Europe is currently the subject of investigation [Dept. Nematology, Wageningen; Britisch Museum; Univ. Milan]

8. CONCLUSIONS

Various aspects of soil and ground water quality standards were discussed. In general, it may be stated that such standards should be indicative of effects on the different functions of soil and ground water. It would therefore be obvious that standards are related to the function of interest. Different functions need not be equally vulnerable to adverse effects of contamination. Dutch soil quality standards, based on total contents, were considered in more detail. Reference values that reflect the mean current background quality appear to be quite different for various heavy metals. Moreover, they depend on soil parameters such as clay and organic matter content. A problem with standards is that they are hardly related to the effects that may be expected. This is the case, because they are based on total contents. A parameter that indicates effect (e.g. via bio-availability, or mobility) is the concentration in solution. To interpret the risks associated with contamination, the concentration in solution (c) may have to be taken into account. Another problem is that quality standards for different compartments may be poorly related. With an example, we showed that currently accepted agricultural production methods, that are in agreement with standards at present, may not be environmentally sustainable. The complications in assessing biological effects and using bio-indicators were illustrated with several other examples.

9. REFERENCES

Barth, H. and P.L. L'Hermite (1987), Scientific basis for Soil Protection in the European Community, Elsevier Applied Sciences Publishers, Essex, England, 630 pp.

Bjerre, G.K. and H.H. Schierup (1985),"Uptake of six heavy metals by Oak as influenced by soil type and additions of cadmium, lead, zinc, and copper", Plant and Soil, 88, 57-69.

Boekhold, A.E. and S.E.A.T.M. van der Zee (1990), "Long-term effects of soil heterogeneity on cadmium behaviour in soil", J. Contam. Hydrol. (in press).

Boekhold, A.E., S.E.A.T.M. van der Zee and F.A.M. de Haan (1990), Prediction of cadmium accumulation in a heterogeneous soil using a scaled sorption model, Proc. Model Care, IAHS publ. 195, 211-220.

Boekhold, A.E., S.E.A.T.M. van der Zee and F.A.M. de Haan (1991), "Spatial patterns of cadmium contents related to soil heterogeneity", Water, Air, Soil Pollution, (in press).

Bongers, A.M.T. (1988), De nematoden van Nederland. Natuurhistorische Bibliotheek van de KNNV, nr. 46. Pirola, Schoorl.

Bongers, A.M.T. (1990), Biologische bodembeoordeling met nematoden. Hoofdstuk H 5100 in Handboek voor Milieubeheer, Deel Bodem, Samsom, Alphen a.d. Rijn.

Bremner, I. (1981), "Effects of the disposal of copper-rich slurry on the health of grazing animals", in P. L'Hermite and J. Dehandschutter (eds.), Copper in animal wastes and sewage sludge, J. Reidel, Dordrecht, 245-255.

Brunekreef, B. (1985), "The relationship between environmental lead and blood lead in children: a study in environmental epidemiology", PhD thesis, Agric. University Wageningen, 185 pp.

Busch, C. (1985), "Der Einfluss der Stikstoff - und Kaliumdüngung sowie der Kalkung auf die Pflanzenverfügbarheit und Löslichkeit des Cadmiums beider Düngung mit Abwasserklärschlam", Dissertation, Univ. Bonn, 203 pp.

Chardon, W.J. (1984), "Mobiliteit van cadmium in de bodem", PhD thesis, Agric. Univ. Wageningen, The Netherlands.

Council European Communities, Directive of the Council of 15 July 1980 relating to the quality of water intended for human consumption, The Official Journal of the European Communities, L 229, August 30, 11-29.

De Boo, W. (1989), "A closer look at cadmium in agriculture" (in Dutch), Meststoffen 1, 36-39.

De Haan, F.A.M. (1989), "Research priorities for soil quality assessment; in De Haan, F.A.M. et al. (eds.), Soil Quality Assessment, State of the art report on soil quality", CEC, Directoral General XII, Brussels.

De Haan, F.A.M. and W.H. van Riemsdijk (1986), "Behaviour of inorganic contaminants in soil", in J.W. Assink and W.J. van den Brink (eds.), Contaminated Soil, Martinus Nijhoff, Dordrecht, 19-32.

De Haan, F.A.M., S.E.A.T.M. van der Zee and W.H. van Riemsdijk (1987), "The role of soil chemistry and soil physics in protecting soil quality: variability of sorption and transport of cadmium as an example", Netherlands Journal of Agr. Science 35, 347-359.

Doelman, P. (1990), "Interaction heavy metals - soil biota; biological availability; special attention paid to abiotic factors and the soil micro-biota", in W.H. van Riemsdijk (ed.), Capita Selecta bodemchemische aspecten van biologische beschikbaarheid, Soil Sci. and Plant Nutrition, Agric. Univ. Wageningen.

Edelman,Th. (1984), Achtergrondgehalten van stoffen in de bodem, Reeks Bodembescherming 34, 's-Gravenhage, Staatsuitgeverij.

Ferdinandus, H.N.M., Th.M. Lexmond and F.A.M. de Haan (1989), "Heavy metal balance sheets as criteria for the sustainability of current agricultural practises" (in Dutch), Milieu 4, 48-54.

Jury, W.A., W.F. Spencer and W.J. Farmer (1983), "Behaviour assessment model for trace organics in soil, I, Model description", J. Environ. Qual. 12, 558-564.

Kiewiet, A. (1989), "Accumulation of lead and cadmium by lumbricus rubellus", Internal report RIN Arnhem.

KNMI/RIVM (1986), "Chemische samenstelling van de neerslag over Nederland" (in Dutch), Annual Rep. 1984, 126 pp, KNMI 156-7, RIVM 218203003.

Kuboi, T., A. Noguchi and J. Yazaki (1986), "Family-dependent cadmium accumulation characteristics in higher plants", Plant and Soil, 92, 405-415.

Lexmond, Th.M. (1980), "The effect of soil pH on copper toxicity to forage maize grown under field conditions", Netherlands Journal of Agric. Science 28, 164-183.

Lexmond, Th.M. (1981), "A contribution to the establishment of safe copper levels in soil", in P. L'Hermite and J. Dehandschutter (eds.), Copper in animal wastes and sludges, 162-183, Reidel, Dordrecht.

Lexmond, Th.M. and Th. Edelman (1986), Voorlopige referentiewaarden en huidige achtergrondgehalten voor een aantal zware metalen en arseen in de bovengrond van natuurterreinen en landbouwgronden, Report 1986-2, Dept. Soil Sci. and Plant Nutrition, Agric. Univ. Wageningen, 59 pp.

Lexmond, Th.M. and Th. Edelman (1987), Huidige achtergrondwaarden van het gehalte aan een aantal zware metalen en arseen in grond. Hoofdstuk D 4110 in Handboek voor Milieubeheer, Deel Bodem, Samsom, Alphen a.d. Rijn.

Ma, C.W. (1982), Regenwormen als bio-indicators van bodemverontreiniging, Min. Volksgezondheid Milieu, reeks Bodembescherming 15, (in Dutch).

Ma, C.W. (1983), Biomonitoring of soil pollution: ecotoxicological studies of the effects of soilborn heavy metals on Lumbrist earthworms, Ann. Rep. 1982, Research Institute for Nature Management, Arnhem, 83-97.

Ma, C.W. (1984), "Sublethal toxic effects of copper on growth, reproduction and litter breakdown activity in the earthworm Lumbricus rubellus, with observations on the influence of temperature and soil pH", Environ. Pollution Series A, 207-219.

Rao, P.S.C., A.G. Hornsby and R.E. Jessup (1985),"Indices for ranking the potential for pesticide contamination of ground water", Proc. Soil Crop Sci. Fla, 44, 1-8.

Rijnaarts, H. and H. Jumelet (1990), "Impact of physical/chemical parameters on aerobic biodegradation of alpha-hexachlorocyclohexane (α-HCH) in soil and soil suspensions", MSc thesis, Soil Sci. and Plant Nutrition, Agric. Univ. Wageningen, Environ. Sci and Technol. (1990).

Stigliano, W.B. et al. (1991), Chemical time bombs: Definition, Concepts, and examples, Executive Report 16, CTB Basic Document 1, Int. Inst. Applied Systems Analysis IIASA, Laxenburg, Austria, 23 pp.

Ten Napel, G.J. (1988), "Influence of pH on toxic effects of copper on the earthworm Lumbricus rubellus", MSc thesis, Agric. Univ. Wageningen, Dept. Soil Sci. and Plant Nutrition, 51 pp.

Van der Zee, S.E.A.T.M. (1988),"Transport of reactive contaminants in heterogeneous soil systems, PhD thesis", Agricultural University Wageningen, 283 pp.

Van der Zee, S.E.A.T.M. and W.H. van Riemsdijk (1987), "Transport of reactive solute in spatially variable soil systems", Water Resour. Res., 23, 2059-2069

Van der Zee, S.E.A.T.M. and J.J.T.I. Boesten (1991), "Effects of soil heterogeneity on pesticide leaching to ground water", Water Resour. Res. (accepted).

Van der Zee, S.E.A.T.M., H.N.M. Ferdinandus, A.E. Boekhold and F.A.M. de Haan

246

(1990a), "Long-term effects of fertilization and diffuse deposition of heavy metals on soil and crop quality", in Van Beusichem, M.L. (ed.), Plant Nutrition, Physiology and Applications, Kluwer Ac. Publ., 323-326.

Van der Zee, S.E.A.T.M., W.H. van Riemsdijk and F.A.M. de Haan (1990b), The protocol phosphate-saturated soils, 1. Fundamental aspects, Dept. Soil Sci. and Plant Nutrition, Agric. Univ. Wageningen, 69 pp, (in Dutch).

Van Driel, W. and K.W. Smilde (1982), "Heavy metal contents of Dutch arable soils", Landwirtschaftliche Forschung Sonderheft 38, 305-313.

Van Genderen, H. (1987), Relatie tussen bodemkwaliteit en effecten, Verslag Studiedag Bodemkwaliteit, Technische Commissie Bodembescherming.

Van Hook, R.I. (1974), "Cadmium, lead and zinc distribution between earthworms and soil: potential for biological accumulation", Bull. Environ. Contam. 12, 509-512.

Van Riemsdijk, W.H. (1990), Introduction on soil chemical aspects of bioavailability, Caput series soil chemical aspects of bioavailability, Soil Science and Plant Nutrition, Wageningen Agricultural University.

Wiersma, D., B.J. van Goor and V.G. van der Veen (1985), Inventarisatie van cadmium, lood, kwik en arseen in Nederlandse gewassen en bijbehorende gronden, Rapport 8-85, Instituut voor Bodemvruchtbaarheid, Haren.

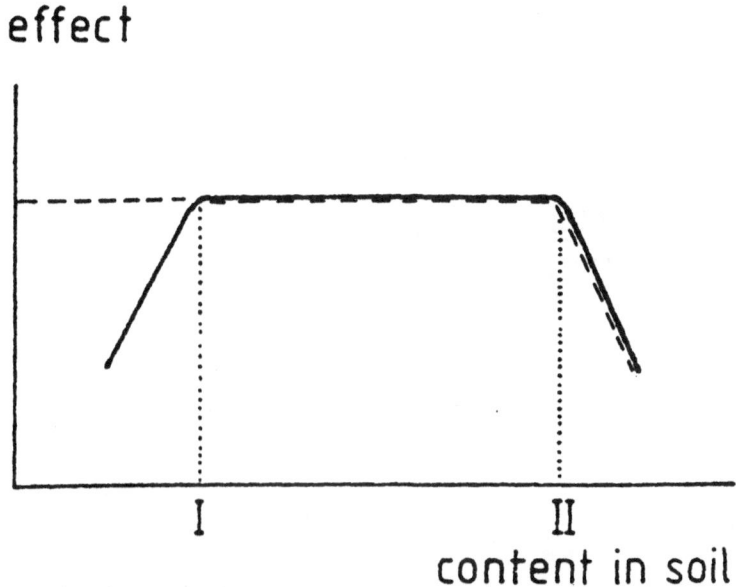

effect

content in soil

Figure 1. Schematic representation of the effect of availability on production quality or yield. Solid line: nutrients, dashed line: non essential solutes. When the slope is positive the effect is positive, when the slope is negative, the quality or the yield decreases.

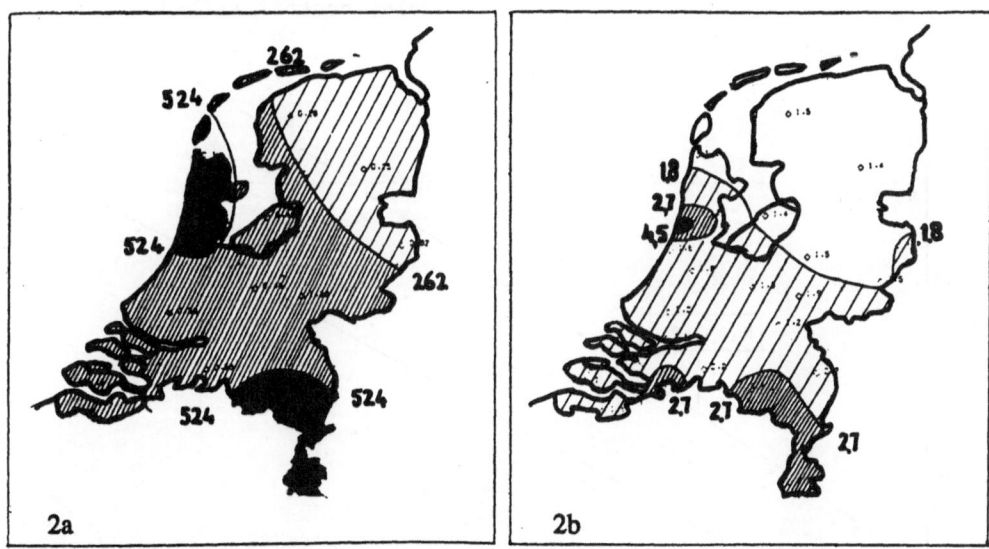

Figure 2. Atmospheric deposition of zinc (a) and cadmium (b) in The Netherlands in g ha^{-1}yr^{-1}. Data from KNMI/RIVM, 1986.

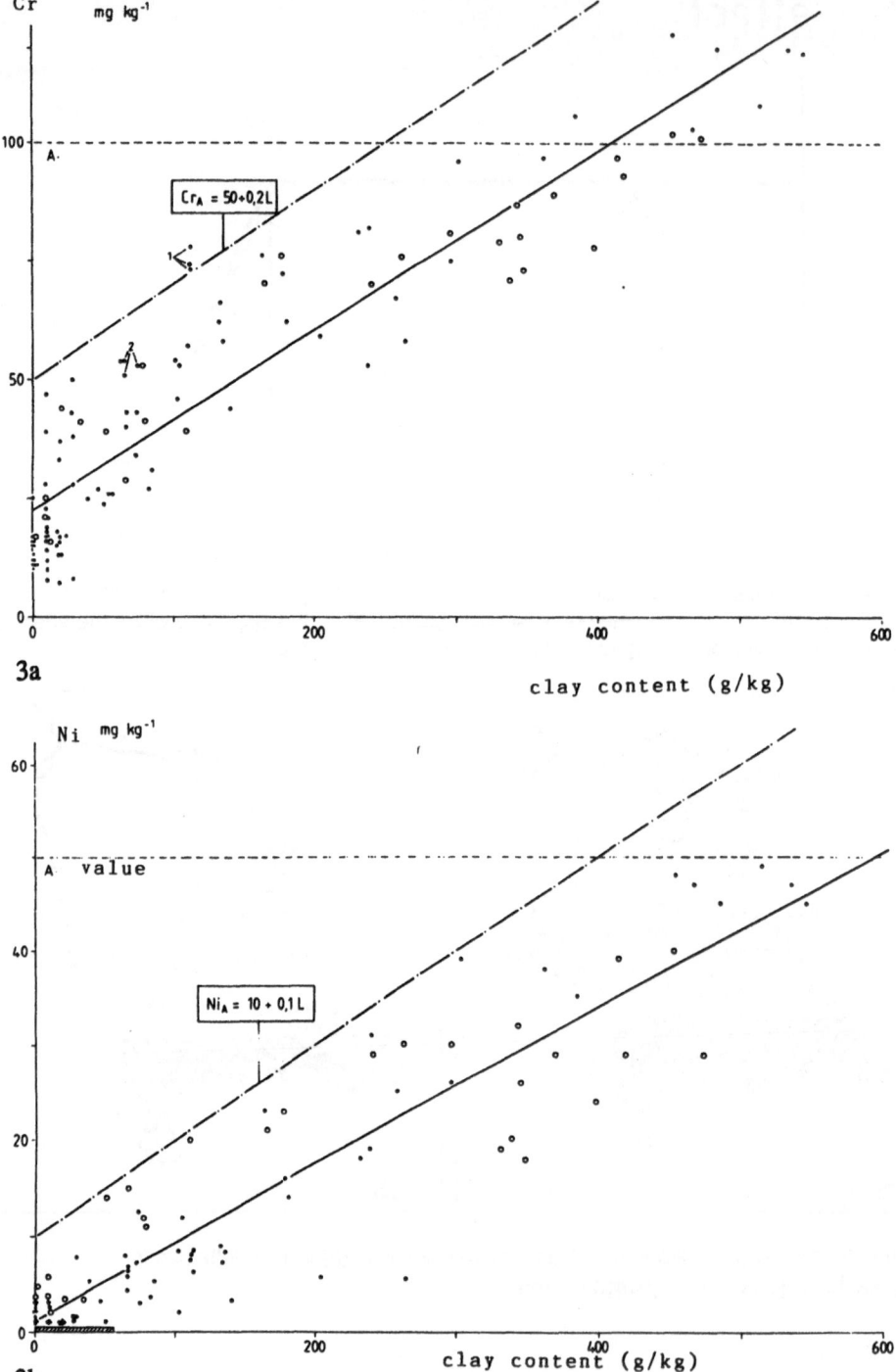

3a

clay content (g/kg)

3b

clay content (g/kg)

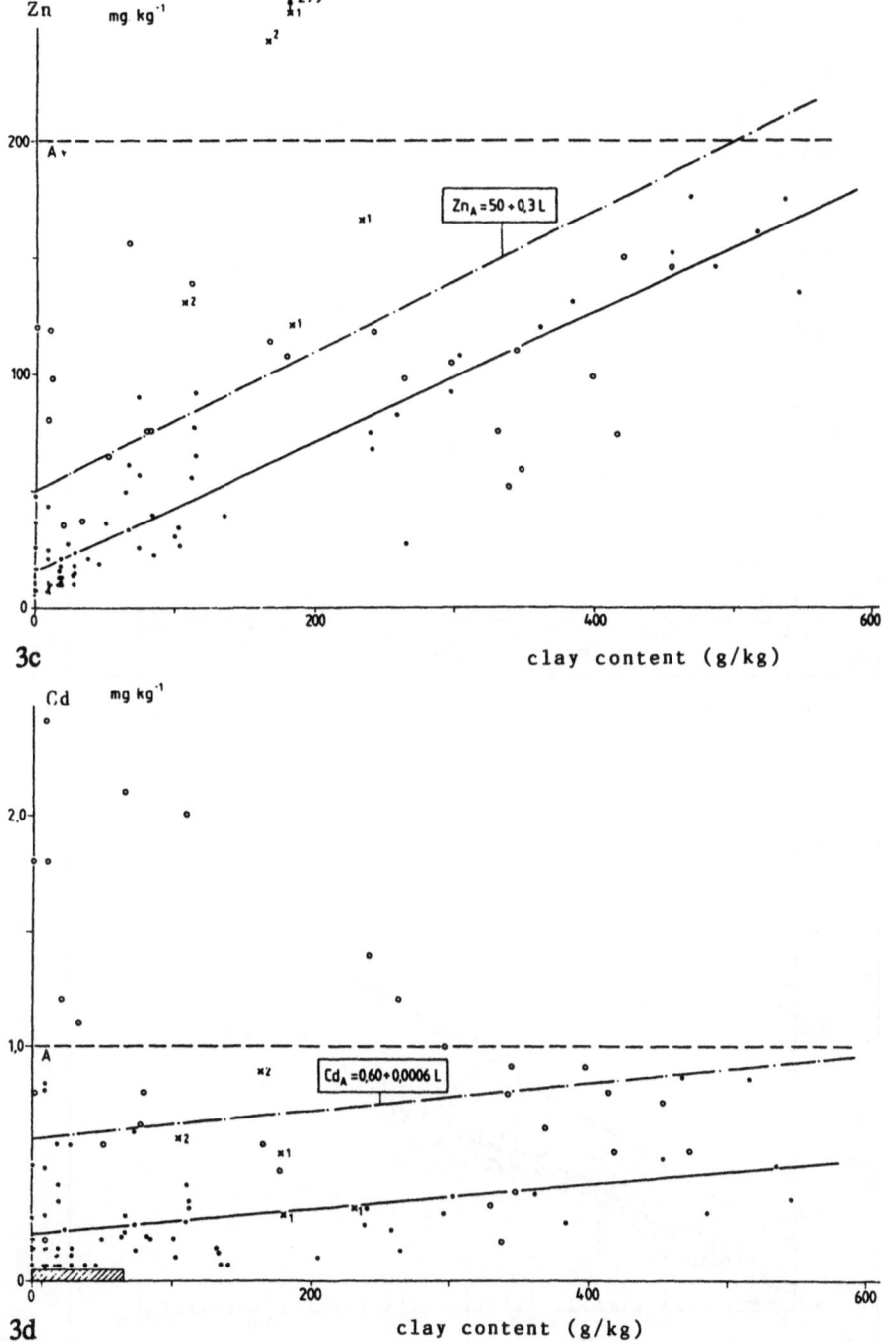

3c

3d

Figure 3. Metal contents (vertical) as a function of clay content(horizontal); explanation given in the text.

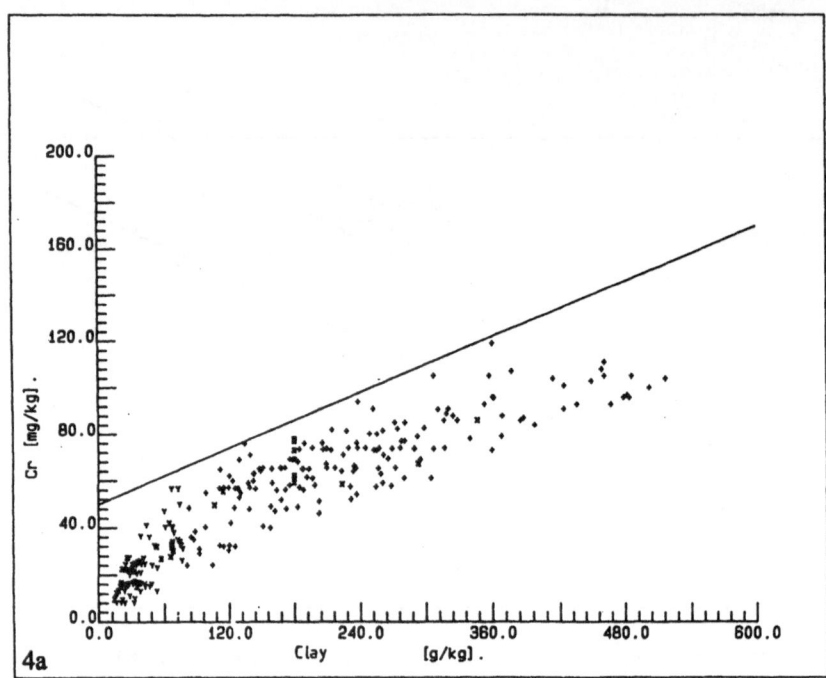

4a

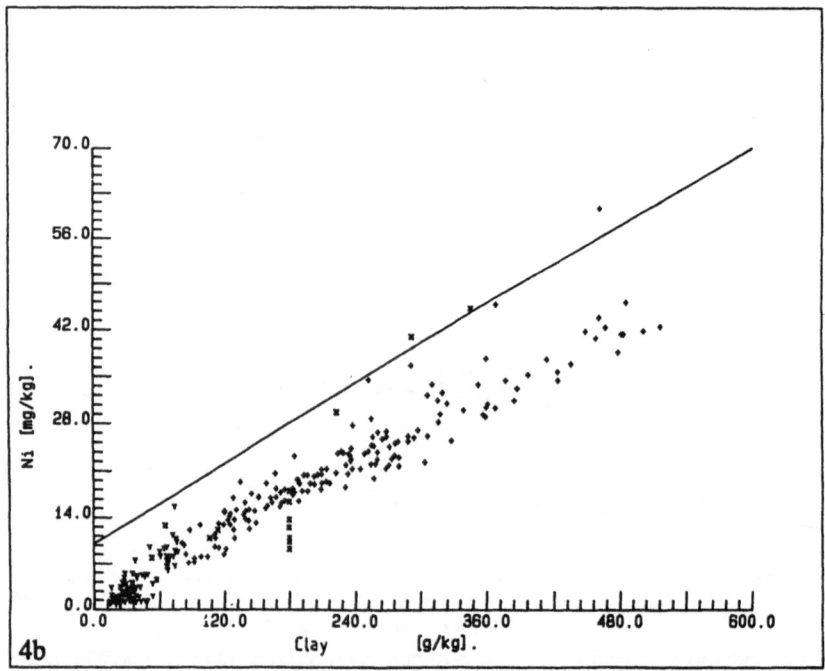

4b

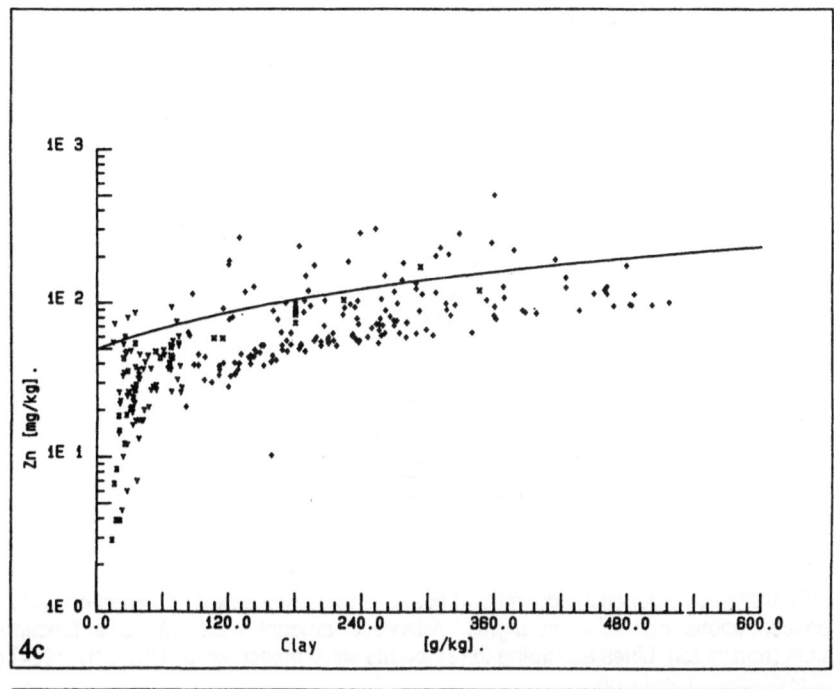

4c

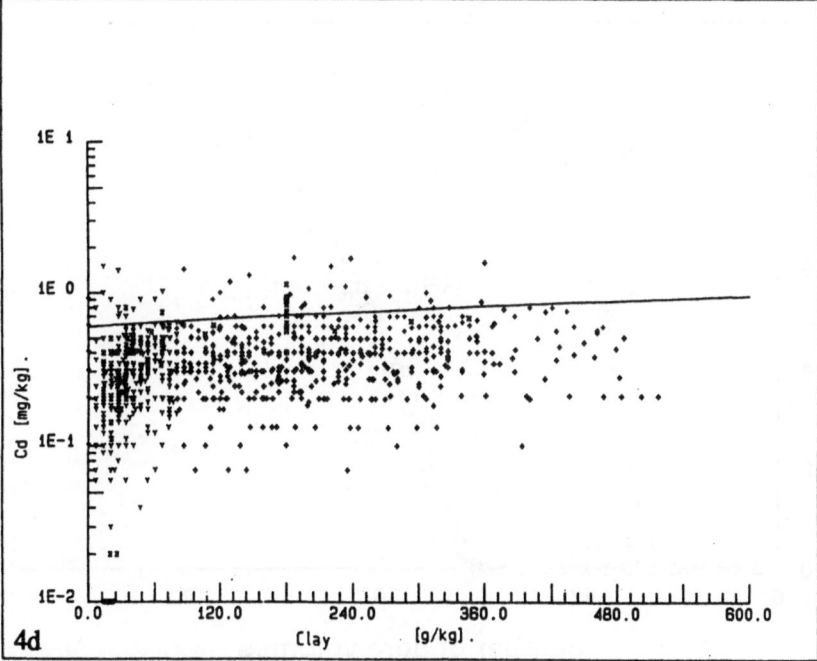

4d

Figure 4. Metal contents (vertical) in arable topsoils as a function of clay content (horizontal); explanation given in the text.

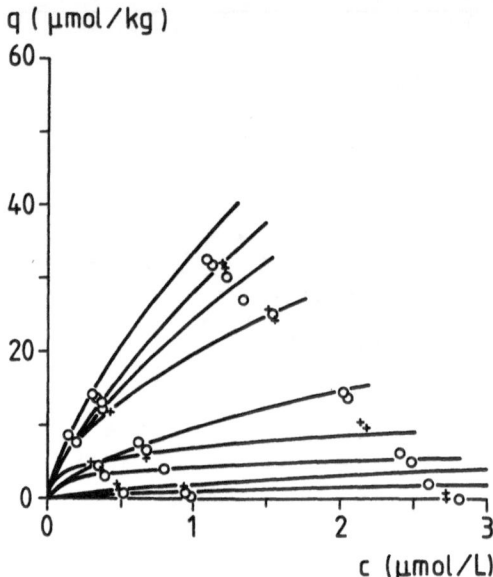

Figure 5. Illustration of Freundlich adsorption equations for cadmium with different chloride and calcium concentrations and ionic strengths. Adsorbed amount (vertical) as a function of concentration (horizontal). Lines according to (3), points were measured by Chardon [1984] and appropriately defined in that work.

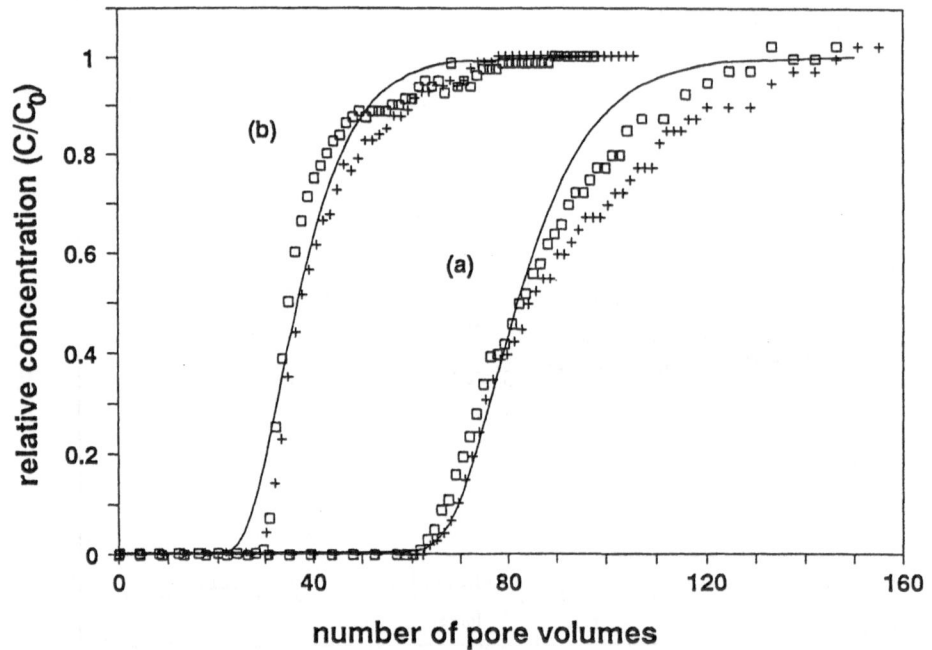

Figure 6. Predicted Cd-breakthrough using adsorption parameter estimated with batch experiments [Boekhold et al., 1990]. For curve (a), $C_O = 2$ mg Cd/l; for curve (b), $C_O = 20$ mg Cd/l.

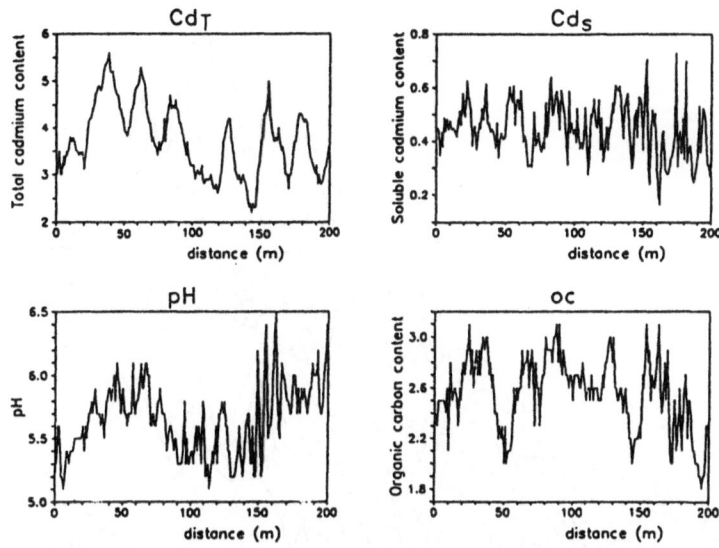

Figure 7. Total Cd (Cd_T), dissolved Cd (Cd_S), organic carbon (*oc*) and pH as a function of position along a transect [Boekhold et al., 1991]. Units were mg/kg for Cd_T and Cd_S, and mass percent (g/g 100%) for *oc*.

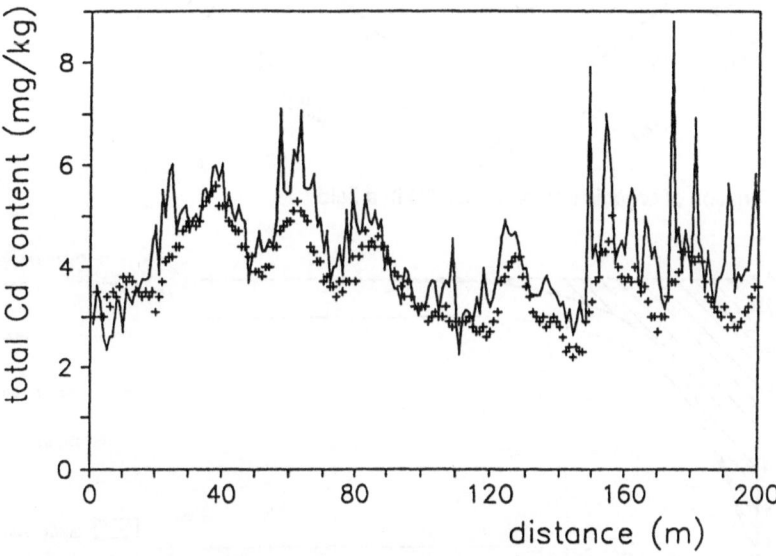

Figure 8. Predicted Cd (points) and measured Cd (line) using local Cd (Cd_S), *oc*, and pH in the Freundlich equation (4); see text.

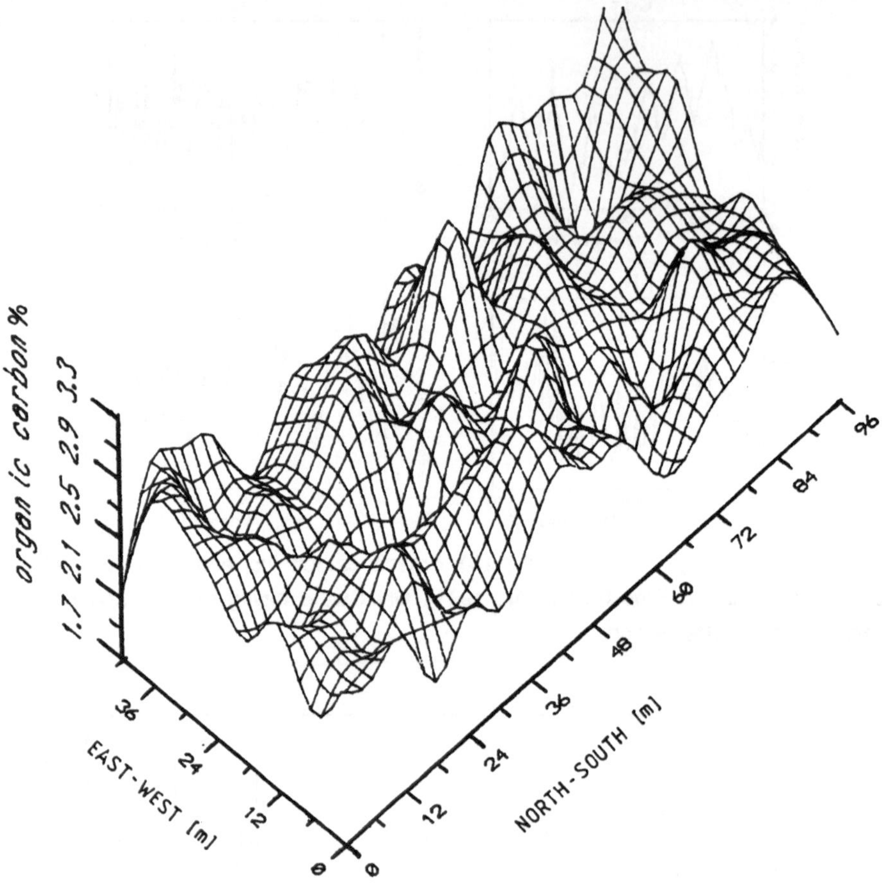

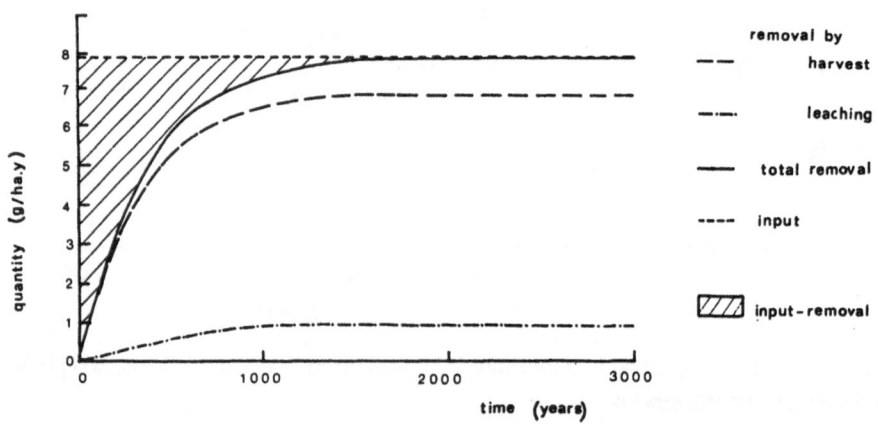

Figure 9. Variation of organic carbon (mass %) in a field.

Figure 10. Approach of Cd-removal and accumulation (for a constant Cd-input) to the steady state situation. Dashed area represents the accumulation in soil [Van der Zee et al., 1990a].

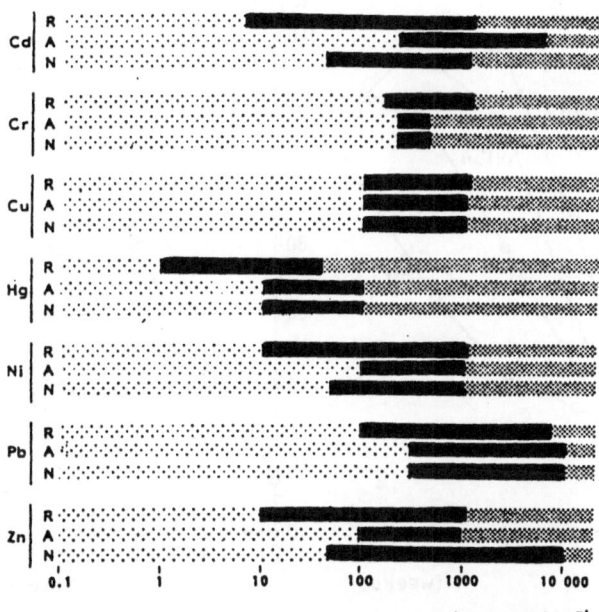

Figure 11. Effects of heavy metals on soil respiration (R), nitrogen mineralization (A), and nitrification (N). Bars denote (left to right) that inhibition was observed never, sometimes, and always, respectively.

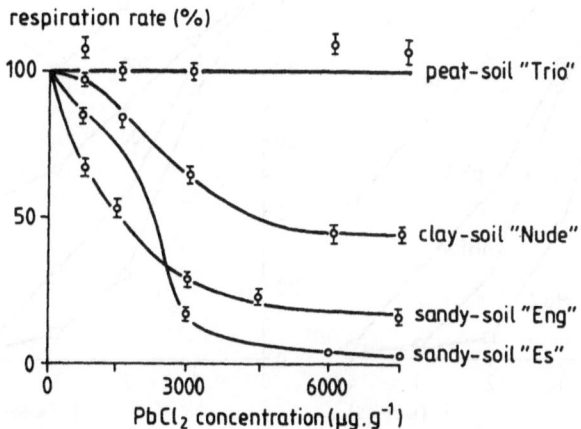

Figure 12. Influence of Pb on respiration rate for four soils, compared to untreated case.

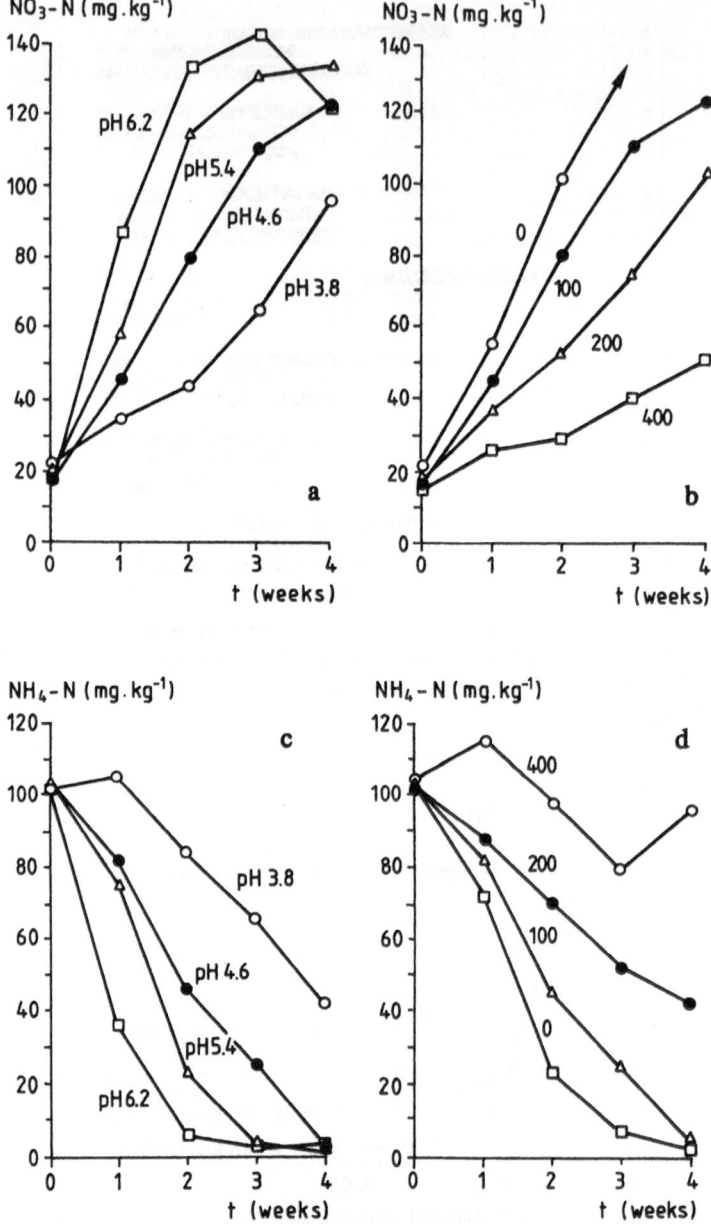

Figure 13. Effect of Cu (in mg/kg) and pH on nitrate and ammonium concentrations when ammonium and Cu were added simultaneously to soil. Figs. 13a and 13c show pH effect for Cu = 100 mg/kg, Figs. 13b and 13d show Cu-effect for pH = 4.5.

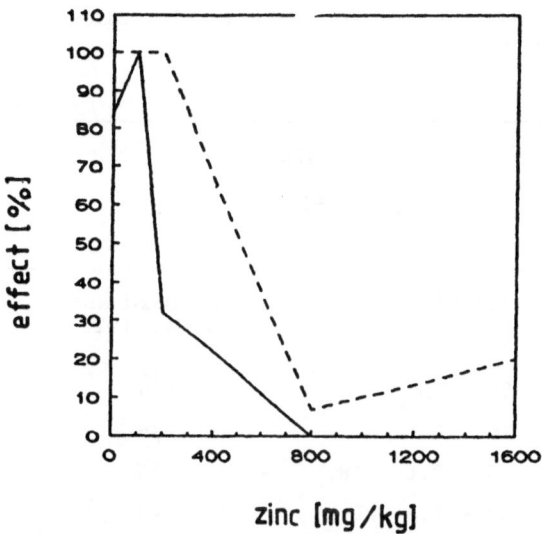

Figure 14. Effect of zinc addition on earthworm survival (dashed) and cocoon production (solid).

Table 1.   Reference values for heavy metals, arsenic and fluoride, with L and H in % by mass.

compound	relationship	SOIL (mg kg^{-1} dry soil) standard soil (H=10, L=25)	GROUNDWATER (µg l^{-1})
Cr	50 + 2L	100	1
Ni	10 + L	35	15
Cu	15 + 0.6 (L+H)	36	15
Zn	50 + 1.5 (2L+H)	140	150
As	15 + 0.4 (L+H)	29	10
Cd	0.4 + 0.007 (L+3H)	0.8	1.5
Hg	0.2 + 0.0017 (2L+H)	0.3	0.05
Pb	50 + L + H	85	15
F	175 + 13L	500	–

Table 2. Calculated input and removal rates (in mg/m² year). Removal rates are in agreement with current Dutch standards for crop quality, and the stand still principle for ground water quality. (After Ferdinandus et al. 1989).

	Cd	Cu	Pb	Zn
Input				
Agricultural crops				
- commercial fertilizer (cf)	0.55	1.7	4.7	15.5
- sewage sludge + cf	0.84	91	55	250
- cattle manure + cf	0.38	6	6.2	21
- pigs manure + cf	0.21	30	5	59.5
- poultry manure + cf	0.15	15.5	4.6	60.5
Horticultural crops (cf)	1.35-1.6	3.6-4.3	5.5-7.5	27.5-330
Precipitation	0.2	3.2	13	20
Removal				
Agricultural crops for				
- human consumption	0.30-0.25	3.6-11	0.1-0.5	36-110
- animal fodder	0.15-0.75	3.6-13.5	7-31	36-135
Horticultural crops	0.43-0.85	3.7-7.5	1.7-4.5	37-75
Groundwater recharge	0.1	1.3	1.3	7.5

ECOTOXICOLOGICAL RISK INDICATORS
FOR ENVIRONMENTAL CHEMICALS

M. Vighi and D. Calamari
Institute of Agricultural Entomology
University of Milan
Via Celoria 2
I - 20133 Milan
Italy

ABSTRACT. The main objective of applied ecotoxicological research is the production of indications useful for hazard assessment and risk evaluation for the management of potentially harmful chemicals. In the present paper, the scheme of an integrated ecotoxicological approach is described. In particular the predictive capability of such an approach and the possibility of applying it to Environmental Impact Studies are discussed. Risk indicators for various kinds of targets (man, ecosystems) and for different time and space scales are proposed and quantified. In particular, numerical indices have been elaborated for: (i) small scale, short term consequences of a direct discharge in air, water and soil; (ii) long term risk for man and the environment deriving from distribution processes in the various environmental compartments on a local scale; (iii) long term environmental risk on a global scale.

1. INTRODUCTION

The Directive 85/337/EEC on the assessment of the effects of certain public and private projects on the environment, states verbatim in Article 3:
"The environmental impact assessment will identify, describe and assess in an appropriate manner, in the light of each individual case and in accordance with the Articles 4 to 11, the direct and indirect effects of a project on the following factors:
- human beings, fauna and flora;
- soil, water, air, climate and the landscape;
- the interaction between the factors mentioned in the first and second indents;
- material assets and the cultural heritage."
It appears that specific care should be taken to assess the damages to various biotic and abiotic components of natural environments and to processes regulating the structure and functioning of ecosystems.

These kinds of damages can derive from the input of potentially harmful chemical substances into the environment. Therefore, it follows that, in the framework of multidisciplinary contributions needed to perform an Environmental Impact Study, ecotoxicological competence can be, in several cases, of paramount importance.

In this paper no attempt will be made to discuss the ecological consequences of physical modifications of an area such as the building of a dam, a motorway, etc, where different ecological approaches such as diversity indices, maintenance of biological diversity and studies on endangered species could be utilized.

261

A. G. Colombo (ed.), Environmental Impact Assessment, 261–275.
© 1992 ECSC, EEC, EAEC, Brussels and Luxembourg.

According to the definition of the SCOPE (Scientific Committee on Problems of the Environment), ecotoxicology "is concerned with the toxic effects of chemical and physical agents on living organisms, especially on populations and communities within defined ecosystems; it includes the transfer pathways of those agents and their interactions with the environment" (Butler, 1978). Thus ecotoxicology deals not only with the effects of chemical substances but also with their environmental distribution and fate.

Moreover, a fundamental characteristic of an ecotoxicological approach to the study of environmental pollution is its predictive capability. Predictive instruments are developed for the evaluation of the effects of chemicals (QSAR: Quantitative Structure-Activity Relationships) and of their environmental distribution and fate (evaluative models) in order to produce useful information not only for the control but also for the prevention of environmental contamination.

This predictive capability is obviously fundamental in an Environmental Impact Study, where the environmental consequences of a structure which is not yet operating must be evaluated.

An example of an integrated ecotoxicological approach is shown in figure 1. Through experimental or predictive studies the environmental exposure and effective levels can be evaluated and, on these bases, a hazard assessment can be performed. Finally, a risk evaluation can be completed by means of studies on exposed populations.

A further aspect to be taken into account is the position of human beings in relation to ecotoxicology. Man can be subject to different kinds of exposure to hazardous chemicals. People can be directly exposed in places of production or intensive use (chemical industry workers, farmers applying pesticides, etc.) or direct exposure can derive from the use of drugs, cosmetics, food additives or preservatives etc. This kind of problem pertains more to occupational health or traditional toxicology than to ecotoxicology.

In contrast, large numbers of people can be subject to indirect exposure to potentially harmful chemicals as a consequence of their environmental distribution and fate on a wide spatial and temporal scale (air pollution, contaminants in drinking waters, bioaccumulation in edible living organisms, etc.). In these cases human exposure can be evaluated or predicted by means of an ecotoxicological approach. In this framework, ecotoxicologists can fruitfully collaborate with epidemiologists or hygienists (Calamari, 1992).

Finally, it should be noted that an integrated ecotoxicological approach like that described, has never been applied till now to official cases of Environmental Impact Studies. Nevertheless, several cases of industrial or agricultural pollution have been studied and risk evaluations have been successfully performed according to this methodology.

2. CONCEPTUAL BASES FOR THE QUANTIFICATION OF ECOTOXICOLOGICAL RISK

Ecotoxicological principles, stated in the previous paragraph, can be applied to an Environmental Impact Study in order to produce a methodology and practical indices suitable for quantifying the ecotoxicological risk deriving from the immission of a chemical substance (or a mixture of chemicals or a complex product such as mineral oils) into the environment due to the setting up of a project.

This quantification should be based on the intrinsic properties of the substance, determining its biological activity and environmental partitioning and reactivity, as well as on extrinsic factors depending on human activities (loads, use patterns, etc.) or environmental characteristics (properties of environmental compartments, environmental processes, biological populations, etc.). In other words, it is necessary to produce, in quantitative terms, a hazard assessment based on the comparison between a predicted environmental concentration and the levels indicating possible harm for living organisms or the ecosystem. Moreover as a function of the exposed biological populations, a risk evaluation should be performed.

In many cases, in this procedure a compromise must be accepted between the need for scientifically sound results and producing suitable answers within the relatively short time of an Environmental Impact Study. Moreover, the information available is often not exhaustive and it is not possible to produce new data through experimental work.

As for the effects on living organisms or ecosystems, the objective should be, ideally at least, the availability of a suitable criterion or objective. By definition, a quality criterion indicates the levels or concentrations of physico-chemical parameters or of potential contaminants in an environmental compartment which are suitable to protect various possible "uses" of the compartment itself, including human utilization of the compartments and the preservation of an ecologically balanced community of living organisms. Quality criteria can therefore be formulated for all the principal environmental compartments (air, water, soil) and for the compartment "biomass", particularly as a function of its utilization as a food source.

According to the Scientific Advisory Committee on Toxicology and Ecotoxicology (SACET) of the EEC, a water quality criterion for the protection of aquatic life "should be such as to permit all stages in the life of aquatic organisms to be successfully completed, should not produce conditions which cause these organisms to avoid parts of the habitat where they would normally be present, should not give rise to the accumulation of substances that are harmful to the biota (including man) whether via food or otherwise, and should not produce conditions which alter the functioning of the ecosystem".

Thus, a quality criterion represents the result of a careful evaluation of the available scientific information on the environmental effects of a chemical, as well of ad hoc experimental research. Traditionally, a complete set of experimental data for the assessment of a quality criterion of a chemical substance should take into account information on:
- acute and chronic toxicity on various living organisms
  and characteristics of different trophic levels;
- behavioural and life cycle effects;
- influence of environmental variables on toxic effects;
- mode of action;
- persistence;
- bioaccumulation and bioconcentration;
- controlled ecosystems and field studies.

Considering the complexity of the problem, it is evident that rigourous environmental quality criteria have been produced for a relatively small number of chemical substances.

Some relatively different approaches to evaluating ecotoxicological hazard and to defining Quality Criteria have been proposed recently by Kooijman (1987) and Van Straalen and Denneman (1989). These methods are founded on statistical principles and requirements. In both methods it is assumed that the LC50 and NOEL values derived from single species present in the community observed may be conceived as mutually independent random trials from a log-logistic probability distribution. Each species tested does not represent any other species but is, in fact, one estimate of sensitivity. With several such estimates, the overall range of sensitivity for all species can then be determined.

Both methods allow a hazardous concentration to be evaluated for the ecosystem on the basis of a relatively small set of toxicity data and statistical processing. The Van Straalen and Denneman method has been applied to aquatic toxicity data by Van Leeuwen (1990), who concluded that the approach is a useful tool in ecotoxicological hazard assessment, although it needs some improvement and care must be taken.

Nevertheless, even these simplified approaches require an amount of information that, in general, cannot be produced by ad hoc experiments during an Environmental Impact Study. In the majority of cases, rougher values must be used, which would be less significant than a precise quality criterion but more realistically achievable.

A NOEL (No Observed Effect Level) can be taken from the literature or extrapolated from toxicity data or, when possible, using toxicity data estimated by means of QSAR equations. As for the effects on human beings, Admissible Daily Intakes (ADI) can be used. These data are produced mainly by the WHO, FAO or other international organizations and are available for several chemical substances. Similarly, referring to the assessment of exposure, a Predicted Environmental Concentration (PEC) could be evaluated on the basis of simple prediction evaluative models requiring a relatively small amount of input information and characterized by easy data management and interpretation, bearing in mind the loss in ecological realism.

3. PROPOSAL FOR ECOTOXICOLOGICAL IMPACT INDICES

Environmental pollution can be studied at different levels and scales in terms of space and time. The scale of distribution of a contaminant in the environment depends, in the short term, on the uses and discharge patterns, and, in the long term, on the mobility and persistence of the substance. In figure 2 some examples of persistence and distribution for different contaminants are shown. As a function of the different space and time scales of the effect of pollution, various levels of evaluation of the ecotoxicological risk must be taken into account (figure 2):
- small scale risk due to direct immission;
- local scale risk for man and the environment;
- global scale risk.

Each level is characterized by different conditions and must be evaluated according to specific criteria. In the following paragraphs, some criteria for the formulation of quantitative indices to evaluate the various levels of ecotoxicological risk will be proposed.

3.1. Data Requirements

From a practical point of view, in order to produce a hazard assessment for a potentially harmful chemical the following information is needed.

a) **Loads and emission patterns.** In general, in environmental studies, this kind of information is relatively difficult to obtain. In contrast, in an Environmental Impact Study, data on the total load and on emission patterns (continuous or discontinuous discharges, air dispersion, effluents in surface waters, etc.) should be easily available as part of the basic information dossier of a project.

b) **Physico-chemical properties.** Physico-chemical properties of the molecules under study are needed to predict environmental distribution by means of evaluative models, to estimate effects on living organisms or to evaluate bioaccumulation by means of QSAR equations. The most relevant parameters (water solubility, vapour pressure, octanol-water partition coefficient, pKa, etc.) should be provided in the information dossier. Otherwise the data needed can be found in the literature or calculated by means of available estimation methods (see for example Lyman et Al., 1982).

c) **Environmental distribution and evaluation of predicted environmental concentrations** (PEC). As was said previously, a contaminant introduced in the environment can produce adverse effects, acute and localized, in the place and time of emission, and consequences on a wider scale, in time and space, as a consequence of transfer and distribution processes in various environmental compartments. In the former case, an evaluation of environmental concentration can be obtained by means of a simple dilution ratio (for example between the water flow of an effluent and those of the receiving water body) or of physical models for the dispersion in air or water. In the latter case the application of an evaluation model to predict partition and distribution in environmental compartments is needed. Examples of these models, which must be characterized by simple application, few input data requirements and easy interpretation of the results, are the fugacity model (Mackay, 1979; Mackay and Paterson, 1981) or those proposed by Cohen et al. (1990). Leaching indices for the potential of groundwater contamination are also available (Rao et al., 1985; Gustafson, 1989).

d) **Effects on living organisms.** Taking into account the difficulty of finding precise quality criteria for the various environmental compartments in the literature, NOEL data for different exposure (inhalation, ingestion, exposure in water) or ADI for man can be used, if available. If reliable toxicity data on different organisms are available, a traditional extrapolation of NOEL can be performed or some of the previously quoted new approaches for quality criteria evaluation can be applied (Kooijmann, 1987; Van Straalen and Denneman, 1989). In the case of the short term risk, from direct immission, threshold levels obtained from acute toxicity data can be used if the immission is periodic or occasional and not continuous. If there is a lack of experimental data, the effects can be estimated by means of QSAR equations. Extensive reviews on the value and limitations of the QSAR approach are available in the literature (Calamari and Vighi, 1990; Hermens,1989).

e) **Bioaccumulation.** Bioaccumulation can be evaluated on the basis of some physico-chemical properties of the molecule, such as the octanol-water partition coefficient (Kow) or Henry's constant (H = vapour pressure/water solubility). Several equations for the calculation of bioconcentration factor (BCF) in aquatic and terrestrial animals have been proposed. These equations are in general of the type:

$$\log BCF = a \log Kow + b \tag{1}$$

(see for example: Mackay, 1982; Travis and Arms, 1988). For terrestrial plants more complicated equations, non linear or biparametric, may be needed to predict bioaccumulation in roots and stem (Briggs et al., 1982, 1983;) while for the prediction of BCF in foliage the following equation can be utilized:

$$BCF = L \, Koa \tag{2}$$

where Koa is the octanol:air partition coefficient and L is the lipid fraction (Paterson et al. 1991) as from the experimental data of Bacci et al.(1990).

f) **Persistence.** Data on persistence are very important for hazard assessment of chemicals but are difficult to obtain, in particularly in a form useful for practical purposes, because of both the intrinsic stability of the molecule and the variability of environmental conditions. Information on transformation constants for various processes (biodegradation, photodegradation, hydrolysis etc.) are rare. Predictive approaches based on QSAR or on other estimation methods are being developed (see for example: Vasseur et Al. 1992; Macalady and Schwarzenbach, 1992; Zepp, 1991). An attempt to evaluate the intrinsic stability of organic chemicals by means of mass spectrometry fragmentation has been proposed recently (Tremolada et al.,1992). All these approaches are very promising, but, at present, are not yet reliable enough. In general, no more than a rough semiquantitative estimate of the persistence (i.e. weeks, months, years) can be derived from all the available information.

g) **Exposed populations.** Quantitative and qualitative information on exposed populations is needed for risk evaluation. Several factors, besides numerical consistency, should be taken into account, for example economic or naturalistic importance, vulnerability, recovery or recolonization capability, etc. Taking into account these and other aspects, some indices of relative importance can be proposed (i.e.: negligible, important, very important). Obviously these schemes are not applicable to human populations, which are always classified as very important.

h) **Mobility.** This property is particularly important in long range, long term risk evaluations. In fact, a chemical substance will produce effects on a wide scale if, besides a certain degree of persistence, it is able to move and circulate in the environment, including by transfers among environmental compartments. A method for a quantitative evaluation of mobility or, at least, for a comparison and ranking among molecules, based on sound and reliable conceptual principles, is not yet available. An attempt at a rough classification can be based on the affinity of a substance for the principal environmental compartments (air, water, soil) and on their role in mass transport. For example, a substance with high affinity for the soil tends to be immobilized in this compartment. In contrast, chemicals with high affinity for air or water will be distributed on a wider scale as a result of transport or advection processes.

A proposal for a mobility index can be based on the percentage distribution in the three principal environmental compartments calculated by means of the standard fugacity model, according to the following scheme:

$$\frac{(\% \text{ distribution in soil})^0 \quad + \quad (\% \text{ distribution in water})^1 \quad + \quad (\% \text{ distribution in air})^2 \quad =}{\text{mobility index}}$$

A numerical index is then produced based on the different and growing role in mass transport for soil, water and air. The index ranges from a minimum value of 1 (molecules present at 100% in soil) and a maximum of 10^4 (100% in air). The range can be arbitrarily divided as follows:

low mobility:	1 - 10
medium mobility:	10 - 100
high mobility:	100 - 1000
very high mobility:	1000 - 10000

3.2. Quantification of Risk Indices

3.2.1. Risk index for direct immission (I_1). This index describes the short range consequences of a continuous or intermittent discharge or of an incidental event. The evaluation can be based on a few parameters.

- Predicted environmental concentration (PEC) in the compartments directly involved. This can be evaluated from information on loads and discharge patterns, applying simple physical models of dilution or mass transport. In the case of the prediction of incidental events the maximum quantities that could be discharged in the environment must be known.
- Effect on living organisms. If a quality criterion (QC) is not available, a threshold level (TL) for acute effects or a NOEL, if continuous emission is involved, can be used. These values must refer to the exposure typical of the particular compartment involved (aquatic toxicity, inhalation, phytotoxicity or bacterial inhibition in soil).
- Evaluation of the importance of the exposed target, on a quantitative and qualitative basis.

A risk index, on a scale from 0 to 10, can be obtained by giving a score to the ratio between PEC and effects (QC or NOEL or TL) and to the exposed target, as shown in table 1.

3.2.2. Environmental risk on a local scale (I_2). This index refers to the impact, over a relatively long time, on the whole territory influenced by the setting up of a project. The extension of the affected area should be evaluated case by case as a function of the characteristics of the project and of the territory. In general it should be in the range of tens to hundreds of km. In comparison with the previous index more parameters and a more complicated evaluation are needed.

- PEC in various environmental compartments must be evaluated by means of the application of a predictive model to the specific territory under study. The fugacity model or other similar evaluative approaches can be used.
- Effects on living organisms should be defined as QC for the various environmental compartments or exposure patterns. If not available, QC can be extrapolated from NOEL values or roughly approximated using suitable application factors.
- A bioconcentration factor (BCF) must be introduced.
- Persistence, not considered in the previous index, plays a relevant role in this evaluation over mean or long times.
- As before, the exposed target must be evaluated.

Separate risk indices would be formulated for the aquatic (I_2w) or terrestrial (I_2t) environment, as before, on a scale from 0 to 10. A calculation scheme is given in table 2.

3.2.3. Risk for man on a local scale (I_3). As previously pointed out, only the risk deriving from an ecotoxicological exposure will be taken into account, i.e. a prolonged exposure to relatively low levels resulting from environmental distribution and fate processes. The risk for man will be evaluated from the comparison between admissible daily intake values (ADI) and the total quantities assumed via air, water and food. For cautionary purposes, the "worst case" will always be taken into account, i.e., for example, food will be assumed as totally produced in the contaminated area. A quantitative index must refer to a hypothetical "average man", characterized by the following values, taken or extrapolated from WHO estimates:

body weight:	70 kg
breath rhythm:	20 m^3/d
drinking water consumption:	2 l/day
animal food consumption:	250 g/day
vegetable consumption:	1 kg/day

Among the quoted values, those dealing with food consumption are to be assumed as roughly indicative. The total daily intake (TDI) will be calculated starting from concentrations in various environmental compartments (including animal and vegetable biomass) according to the following algorithm:

air concentration (mg/m^3)	x 20 m^3	+
water concentration (mg/l)	x 2 l	+
animal biota concentration (mg/kg)	x 0.25 kg	+
animal biota concentration (g/kg)	x 1 kg	=

TDI: Total daily intake (mg)

Concentrations in various compartments could be evaluated as for the previous index, by means of data on loads and applying evaluative models or bioconcentration equations. The TDI will be divided by 70 kg and expressed as mg/kg, so that it can be compared to the ADI.

A risk index for man (I_3) will be evaluated on the basis of the TDI/ADI ratio and persistence, as shown in table 3.

3.2.4. Environmental risk on a global scale (I_4). For a complete assessment of the environmental impact of chemicals, in particular when highly persistent or mobile molecules are concerned, the evaluation should go beyond local boundaries and the possibility of unwanted effects on a wide spatial and temporal scale should be taken into account. These effects cannot be predicted easily, but some general criteria can be proposed:

- On a global scale, the chemicals of maximum concern are not the most toxic but those for which there is high probability of a loss of control. Thus, persistence and mobility are, in this case, the most important parameters. In particular, persistence is the main factor, in fact highly persistent chemicals can be globally distributed even if the mobility is relatively low (see for example the case of DDT); in contrast, rapidly degradable molecules are of relatively low global concern even if characterized by high mobility.
- The toxicity is, on the global scale, less important than on the local one. In fact, persistent and diffused chemicals can produce serious environmental modifications even if the toxicity is negligible (see for example, CO_2 and the greenhouse effect, CFCs and the ozone layer reduction). Moreover, on a global scale, when the effects on all kind of living organisms, environmental compartments and ecosystems are considered, it is difficult to elaborate a reliable quantitative toxicity index. Thus a rough qualitative index could be proposed, based on the review of the information available and on the operator's experience. Obviously this is a subjective answer but more objective solutions are not easy to propose.
- On the other hand, bioaccumulation plays a relevant role in the evaluation of the environmental impact on a large space and time scale.
- Another important factor is the load introduced into the environment, taking into account that the contribution of a single source of emission to the global load is yet to be evaluated. Thus, a load of more than 100 metric tons per year, that, roughly speaking, is around the 0.1% of the global emission of an industrial chemical produced at a relatively high rate, can be assumed as a very high load.
- Evaluation of the target is not important, because, in this case, the target is the global ecosystem, which is always present.

On these bases a global risk index can be proposed according to table 4.

4. CONCLUSIONS

The schemes described in the previous paragraphs are an attempt to propose quantitative indices practically applicable in an Environmental Impact Study for environmental chemicals. Nevertheless, they must not be assumed as a precise methodology or a "cookbook" for the ecotoxicological evaluation of chemical substances. The conceptual framework that led to the proposed indices is certainly acceptable on a qualitative basis but it still presents a high level of uncertainty on a quantitative basis. Improvement can be achieved by means of applications to scenarios and case histories.

Many difficult aspects, requiring particular care and experience, have only been mentioned in this paper. The criteria for the selection of reliable input data, the application of suitable predictive models for the evaluation of a PEC and the methods for the extrapolation of environmental quality criteria from a set of toxicity data, are important problems and need to be described in ad hoc detailed papers.

Moreover, for some aspects, it is difficult to propose a standardized procedure due to the frequent lack of available information. Thus, in these cases, relatively subjective decisions based on the operator's experience are needed. This is a critical point, because, in regulative procedures, subjective opinions should in general be avoided. The solution of this aspect lies in the production of more data or in the development of suitable predictive models. For example, the development of methods able to predict the environmental stability of a molecule would be of help in the formulation of less subjective persistence indices.

In any case, even if all aspects are more precisely defined, it goes without saying that the direct experience of an ecotoxicologist cannot be neglected in the evaluation of the environmental impact of potentially dangerous chemicals.

REFERENCES

Bacci, E., Calamari, D., Gaggi, C. and Vighi, M. (1990), "Bioconcentration of organic chemical vapors in plant leaves: experimental measurements and correlation", Environ. Sci. Technol. 24, 885-889.

Briggs, G.G., Bromilow, R.H. and Evans, A.A. (1982), "Relationship between lipophilicity and root uptake and translocation of non-ionized chemicals by barley", Pestic. Sci. 13, 495-504.

Briggs, G.G., Bromilow, R.H., Evans, A.A. and Williams, M. (1983), "Relationship between lipophilicity and the distribution of non-ionized chemicals in barley shoots following uptake by roots", Pestic. Sci. 14, 492-500.

Butler, G.C. (1978), Principles of Ecotoxicology, John Wiley and Sons, New York.

Calamari, D. (1992), "The role of ecotoxicology in the assessment of human exposure to chemical substances", Human and Experimental Toxicology (in press).

Calamari, D., Vighi, M. (1990), "Quantitative Structure-Activity Relationships in Ecotoxicology: value and limitations", Reviews Environ. Toxicol. 4, 1-112.

Cohen, Y., Tsai, W., Chetty, S.L. and Mayer G.L. (1990), "Dynamic partitioning og organic chemicals in regional environments: a multimedia screening-level modeling approach", Environ, Sci. Technol. 24, 1549-1558.

Gustafson, D.I. (1989), "Groundwater Ubiquity Score: a simple method for assessing pesticide leachability", Environ. Toxicol. Chem. 8, 339-357.

Hermens, J. (1989), "Quantitative structure-activity relationships on environmental pollutants", in O. Hutzinger (ed.), Handbook of environmental chemistry, vol. 2E, Springer Verlag, Berlin, pp. 111-162.

Kooijman, S.A.L.M. (1987), "A safety factor for LC50 values allowing for differences in sensitivity among species", Water Res. 21, 269-276.

Lyman, W.L., Reehl, W.F. and Rosenblatt, D.H. (1982), "Handbook of chemical property estimation methods", McGraw-Hill Book Company, New York.

Macalady, D.L. and Schwarzenbach, R. (1992), "Predictions of chemical transformation rates of organic pollutants in aquatic systems", in D. Calamari (ed), Chemical Exposure Prediction, Lewis Publishers, Chelsea, Michigan (in press).

Mackay, D. (1979), " Finding fugacity feasible", Environ. Sci. Technol. 13, 1218.

Mackay, D. (1982), " Correlation of bioconcentration factor", Environ. Sci. Technol. 16, 274.

Mackay, D. and Paterson, S. (1981)," Calculating fugacity", Environ. Sci. Technol. 15, 1006.

Paterson, S., Mackay, D., Bacci, E. and Calamari, D. (1991), " Correlation of the equilibrium and kinetics of leaf-air exchange of hydrophobic organic chemicals", Environ. Sci. Technol. 25, 866-871.

Rao, P.C.S., Hornsby, A.G. and Jesup, R.E. (1985), " Indices for ranking the potential for pesticide contamination of groundwater", Soil Crop Science Society of Florida 44, 1-8.

Travis, C. and Arms, A. (1988), "Bioconcentration or organics in beef, milk and vegetation", Environ. Sci. Technol. 22, 271.

Tremolada, P., Di Guardo, A., Calamari, D. and Fanelli, R. (1992), "Mass spectrometry derived data as possible predictive method for environmental persistence of organic molecules", Chemosphere (in press).

Van Leeuwen, C.J. (1990), " Ecotoxicological effects assessment in the Netherlands: recent developments", Environ. Management 14, 779-792.

Van Straalen, N.M. and Denneman, C.A.J. (1989), " Ecotoxicological evaluation of soil quality criteria", Ecotoxicol. Environ. Safety 18, 241-251.

Vasseur, P., Kuenemann, R. and Devillers, J. (1992 , "Quantitative structure biodegradability relationships for predicting purpose", in D. Calamari (ed), Chemical Exposure Prediction, Lewis Publishers, Chelsea, Michigan (in press).

Zepp, R.G. (1991), " Photochemical fate of agrochemicals in natural waters", in H. Freshe (ed.), Pesticide chemistry, VCH Verlaggesellschaft, Weinheim, 229-246.

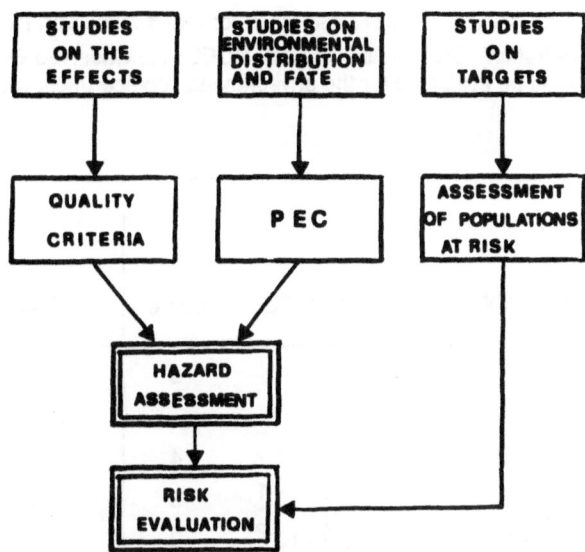

Figure 1. Ecotoxicological approach for hazard assessment and risk evaluation (PEC: Predicted Environmental Concentration).

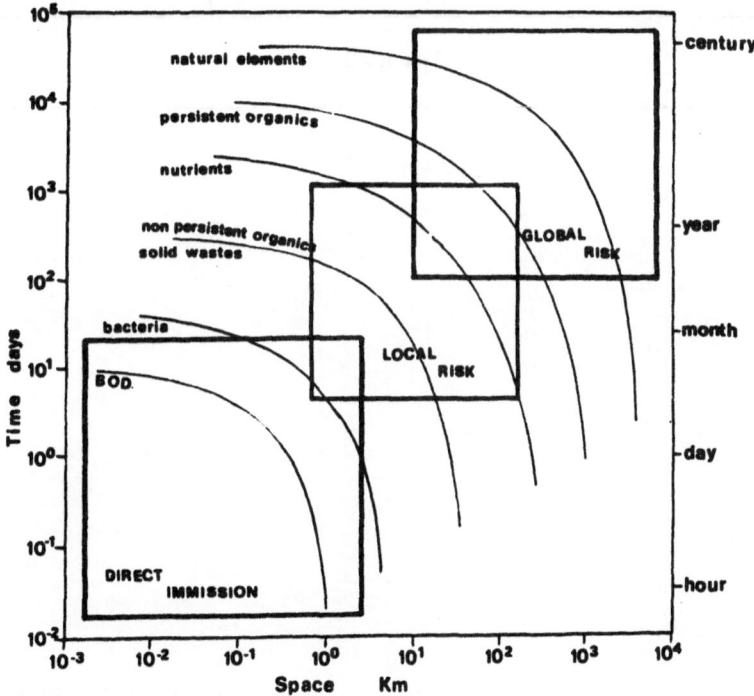

Figure 2. Persistence and distribution of various environmental contaminants and boundaries of the three levels of risk evaluation.

Table 1. Scheme for the calculation of the risk index for direct immission (I_1). PEC (Predicted Environmental Concentration), QC (Quality Criteria), NOEL (No Observed Effect Level) and TL (Threshold Level) should be evaluated for the environmental compartment where the immission takes place.

A				B	
PEC / QC or	PEC / NOEL or	PEC / TL	Score	Target	Score
$> 10^2$	> 10	> 1	5	very important	2
$10 - 10^2$	$1 - 10$	$10^{-1} - 1$	4		
$1 - 10$	$10^{-1} - 1$	$10^{-2} - 10^{-1}$	3	important	1
$10^{-1} - 1$	$10^{-2} - 10^{-1}$	$10^{-3} - 10^{-2}$	2		
$< 10^{-1}$	$< 10^{-2}$	$< 10^{-3}$	1	negligible	0.5
$I_1 = A \cdot B$					

Table 2. Scheme for the calculation of the risk on a local scale for the aquatic (I_{2w}) and terrestrial (I_{2t}) environment. PEC and QC must be calculated for all the main environmental compartments: water (w), air (a), soil (s), sediments (sed). For the two different indices the role of the target (D) must be evaluated respectively for the aquatic or terrestrial environment.

A		B		C		D	
PEC / QC	Score	BCF	Score	Persistence	Score	Target	Score
> 10^2	5						
10 - 10^2	4	log Kow > 3.5	1	months	1	very important	1
1 - 10	3	2.5 < log Kow < 3.5	0.75	weeks	0.8	important	0.75
10^{-1} - 1	2	log Kow < 2.5	0.5	days	0.6	negligible	0.5
< 10^{-1}	1						

$$I_{2w} = (Aw + Ased) \cdot B \cdot C \cdot D$$

$$I_{2t} = (Aa + As) \cdot B \cdot C \cdot D$$

Table 3. Scheme for the calculation of the risk for man on a local scale (I_3).

A		B	
TDI / ADI	Score	Persistence	Score
> 1	5		
10^{-1} - 1	4	years	2
10^{-2} - 10^{-1}	3	months	1.5
10^{-3} - 10^{-2}	2	weeks	1
<10^{-3}	1		
I_3 = A • B			

Table 4. Scheme for the calculation of the risk on a global scale (I_4).

A		B		C		D		E	
Persistence	Score	Mobility	Score	Toxicity	Score	BCF	Score	Load t / year	Score
years	1.5	very high	1	very high	1	log Kow > 3.5	1.5	> 100	2
months	0.8	high	0.6	high	0.7	3.5 > log Kow > 3	0.9	1 - 100	1
weeks	0.4	medium	0.35	medium	0.45	3 > log Kow > 2.5	0.5	10^{-2} - 1	0.5
days	0.2	low	0.2	low	0.3	log Kow < 2.5	0.3	10^{-4} - 10^{-2}	0.25
								< 10^{-4}	0.125

$$I_4 = (A + B) \cdot (C + D) \cdot E$$

275

INDUSTRIAL PLANT RISK INDICES

A. Romano and V. Rossini
TRR
P.za Papa Giovanni XXIII
I - 24046   Osio di Sotto   (Bg)
Italy

A. Servida
TECSA, R & D Division
Via Caravaggi
I - 24040   Levate   (Bg)
Italy

ABSTRACT. The characterization of the level of risk of an industrial plant can be achieved by means of either qualitative or quantitative criteria, and according to both the objectives of the analysis being performed and the available resources. Generally speaking, the use of indices is a suitable method which allows the analyst to produce a consistent ranking of industrial installations/units based on theirs quantified levels of hazard easily. The risk index method provides the analyst with procedures both to identify in a plant those units or sections which are potentially at risk, and to evaluate their level of hazard. The aim of the paper is to introduce the reader to the use of risk indices, commonly used in risk analysis, and to show how they work in screening and ranking plants on the basis of their hazard. The features of the most widely used risk indices, namely the Dow Index, Mond Index and the Italian Risk Index, are presented with regard to the scope of the analysis they allow the analyst to perform.

1. INTRODUCTION

In the paper we will present the following indices, which are those most widely used in the risk analysis community:
- Dow Index, developed by Dow Chemical [3];
- Mond Index, developed by ICI (Imperial Chemical Industries) [5];
- the Italian Risk Index "Metodo Indicizzato di Riferimento", as is stated in Annex 2 to the Italian DPCM 31/03/1989 [6,7].

All the above methods provide guide-lines aiming at supporting the assessment of the criticality of a given installation as the hazard would result just from both the presence of certain quantity of hazardous material and the type of technological process being performed and operated in the plant. Thus, we may say that the risk index methods belong to the so-called qualitative techniques for characterizing hazard in plants. We refer the interested reader to other papers, included in this book, for a complete review of the quantitative techniques [1] in use in the risk analysis community, as well as for the formal treatment of the fundamental theory of indicators and indices [2]. In this regard, we would like to point out that the named risk index methods were developed mainly on the basis of experience acquired from a large number of accidents and near misses, and not by formalising a sound mathematical theory. Although, we will not investigate this aspect of the problem any further, because it is out of the scope of the paper, we would like to draw the reader's attention to it, as it would make the understanding of these index procedures much more valuable.

The risk index methods will be described in the paper and their applicability discussed. Differences and commonalities would be highlighted for a proper use of the methods. Moreover,

A. G. Colombo (ed.), Environmental Impact Assessment, 277–286.
© 1992 *ECSC, EEC, EAEC, Brussels and Luxembourg.*

as a benchmark, two of the above indices will be applied separately to a typical petro-chemical plant layout. The outcomes of each method will be compiled in a table which then will be used for comparison. The actual extent of the results will be emphasised. The final analysis of the benchmark results may help the reader to grasp the actual aim and scope of the risk index methods.

2. METHODS FOR THE COMPUTATION OF RISK INDICES

In this chapter we will describe the main features of the most important existing methods for ranking hazard of substances and plants. Precisely, we will present the Dow Index, the Mond Index, and, finally, the index adopted in Italy for the preliminary analysis of the critical areas, as stated in the Italian regulation.

From a procedural view point, we may say that all the methods named start from the subdivision of the plant into process units. Commonly, a unit is defined as a physical part of the plant which is (or could be) physically separated from the remaining parts of the plant and which is causing a specific hazard, due to either the chemical process that is performed in the unit, or the nature and properties of the materials used. By specific hazard is meant that the plant unit should actually induce an hazard different or separate from those ascribable to the other remaining units close to it.

The characterization of the plant in terms of plant units allows the analyst to break down the problem of evaluating the level of hazard of a plant into pieces which individually correspond to computation of hazard indices for smaller areas, the units. In fact, once the units are identified, the level of hazard given by each unit is evaluated through a numerical classification of the unit sections, and according to numerical values which are related to the nature of the materials involved in the process, the existing quantities of materials, the process operating conditions, the type of process and, lastly, the presence and the performance of the safety and protection systems.

The last step in the computation of the plant risk index is the combination of all the unit indices to produce an overall picture of the plant hazard. The combination is usually performed by means of arithmetic functions that are designed to weigh somehow the contribution of the individual unit and material indices to the plant risk index. The key point to be realised is that the quantification of plant risk index is approached as if it were possible to break it down in smaller but self-consistent subproblems.

2.1. The Dow Index

The Dow Index allows one to estimate the degree of potential hazard of a plant in which flammable and/or reactive materials are used. Moreover, the Dow index is capable of providing an estimation of the economic risk, based on the evaluation of the area that might be damaged and the worth of the plant equipment in that area. In the following, the different phases of the Dow index calculation method are discussed.

The Process Units are identified on a detailed plot plan of the plant, according to the greatest impact they may have on the environment. Then, for each unit, the estimation of the level of hazard is carried out through the following steps:

a) Identification of the "Material Factor" which takes into account the characteristics of the materials involved, mainly reactivity and flammability. Although directly determined on the basis of the net heat combustion value, the MF value is found in predefined tables;

b) Identification of the "General Process Hazard" and the "Special Process Hazard" in terms of factors which express the potential hazard of the unit. The General Process

Hazard takes into account the characteristics of the plant which might increase the consequence of a possible accident in the process unit. The Special Process Hazard is, on the other hand, related both to those characteristics of the process unit and to the operating conditions which could increase the probability of occurrence of an accident (fire or explosion). The combination of the General Process Hazard and the Special Process Hazard characterises the Unit Hazard Factor that indicates the degree of hazard exposure of the process unit;

c) Calculation of the Fire and Explosion Index which is related to the potential hazard of the process unit. This index is then used to determine the area of exposure surrounding the process unit. In this step of the procedure the presence of protection and safety systems is completely disregarded;

d) Calculation of the Damage Factor that tries to assess the percentage of the whole process unit that is likely to be damaged by an accident, if the protection and safety systems do not perform. As the Fire and Explosion Index allows determination of the exposure area, it is possible, from the knowledge of the economic worth of the unit in the exposure area, to compute the maximum economic damage induced by an accident (Base Maximum Probable Property Damage - MPPD);

e) Taking into account the protection and prevention systems (Credit Factors) it is possible to compute the damage if the protection systems work successfully (Actual MPPD);

f) From the knowledge of the actual MPPD, the average outage time of the plant (called Maximum Probable Day Outage - MPDO) and the probable economic damage due to the production interruption (Business Interruption - BI) are then computed.

The Dow Index does not provide a procedure to produce an overall assessment of the degree of hazard that could take into consideration the whole plant. Moreover, the method is not capable of dealing with those cases in which toxic materials are involved in the process.

Lastly, we point out that, because of the method's limitations, the Domino effect is not considered in the evaluation of the hazard.

2.2. The Mond Index

The Mond index is used for the identification of the hazard of industrial plants where flammable, explosive and toxic materials are involved in the plant process.

Conceptually, the analysis procedure is very similar to that presented in the previous section, as the Mond Index was derived from the Dow index. Thus, in the following we will discuss only the differences between the two indices. The Mond Index differs from the Dow Index in the following aspects:

i) Beside the material factor, other specific characteristics of the material are considered in order to assess the so-called special material hazard. Moreover, the evaluation of the general process hazard and the special process hazard is carried out according to more detailed procedures;

ii) This index allows one to consider in the analysis also the layout of the unit as significant factor for the estimation of the hazard in the unit;

iii) The toxicity of the material is considered in assessing the material factor.

Thus, the Mond index is more descriptive as it also takes into account the structural aspects of the plant by embedding in the analysis those elements which pertain to the actual plant layout. From the point of view of expressiveness, the Mond index is much richer as it includes the calculation of the following risk sub-indices:

- the general Dow index,
- the fire load (thermal power and duration),
- the internal explosion index,

- the aerial explosion index,
- the toxicity index, and
- the overall hazard index.

The above sub-indices are calculated twice. First, the calculation is carried out without taking into account the existence of the protection systems. Thus, the outcome of this calculation represents, to some extent, the maximum hazard which might be expected from the plant and it is referred to as the intrinsic hazard. Then the actual hazard is calculated, taking into account the existing protection/mitigation systems.

Hence, a comparison of the outcomes of the two above computations can give a preliminary or qualitative measure of the performance of the installation of the protection and safety systems. Moreover, it is also possible to proceed with the sensitivity analysis of the hazard indices as function of the protection systems being adopted. In this respect, the Mond Index is a tool, although still very approximate, of much wider use than the Dow index.

Lastly, we should emphasise the fact that the Mond index does not give any estimation of the economic damage or of the areas affected by the accidents.

In table 1, the main characteristics of the two indices treated so far are compared. We would like to draw attention to the individual capabilities of the two indices with regard to the insight on the plant hazard nature and degree that possibly the analyst might gain in working with each of the two indices.

Table 1. Comparison between Mond and Dow indices.

	Mond Index	Dow Index
Material factor	X	X
Specific material hazards	X	
General process hazards	X	X
Special process hazards	X	X
Quantity hazards	X	X
Layout hazards (DOMINO effect)	X	
Toxicity hazard	X	
General hazard indices	X	X
- fire	X	X
- air explosion	X	
- process explosion	X	
- toxicity	X	
Protection measures	X	X
Hazard factor to the off-site environment	-	-
Estimation of the damaged areas		X
Economic damage		X
Operator experience	X	

The very first objectives of the Mond index method are both to provide the analyst with a simple tool for the identification of the plant areas where the hazard levels can be high, as well as to assess the possible causes of such a hazard. The method is designed to achieve this objective in such a way that, for each plant unit, firstly all the different indices are computed and secondly they are converted to numerical indices, for the aim of composition and comparison.

Once the unit indices are computed, it might be the case that the process is so safe that there is no need for any further analysis.

However, the Mond method foresees two further phases. The former consists of a kind of basic review that allows the analyst to reduce the value of some factors.

The second is based on the concept of the offset values, which are considered when certain types of credit factors can be considered because of very particular software aspect of the unit. In this context, software refers to all those aspects related to the plant operations, such as quality of the procedures, level of the personnel training, periodic inspections, and so forth. On the other hand, hardware, in this context, refers to the equipment as a whole, also taking into account the design criteria, the construction criteria, the installation and licensing.

The basic review is very easy to apply and does not require a long time. Thus, it is usually adopted as the very first screening procedure.

The offsetting procedure is more complex as it discounts the hazard effect according to the existing procedures implemented to decrease either the accident occurrence probability or the damage caused by an accident. An example of the first type is the monitoring system; an example of the second type is the fire fighting system for the protection of vessels against fire. The Mond method manual [4] gives all the details for evaluating the proper credit factors and it describes how these credit factors are to be used to reduce the computed index values and finally to reduce the computed level of hazard.

The offsetting procedure is useful in many cases. The most important one is the case of an existing plant for which the application of the offsetting procedure can assess the safe impact of implementing new safety procedures or of installing new safety systems. Thus, the analyst can, by the offsetting calculation, measure the extent to which the new safety procedures reduce the hazard level of the plant layout hardware and software setting.

Another possible application of the offsetting procedure is to compare the performance (with respect to the hazard level reduction) of alternative safety systems. Hence, in this context it might be used to optimise the choice of the cheapest system with the highest safety performance.

The Mond index is also suitable to verify the layout of a plant during the design and to identify the best solution. In this regard, the numerical values of the units are used in combination with the general criteria of the design which performs to the aim of locating at the correct distances the individual unit. The Overall Risk Rating allows the analyst to approach this problem, as it characterises the level of hazard of a unit taking the main indices that are computed for a single unit.

The application of the Mond index in the plant design phase, for the purpose of evaluating alternative hardware unit layouts, can start when the design of the plant has reached the state of knowing at least the flow diagram and the preliminary quantities of the materials, and preliminary dimensions of the vessels. Thus, the Mond index may give a very early evaluation of the hazard level for the different processes. Applying the method at the early stage of the design can highlight the potential hazard which possibly may be reduced by modifying the plant layout properly and by enhancing the safety and protection system capabilities.

Finally, we emphasise that the Mond index is very useful in the analysis of both existing plants and plants under construction. It points out whether measures to improve the safety of the plant are to be taken.

Let us now say a little about the procedure which is implemented in the Mond index. As we pointed out above, for the Mond Index method also the first step is to identify the different plant units, which are parts of the plant that can be readily and logically characterised as separate physical entities. For each units the following should be performed:

   a) The identification of the dominant (key) material, which is meant to represent the most hazardous material in the unit. The key material is that compound or mixture in the unit that, due to its inherent properties and the quantity present, provides the greatest potential for an energy release by combustion, explosion or exothermic reaction;

b) The calculation of the Material Factor (B) for the key product, i.e. the measure of the potential hazard for fire or explosion. Procedures are provided for the calculation of B for: normal flammable materials, mixtures, marginally flammable materials, material of unspecified composition, reactive combination of materials and, lastly, materials capable of condensed phase explosion;

c) The evaluation of the Special Material Hazard (M), related to the chemical properties of the material itself. In fact, M takes into account any special properties of the key material which may affect either the nature of an accident or the likelihood of the occurrence;

d) The evaluation of the General Process Hazards (P), induced by the type of process and the other operations being carried out in the unit;

e) The evaluation of the Special Process Hazards (S) which are related to features of the process operation which increase the overall hazard (i.e. low and high pressure, low and high temperature, corrosion and erosion hazards, and so forth);

f) The evaluation of the Quantity Factor (Q), which is allocated for the additional hazards associated with the use of large quantities of combustible, flammable, explosive materials;

g) The evaluation of the Layout Hazard (L), related to the hazards which may be introduced by the location of the unit compared to the others;

h) The evaluation of the Acute Health Hazards (T) which takes account of the influence of acute toxic hazards on the overall assessment of the unit.

Once the above factors are characterised, the Mond Index method provides a number of combination rules which define the following indices:

Equivalent Dow Index "D". It provides a global measure of the hazard in the unit and can be rated according to a fixed number of classes. This index is equivalent to the former Dow index to the extent that it does not explicitly takes account of the fire and explosion hazards.

Fire Index "F". It provides a measure of the duration of the fire and of the induced hazard in a unit. It takes into account the amount of flammable material in the unit, its energy release potential and the area of the unit. Thus, the index F is a function of B and the working area.

Internal Explosion Index "E". It gives a measure of the potential hazard for explosion within the unit.

Aerial Explosion Index "A". It measures the hazard due to the explosion of unconfined vapour clouds. Thus, the important features in assessing A are Q, E, B, and the rate and height of the release. There are five corresponding descriptive categories to describe the index.

Toxicity Unit Index "U". It represents the hazard due to toxicity of the materials.

Overall Risk Rating "R". It is an index that characterises the level of hazard of a unit taking into account all the above indices (F, D, E, A). Thus, this index considers the overall unit risk by including the fire and explosion hazards also.

We point out that some of the above indices, precisely F, E, A and R, may be reviewed in order to take account of the effect of the protection and the safety systems. Generally speaking, we may identify two broad classes of safety measures:

a) Measures to reduce the frequency of occurrence of an accident (i.e. monitoring systems, alarm system, emergency system, training, fire protection system, and so forth);

b) Measures to reduce the effects of an accident (i.e. product confinement, fire fighting, plant sectioning).

Concerning the operational contents of the Mond index procedure we refer the reader to [5].

2.3. The Italian Risk Index

The method is based on a proposal of the Italian ISPESL (Istituto Superiore per la Prevenzione e la Sicurezza del Lavoro) and ISS (Istituto Superiore di Sanità). The method aims at evaluating the hazards which are induced by the plant in all the phases of the plant cycle (design and operation). Concerning fire and explosion hazards the method does make reference to the Dow and Mond methods, although some improvements have been introduced by ISPESL. Concerning toxicity, the method is very original because of ISS's contribution. We will not spend long on this index method as conceptually and also procedurally is quite similar to those presented earlier in the paper. Nevertheless, we will highlight those features and characteristics that are specific of this method.

The method can be applied to a number of plant classes (chemical, petro-chemical, etc.) and to a variety of processes. Like the Dow and Mond indices, this risk index method also produces a numerical hazard rating of the units in the plant. Moreover, It introduces the following new and particular factors:
- The factor "S" (hazard to health), that takes into account the effect of the toxicity of the materials in the global evaluation of the unit. Specifically, it considers the delay that might occur in confronting the developing accident due to the toxicity of the materials involved;
- The IIT (intrinsic toxicity index) of each material in the plant, and it expresses the toxicity as given by the properties of the material and the exposure time;
- The toxicity index T, that is calculated once for the industrial activity as a whole and according to the quantities of each materials in the plant.

Once the indices are computed, combination formulae for the calculation of global indices are given. We highlight the fact that the method eventually produces two basic indices: G (Overall Hazard Index) and T (Toxicity Index of the activity). These two indices are then combined in a risk matrix to characterize the specific plant. In this method discounting factors are also included to tune the results of the analysis.

In the following section we will consider an example which aims to give the reader the feeling of the actual extent of the applicability of the Mond and Italian Risk Indices.

3. APPLICATION OF RISK INDICES TO A PLANT

Figure 1 is a drawing of plant which will be considered as case study for the application of the index methods. The aim of this case study is to show the difference in the insight that can be acquired in performing the index analysis with the Mond and the Italian risk index methods.

The first step taken is to identify single plant units according to the nature of the process, the material involved and the operating conditions. The units are to be identified in collaboration with the plant operators as their support and plant knowledge are necessary to subdivide the plant effectively.

The plant we are considering is a typical desulphurization plant. We have identified the following units:
- **Furnace/reactor** (Unit 1). It is made up of the preheating furnace operating in the reactor and of the reactor itself.
- **Thermal exchange** (Unit 2). It is made up of the charge equipment of the plant and of the heat exchangers downstream of the furnace.
- **Compressors/drum** (Unit 3). The unit is made up of the recycle compressors and the separators.
- **Stripping** (Unit 4). The unit contains the stripping column.
- **Absorption** (Unit 5). The unit represents the fuel washing column.

We are not going to give the tedious calculations of the individual factors in detail as it would not add much to the real aim of the example, but we will summarise the main results below.

3.1. Summary of the Results

Table 2 gives the outcomes of the analysis, performed according to the Mond Index and the Italian risk index methods, for units 1 and 5 which, for the scope of the paper, are the most representative.

Table 2. Application of the Mond and Italian risk indices to two units of a plant.

		Mond Index		Italian Risk Index		
	Index	Value	Categ.	Index	Value	Categ.
Unit 1	Ri	2077	high 2	G	1283	G3
	Rf	46	low	G'	28	G1
	U	0	light	Tu	0.162	T1
Unit 5	Ri	70	low	G	307	G3
	Rf	1.55	mild	G'	6,7	G1
	U	10	very high	Tu	0.06	T1

where:
Ri   Overall hazard index (initial)
Rf   Overall hazard index (final)
U   Unit toxicity index
G   Intrinsic general hazard index
G'   Intrinsic general hazard index after compensation
Tu   Unit toxicity index.

By considering both the initial and final indices we may say that both Rf and G' deal with the same type of hazard. Concerning the unit toxicity, the value U=10 puts unit 5 in the very high class; whereas the value T=0.06 puts the same unit in class T1. In this case the values are in disagreement as the Italian risk index completely disregards the presence of H_2S in the unit.

4. CONCLUSIONS

In the paper we dealt with the problem of use of risk indices as operational tools in the risk analysis. The widely used risk indices were reviewed aiming at emphasising their individual characteristics as well as the scope of their possible applications.

At the methodological level, we may say that the risk indices allow the analyst to break down the problem of assessing the level of hazard in a plant into a number of specific problems that may shed some light on the nature and types of the existing hazards. Moreover, we may appreciate the fact that, although the use of indices is just a qualitative method, the effort and resources to be allocated for this type of analysis are not as great as those required for a complete quantitative risk analysis. Nevertheless, the understanding of risk level in the plant that can be achieved by the correct application of the risk index method is undeniable.

None of the methods presented allows consideration of the effects of external environmental events, such as earthquake, fire, flooding, as they are concerned just with the assessment of the intrinsic plant hazard.

The effective application of the methods is very dependent on the skill and experience of the analyst in charge of the study. However, the analyst should always collaborate with the plant personnel in order to characterise the plant unit hazards properly.

Finally, we should highlight the fact that the methods also try to include in the analysis those aspects related to the software aspects of the plant which may be responsible for increasing or decreasing the hazard level in a plant.

REFERENCES

[1] Contini, S. and Servida, A. (1991), "Risk Analysis and Environmental Impact Studies", this book.
[2] Volta, G. and Servida, A. (1991), "Environmental Indicators and measurement scales", this book.
[3] Dow Chemical (1986), "Fire and Explosion Index. Hazard Classification Guide", 5th edition.
[4] Lewis, D.J. (1980), "The Mond fire, explosion and toxicity index applied to plant layout and spacing", 13th Loss Prevention Symposium.
[5] AIChE, "The Mond fire explosion and toxicity index, a development of the Dow index", 2nd Edition.
[6] Binetti, F. et al.(1990), "Metodo indicizzato per l'analisi e la valutazione del rischio di determinate attività industriali", Tech. Report ISPESL.
[7] Decreto del Presidente del Consiglio dei Ministri 31/3/1989: "Applicazione dell'art. 12 del decreto del Presidente della Repubblica 17 maggio 1988, n. 175, concernente rischi rilevanti connessi a determinate attività industriali.", Suppl. Gazzetta Ufficiale n. 93, 21/4/1989.

286

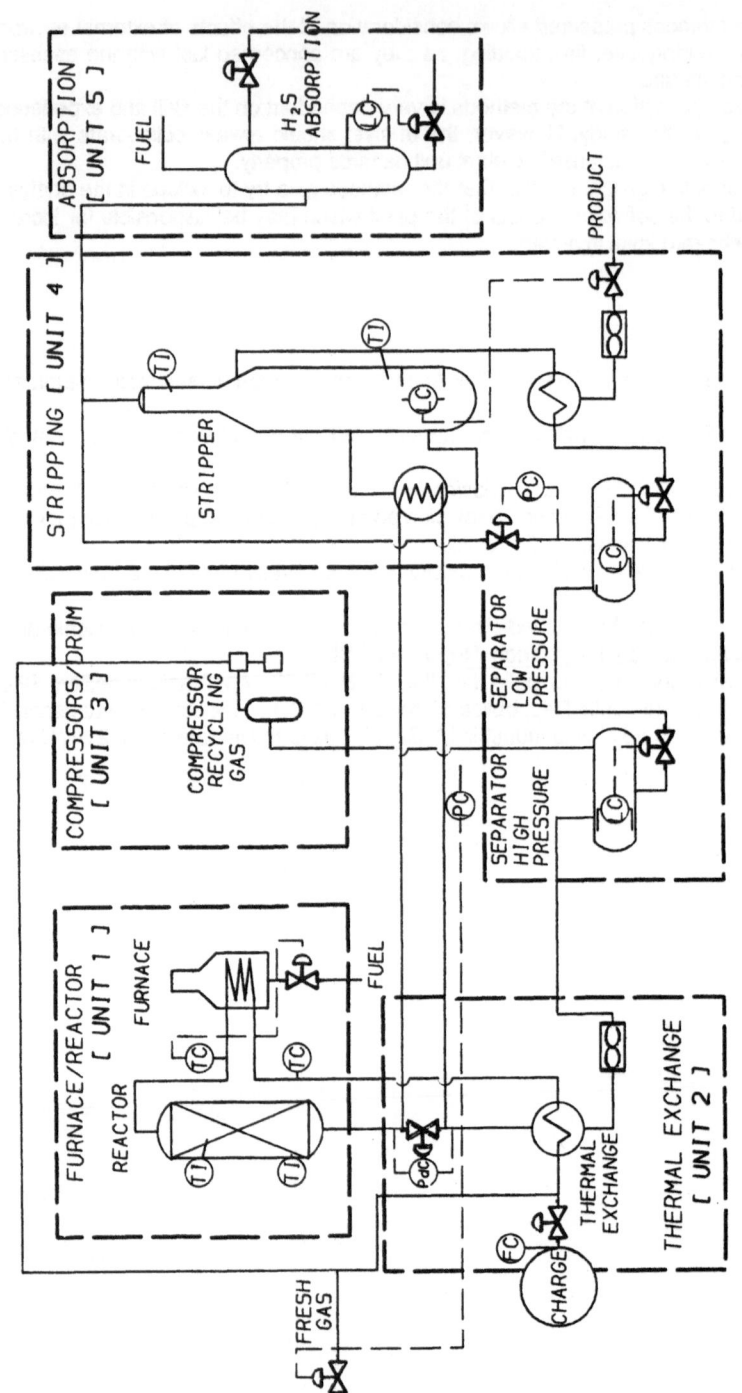

Fig. 1. DESULPHURIZATION PLANT LAYOUT.

SUSTAINABILITY, EFFICIENCY AND EQUITY: PROJECT APPRAISAL IN ECONOMIC DEVELOPMENT STRATEGIES

M. J. F. van Pelt and A. Kuyvenhoven
Dept. of Development Economics
 Wageningen Agricultural University
PO Box 8130
6700 EW Wageningen
The Netherlands

P. Nijkamp
Dept. of Economics
Free University of Amsterdam
PO Box 7161
1007 MC Amsterdam
The Netherlands

Abstract. This paper focuses on conflicts between short-term economic development and long-term environmental problems. It starts with a brief historical overview of the role of environmental externalities and related social costs in economic theory. Following the international debate after the publication of the Brundtland report - Our Common Future - the paper will elaborate on the notion of ecologically sustainable development. The main question tackled in this paper is how to use the sustainability concept in project appraisal in addition to the traditional criteria of efficiency and equity. Issues in the triangular relationship between these three key development objectives will be discussed in great detail, with specific reference to developing countries. Definitions of the key criteria of sustainability, efficiency and equity will be proposed, including issues such as risk assessment and intergenerational trade-offs. Identification of sustainability-oriented project profiles and impact matrices are the logical follow-ups of the previous approach. In this context, considerable attention will be paid to problems inherent in measuring scores on the environment and on sustainability. Finally, comparison and integrated evaluation of projects in sustainability-oriented project appraisal will be dealt with.

1. INTRODUCTION

Environmental decay is not exclusively a problem of the post-war period. In ancient times, the Greek philosopher Plato complained about human actions which had turned the landscape of Attica into a skeleton. Also in the medieval period we find many examples of environmental externalities, e.g. in cities where horse-driven carriages were forbidden during parts of the night. A well-known example is the prohibition on burning certain types of smokey coal in London.

In the economic literature, environmental deterioration has often been regarded as a peculiarity which did not belong to the heart of economics. With the exception of Marx (who recognized the poor quality of living conditions of the working class), environmental externalities were mainly treated as an interesting example of social costs (Marshall, Pigou). The environment only became a focal point of economic research in the post-war period, and in particular since the 1970s. This new interest concentrated attention on both the (individual and social) welfare aspects of environmental decay and the empirical-analytical assessment of the social costs involved (e.g. via extended input-output analysis, material balance models, etc.). The incorporation of environmental costs in social cost-benefit analysis and in project appraisal received some attention, but did not develop into a mature methodology.

In the meantime the scene has changed. Next to the recognition of the important socio-economic consequences of the environmental problem, two new developments have taken

287

A. G. Colombo (ed.), Environmental Impact Assessment, 287–312.
© 1992 ECSC, EEC, EAEC, Brussels and Luxembourg.

place. First, the intensity and threats of environmental pollution have increased dramatically, especially because of the emission of non-biodegradable pollution (e.g. toxic substances, persistent micro-pollutants such as pesticides and herbicides, etc.); these pollutants may also endanger human health. Secondly, there is also the awareness of global environmental changes (e.g. desertification, acidification, deforestation, climate change, ozonization). These changes will have a long-term impact on environmental conditions on earth over a time span which goes far beyond the time horizon in conventional economic models.

These two developments imply that standard economic tools based on efficiency and allocation are in various cases no longer applicable. This has provoked in recent debates in international agencies the notion of ecologically sustainable economic development as a central element for environmentally benign policy strategies. Especially following the Brundtland report "Our Common Future", (WCED, 1987), interest in the question of how to treat the natural environment in economic theory has increased considerably. An important new element in recent contributions in comparison to the literature published particularly in the 1970s and 1980s (see for instance Mäler, 1985; Seneca and Taussig, 1984; Nijkamp, 1979; Hueting, 1980) refers to the notion of sustainable development. The number of definitions is overwhelming (for an excellent overview see Pezzey, 1989), but the interpretation in the Brundtland report is still one of the clearest. It says that sustainable development is development that meets the needs of the present without compromising the ability of future generations to meet their own needs. Whereas "needs" may be translated into social welfare, "ability" is especially concerned with the availability of ecological resources ("ecological sustainability"). Sustainable development requires that the use of such resources by the present generation remains below certain levels. The definition of sustainable levels of resource use is a normative affair (Opschoor and Reijnders, 1989).

This paper is concerned with the analytical consequences which the growing emphasis on ecological sustainability may have for project appraisal, not only for developed but also for developing countries (or regions). Traditionally, development policies, and hence project appraisal, started from two independent objectives, viz. aggregate welfare (income) maximization implying an optimal allocation of scarce resources (efficiency), and a "fair" distribution of income (equity).

Economic theory gives unequivocal guidelines on how to achieve and measure efficiency. Various, basically similar indicators show the extent to which growth and development activities through their use of resources contribute to aggregate welfare improvement. In many developing countries (and regions), the existing distribution of income (as well as productive assets) is relatively unequal. This explains why the scope of project appraisal, especially for developing countries (or regions), has been broadened to incorporate ethical judgements on a just distribution of income generated by development activities among contemporaries (intratemporal equity). Value judgements on the fairness of the distribution of net income flows over time (intertemporal equity) have received much less attention. Efforts have been made, however, to account for growth objectives through a distinction between income used for consumption and for savings (i.e. investment). Again this particularly refers to project appraisal for developing countries.

Daly (1990) argues that scale (ecological sustainability) should become the third macroeconomic objective in addition to the existing objectives of optimal allocation (efficiency) and a fair distribution (equity). As a consequence, it may be a real challenge to incorporate sustainability as a third key criterion - in addition to efficiency and equity - in (microlevel) project appraisal for both developed and developing countries. This paper explores how the structure of project appraisal, i.e. the kind of issues to be tackled, may be affected. Such an overview of issues is a prerequisite for investigating the applicability of appraisal methods to sustainability-oriented project appraisal [1].

Project appraisal is a multi-stage process, and the impact of the sustainability criterion will be explored phase-by-phase. Our classification of project appraisal stages is summarized in diagram 1 [2].

Section 2 treats the incorporation of the sustainability criterion in the first phase, viz. the description of the policy framework. Starting from a social welfare function, appraisal criteria and criteria weights can be derived. Criteria weights show the relative priority of criteria, and may be quantitative (including (shadow) prices) or qualitative. Special attention is paid to policies regarding possible conflicts between the "old" criteria of efficiency and (intratemporal) equity and the "new" sustainability criterion. An important question is whether sustainability is considered a precondition for approval of projects, or whether trade-offs between, on the one hand, sustainability and, on the other hand, efficiency and equity are allowed.

Section 3 deals with the role of the sustainability criterion in the second phase, viz. the definition of the project alternatives and particularly the project setting. Gaining insight into the economic and environmental context of a project is a prerequisite for estimating its effects. Sustainability being a variable linking environmental and economic factors, the project setting should provide insight into welfare patterns, socio-economic and environment systems in the project area, and relations in these fields between the project area and supra-project levels.

The third phase (section 4) comprises impact assessment, viz. the estimation of the effects of project alternatives on the criteria of efficiency, equity and sustainability. Incorporation of the sustainability criterion may drastically affect the treatment of time. Whereas the focus used to be on short and medium term effects, sustainability-oriented project appraisal requires an (additional) analysis in terms of generations. In view of the important role of risk strategies in sustainability policies, assessments of risk and uncertainty should be integrated in impact matrices. Possible measurement scales for efficiency, equity and sustainability are discussed. Finally, practical difficulties in assessing scores on the sustainabilty criterion are outlined.

The final phase in project appraisal concerns evaluation, the basis for decisions to reject or approve project alternatives (section 5). A first evaluation step assesses whether alternatives satisfy constraints on individual criteria, particularly sustainability. If a project does not comply with the sustainability constraint, i.e. involves resource use in excess of normative levels, the project may be reformulated in such a way that it does satisfy this constraint. The nature and various consequences of such constraint-satisfying activities are discussed. In the second evaluation step, the set of project alternatives that satisfy constraints are subjected to an integrated appraisal on the basis of their performance on all remaining criteria and the relevant criteria weights.

Section 6 contains some concluding remarks.

2. SUSTAINABILITY AND THE POLICY FRAMEWORK

2.1. Defining Efficiency, Equity and Sustainability

A policy framework is based on the objectives of development (and hence appraisal criteria) and on how conflicts between objectives should be treated (criteria weights). An analysis of the impact of sustainability on project appraisal starts with understanding the difference this criterion will make for policy frameworks. Differences between policy frameworks in traditional

---

[1] For a first analysis of the usefulness of two groups of appraisal methods, viz. cost-benefit analysis and multi-criteria analysis, see van Pelt, Kuyvenhoven and Nijkamp (1990).

[2] A more detailed note on project appraisal stages is available on request.

and sustainability-oriented project appraisal are dealt with in two steps:
- definition of the key criteria of efficiency, equity and sustainability,
  and the weighting of respective sub-criteria (this section),
- possible conflicts between the key criteria, and basic weighting policies
  (section 2.2.).

Diagram 2 contains an overview of the key criteria of efficiency, equity and sustainability, as well as their attributes (i.e. subcriteria). It shows that the definition of the sustainability criterion is related to several efficiency and equity attributes. These links will be outlined below. The following observations serve as a starting point for this analysis.
- Sustainability is expressed in: a) acceptable levels of natural resource use, and,
  b) acceptable long-term ecological risks, both at a certain level of decision-making
  (global, national, regional, local/project).
- Sustainability depends on views on: a) the direct impact of the environment on
  social welfare, b) substitutability between man-made and natural capital as factors
  of production, and, c) the responsibility of the present generation to future
  generations (intergenerational equity).

The choice of efficiency, equity and sustainability as main appraisal criteria can be illustrated by the Netherlands policy for development co-operation. Since the early 1980s the overall objective has been "structural combat of poverty". It combined the objectives of increases in production and income (efficiency) and a fair distribution (intratemporal equity). In 1990, without altering the overall objective, ecological sustainability was added as a third attribute (Ministry of Foreign Affairs, 1990).

Of course, there may be several other development goals besides efficiency, equity and sustainability. Dutch development policy, for example, aims at strengthening the position of women, institutional development, promoting appropriate technology and several other attributes of development. In the remainder of this paper we will either show that these goals may be considered an attribute of one of the key criteria (women - intratemporal equity) or assume that they are of a lower order than efficiency, equity and sustainability.

2.1.1. Efficiency. The attributes of aggregate welfare correspond directly with the efficiency criterion. Efficiency has been a key criterion in policy frameworks in conventional (economic) project appraisal for developing countries (milestones have been Little and Mirrlees, 1974; UNIDO, 1972; Squire and van der Tak, 1975; Squire, 1989). Efficiency constitutes the difference between gross aggregate welfare changes (benefits) and all use of scarce resources (costs)[3]. In the past, welfare benefits tended to be equated with availability of material goods and services produced in the socio-economic system (maximization of material consumption or income). Such goods are partly traded in markets, partly non-traded (social overhead, public goods). Increasingly, shortcomings of the narrow welfare concept are being acknowledged (see for instance Hueting, 1980; van Pelt, Kuyvenhoven and Nijkamp, 1990). Assuming a broader interpretation of welfare, the availability of environmental amenities with a direct impact on the well-being of men has also been considered a welfare attribute.

On the cost side, basic resources comprise man-made capital and natural capital. Each of these categories can be further subdivided. Natural capital, for instance, may be subdivided into classes related to its functions: assimilation of waste, provision of renewable and non-renewable resources, supply of environmental amenities essential to production processes. It has been proposed that there should be a differentiation between objectives regarding

---

[3] No agreement exists on how welfare losses and benefits accruing to individual households are to be aggregated. The authors referred to all argue in favour of differential weighting on the basis of the individual's pre-project income. Thus, equity concerns are explicitly introduced into the selection criterion (see Squire, 1989).

irreversible vis-a-vis reversible environmental problems (Hedman, 1990).

The policy framework should develop the weighting function converting individual classes of costs and benefits (sub-criteria) into overall efficiency performance. Traditionally, cost-benefit analysis (CBA) techniques have been applied, whereby prices serve as weights. If available and a true reflection of the value to society, market prices are applied. If markets are imperfect, generate external effects, or are considered distorted, shadow prices may be applied. The latter approach, among other things, sets project appraisal for developing countries apart from approaches for developed countries. Problems occur if no price can be determined for one or more efficiency attributes, particularly environmental amenities (benefits) and environmental resource use (costs). In such circumstances there are two possible approaches. The first is to replace the (comprehensive) efficiency criterion by at least two other criteria, viz. a partial-efficiency criterion (covering all monetarized efficiency attributes) and other criteria covering the remaining efficiency attributes (such as the environment). An alternative approach would be to break down the comprehensive efficiency criterion fully into its respective attributes: contributions to material welfare, contributions to environmental amenities, use of man-made capital, use of natural capital, etc. Both approaches require weighting mechanisms other than prices to arrive at conclusions regarding efficiency. The emphasis is then likely to be on weights derived from views of policy-makers (see Nijkamp, Rietveld and Voogd, 1990).

2.1.2. Equity. The policy framework should provide insight into preferences regarding intertemporal equity, viz. the distribution of welfare over time. In traditional project appraisal, usually, a time horizon encompassing not more than one generation is assumed, which in diagram 2 is considered a short-term approach. The frequently applied discounting technique implicitly assigns the consequences of projects affecting future generations a negligible or zero weight.

In view of the long-term focus implied by sustainability concerns, the diagram also emphasizes intergenerational equity, the distribution of welfare among successive generations. In other words, how important is the welfare of the present generation compared to the welfare of future generations? How much welfare are those who are living now willing to sacrifice in order to safeguard the interests of future generations? Moreover, what are views on the possibility of compensating future generations for a lower level of environmental amenities by higher material welfare levels?

Sustainability concerns draw particular attention to long term ecological risks. Such risks may have various specific characteristics (Quiggin and Anderson, 1990). One is that "surprises" may occur, events that cannot be predicted, and particularly unpleasant surprises with potentially disastrous effects for future generations. Often, probabilities associated with various possible events cannot be estimated. Judgements on the present generation's responsibility to future generations should therefore contain a risk strategy, describing subjective attitudes towards risks and associated extreme events. Risk-aversive strategies imply a larger willingness-to-sacrifice present welfare than optimistic views on future possibilities to respond to any harmful events, for instance through technological progress. One approach is to follow the maximin strategy, whereby the alternative is preferred of which the worst possible outcome is better than the worst possible outcomes of other alternatives. An alternative approach consists of assigning weights to risks and their possible consequences. Reijnders (1990), for instance, argues that long-term ecological risks are unacceptable. This implies that he assigns a weight of 1 to the environmental risk criterion. "No-regret" strategies aim at avoiding highly uncertain but potentially disastrous events and surprises by embarking on measures that also can be justified on the basis of their impact on related, but more predictable fields.

A main factor in the present generation's willingness to avoid risk possibly affecting future generations is views on future possibilities for substitution in two fields. First, to what extent may an increase in man-made capital compensate for a loss of environmental capital in the production of goods and services? Second, to what extent is it possible to compensate for a reduced availability of one type of environmental resources by enhancing the quality or quantity of another environmental stock attribute? According to some, ecological decay may not be unacceptable because technological progress and increases in man-made capital may provide compensation, or because of optimism regarding the environment's capacity to recuperate over time. Others may prefer a much more cautious approach, arguing that compensation possibilities are very limited.

In the 1950s and 1960s benefits of economic growth were widely assumed to trickle down to the poorest groups. Consequently, no particular need was felt to integrate the distributional impact of development activities in project appraisal. In reality, however, economic growth often showed a biased distribution. It was observed that welfare benefits often did not accrue equally to all population groups and that central governments were unable to redistribute income. Efficiency and (intratemporal) equity turned out to be potentially conflicting goals. In the late 1960s, redistribution of income to the benefit of low-income groups became a second key objective in project appraisal for developing countries. Moreover, it was recognized that not only the direct redistribution of income should be tackled, but also the question of who owns or has access to productive assets.

In view of links between poverty, distribution and environmental problems (WCED, 1987), intratemporal equity will continue to be a central issue in sustainability-oriented project appraisal. To account for possibly conflicting interests of different social groups, the policy framework should include, as in the past, value judgements (weights) regarding the distribution of material welfare among contemporaries, particularly among higher-income and lower-income groups. Moreover, views should be developed on the question of who owns or has access to environmental resources upon which income-generating activities depend.

Another dimension of intratemporal equity, viz. the distribution of welfare between spatial levels, used to be reflected in project appraisal from a national point of view. Supra-national effects were implicitly assigned a weight of zero. In sustainability-oriented project appraisal, views on trade-offs between welfare at the project level, the national level and the supra-national (continental, global) level need to be tackled. Are welfare objectives defined at higher levels, implying that welfare trade-offs at and between lower levels are allowed? Or do welfare objectives show a spatial disaggregation?

2.1.3. Sustainability. In the traditional approach, no constraints used to be imposed on the use of environmental resources, one of the efficiency sub-criteria. Implicitly, any use of natural resources is permitted provided compensation is offered in the form of a larger production of man-made goods and services. The environment sub-criterion continues to play this role in sustainability-oriented project appraisal. Through the sustainability criterion, "environment" is given a second function. Sustainability, whatever its definition and operationalization, always refers to a certain threshold level regarding the use of environmental capital (or the total stock of capital, comprising man-made capital as well; this approach is taken up later). In its basic form (actual resource use is either lower or higher than sustainable levels), this third key criterion is therefore of a different nature from efficiency and equity, which are expressed in terms of desired directions of change (maximization of aggregate real income, enhancement of the part of income accruing to target groups). More data-demanding forms of sustainability criteria involve measurement of the degree of sustainability on a cardinal or ordinal scale expressing the relative difference between normative threshold levels and actual resource use (see Opschoor and Reijnders, 1989).

The choice of threshold levels for sustainable resource use depends to a great extent on how the present generation judges its responsibilities to future generations, including assessment of risks and possibilities for substitution in production functions (intergenerational equity).

Two important normative interpretations of sustainability can be illustrated using the two types of resources shown in diagram 2, viz. man-made capital (M) and environmental capital (N). The two approaches, termed strong sustainability (sS) and weak sustainability (wS; see Foy and Daly, 1990), are summarized in table 1.

Table 1. Strong sustainability and weak sustainability.

definition	sustainability condition
wS	$d(N+M) \geq 0$
sS	$d(N) \geq 0 \cap d(M) \geq 0$

The wS approach (see for instance Bojö, Mäler and Unemo, 1990) puts a non-negative constraint on the total of man-made and natural capital. This approach may be explained by the view that especially thanks to technological progress, man-made capital may increasingly substitute for natural capital in the production of material goods, and the opinion that the loss of one type of environmental resource may be compensated for by increasing the supply of another type. The sS definition (advocated by among others Pearce, Barbier and Markandya, 1990), involving a non-negative constraint on the two stocks separately, is much more cautious in these respects [4].

An important question concerns policies regarding the weighting of attributes of particularly the environmental stock. This stock may be disaggregated into types of environmental resources (for instance, renewable and non-renewable resources) and environmental functions (for instance, waste assimilation and life support systems, such as the ozone layer). Following the sS approach, the question arises to what extent a decline in one environmental attribute, for instance the ozone layer, may be compensated for by an improvement in another attribute, for instance the number of species. One weighting strategy would be to impose non-negative constraints on each attribute. Sustainability would then require that these constraints on individual attributes are complied with. Alternatively, trade-offs between attributes may be allowed. Scores on attributes then need to be standardized. In principle, prices could be taken as weights, but valuation problems are likely to be significant. The alternative is to use other willingness-to-pay indicators or policy weights (Opschoor and Reijnders, 1989).

If the wS approach is applied, even more difficult weighting problems occur. How does one compare changes in man-made capital (attributes) with changes in environmental capital (attributes)?

Besides specifying normative limits to resource use, the sustainability criterion should specify a certain spatial level as a point of reference. Is sustainability defined and to be achieved at the project level or at a supra-project level? Hence, are limits to resource use defined at the project level, the programme level, the national level, or the global level?

Starting from the "strong sustainability" requirement, (viz. a non-declining stock of natural capital), Klaassen and Botterweg (1976) proposed to apply the sustainability constraint at the project level. Hence, no individual project should negatively affect the size of the stock of environmental resources. Consequently, the sustainability constraint is also adhered to at

---

[4] For a more wideranging discussion on sustainability constraints and underlying assumptions, see van Pelt, 1991.

higher levels (as far as projects are concerned). Pearce, Barbier and Markandya (1990) consider this approach not to be feasible. They argue that the sustainability condition should be applied at the "programme level", i.e. across a set of projects. In this case, individual projects may use environmental resources as long as this is compensated elsewhere in the programme (see section 5.1.).

In view of practical problems in impact assessment (see section 3), it is desirable that global sustainability levels are at least expressed in national parameters. Winpenny (1990) elaborates on approaches to translate global climate policies into national targets for emissions of greenhouse gases or energy efficiency. When a project uses up a part of that target, emissions elsewhere in the country would need to be decreased or abatement measures would need to be undertaken.

The impact of the sustainability criterion on the outcomes of project appraisal has an inverse relationship with the level at which sustainability is defined. Global sustainability may be commensurate with unsustainable development at some places and individual projects that do not satisfy overall constraints. When sustainability is defined at the project level, however, its impact on the design of individual activities is much larger.

In summary, ecological considerations play a dominant role in policy frameworks if:
- the strong sustainability approach applies instead of the weak sustainability approach,
- non-negative constraints on overall stocks are applied to attributes as well,
- extreme risk-aversive strategies are followed,
- sustainability is defined at the lowest spatial level.

2.2. Weighting Efficiency, Equity and Sustainability

Policy frameworks would not need to include relative priorities of the key criteria sustainability, efficiency and equity if one alternative may be expected to outrank all other alternatives in every field. In reality, however, conflicts between, on the one hand, efficiency and equity, and, on the other hand, sustainability are likely to prevail.

Possible conflicts between efficiency (real income increases in a narrow sense) and sustainability are at the core of the public debate on the Brundtland report. The WCED (1987) holds the fairly optimistic view that economic growth need not be at the expense of the environment. Economic growth is even considered a prerequisite for sustainable development in developing countries. The WCED emphasizes that the nature of growth patterns would need to be adjusted. This view has raised fundamental criticism, for instance by Hueting (1990), who has repeatedly argued that economic growth and preservation of ecological resources cannot go together. The present debate appears to give little support for the assumption that trade-offs between short-term efficiency and long-term sustainability can be ruled out at the project level.

Similar questions refer to possible conflicts between sustainability and intratemporal equity. Pearce, Barbier and Markandya (1990) have discussed the consequences of this problem following the "strong sustainability" approach. They claim that maintaining the present stock of natural resources over time in low-income countries is "likely to serve the goal of intragenerational fairness, i.e. justice to the socially disadvantaged both within one country and between countries at a given point in time". The argument is unconvincing for several reasons. First, without a proper definition of ethical notions such as "fairness" and "justice", one cannot assess whether the sustainability goal would be beneficial in this respect. Second, if it is assumed that the goal would be redistribution of welfare towards the poor at various levels, it is doubtful whether optimism regarding the effect of the sustainability constraint on these groups in the present generation is justified. Possible trade-offs between long-run benefits of ecologically sound policies and short-run economic costs are ignored. These trade-offs, however, may be particularly strong at low-income levels (see van Pelt, Kuyvenhoven and Nijkamp, 1990).

In many developing countries the poor are extremely dependent on natural resources. Reducing their use of natural resources may often be difficult without unacceptable income sacrifices (opportunity costs). One reason is that as long as market prices do not (fully) account for ecological costs and benefits, differences between private and social valuation can be large. Investments aimed at improving the efficiency of resource use in existing activities may therefore be costly. Moreover, income-generating alternatives for environmentally problematic activities may simply be absent. Hence, at least in the short and medium run, and before market prices fully incorporate ecological costs and benefits, a key question is who will pay for the transition from non-sustainable to sustainable practices. From the viewpoint of the poorest countries and social groups, for whom fighting poverty is a primary objective, it is imperative that sustainability concepts focus not only on the interests of future generations but also on those of the present generation, especially the poorest groups.

Given the possibly conflicting nature of efficiency, equity and sustainability, policy frameworks should elaborate on their relative priority. Particularly interesting cases involve at least one criterion with a negative score. Through weighting, criteria may be assigned specific roles in an appraisal. A key question is whether values of weights are made dependent on (ranges) of values of a criterion. Depending on the answer to this question, criteria may be converted to objectives, goals and veto criteria (this terminology is a mixture of approaches of Zionts, 1989, and Voogd, 1983):
- objectives: weights are independent from scores on criteria,
- goals: weights vary with specific intervals of possible scores of criteria, but do not take the maximum value of 1, and
- veto criteria: weights take the maximum value of 1 above or below a threshold for a criterion. In the relevant range, veto criteria (constraints) overrule all objectives and goals.

Threshold levels for efficiency may be used to divide sets of alternatives into efficient and inefficient activities. For instance, a project is efficient (inefficient) if its net present value is positive (negative). If a positive net present value is considered a prerequisite for accepting a project, efficiency is a veto criterion. Efficiency, however, may also be treated as a goal. Whereas a strong preference to avoid inefficient alternatives might exist (i.e. a high efficiency weight in the range of negative net present values), the option of compensating inefficiency by positive scores on equity or sustainability need not be ruled out.

No straightforward threshold level can be defined for (intratemporal) equity. This is a reflection of the fact that a pure value judgement is involved. In some cases, equity may therefore be considered an objective, implying a constant equity weight. It is possible, however, to define thresholds, for instance a minimal part of net benefits that should accrue to specified target groups. Then, equity may be expressed as a goal or even a veto criterion.

Almost by definition, sustainability is either a goal or a veto criterion. Sustainability as a goal criterion implies a strong preference for alternatives, satisfying normative levels of natural resource use. It is not ruled out, however, that non-sustainability is compensated for by sufficiently large efficiency and/or equity gains. The weight assigned to sustainability is lower in the range of sustainable resource use than in the range of non-sustainable resource use. If sustainability is converted into a goal, it is preferably measured on a cardinal (quantitative) or ordinal (ranking) scale. Measurement on a binary (0/1) scale is sufficient if sustainability is a veto criterion. Compliance with sustainable resource use levels is a prerequisite for accepting a project, whatever the scores on efficiency and equity.

3. SUSTAINABILITY, THE PROJECT AND THE PROJECT SETTING

There is no major difference between traditional and sustainability-oriented project appraisal regarding the definition of alternatives. In view of the emphasis on environmental issues in the new approach, and to facilitate subsequent analysis, however, projects may be classified according to their potential environmental relevance. Project profiles may, for instance, give information about resource use: changes in the use of renewable resources, changes in farming and fishing practices, exploitation of water resources, infrastructure, industrial activities, extraction industries and waste management disposal (OECD, 1990).

A sustainability-oriented analysis of the project setting should provide insight into welfare (development) patterns in the project area, and into links between the project area and supra-project levels affecting welfare potentials both in the project area and at higher levels (diagram 3).

Project setting profiles provide insight into attributes of the three key criteria of efficiency, equity and sustainability. They are built upon four cornerstones: welfare patterns, the socio-economic system, the environmental system and links between the socio-economic and environmental systems. In every field, questions of intratemporal distribution (what is the position of specific social groups with respect to the issue concerned?) and of intertemporal distribution (how do variables change over time, and particularly what are the positions of present and future generations?) should be dealt with. Especially where ecological variables are involved, shortcomings in knowledge in terms of risk and uncertainty should be identified.

The project setting profile would focus on the following issues:
- The analysis starts with a description of welfare levels, and its attributes (consumption of material goods and environmental amenities). What are income and consumption levels; what is the extent of poverty? How is income distributed? What are expected changes in welfare levels? In what way does the environment directly affect the well-being of people?
- The description of the socio-economic system should provide insight into the economic structure and production and consumption processes. What is the state of man-made capital? How is it distributed? What is the level of economic efficiency? Which part of the output is marketed? How do economic policies affect these variables?
- With respect to the environmental system, an environmental profile is prepared which give information about the natural resource base, i.e. the type of prevailing ecosystems and on the problems in these systems. The analysis of the environmental system should provide insight into the stock of environmental resources, with specific attention to sensitive areas, such as: soils and soil conservation areas, areas subject to desertification, arid and semi-arid zones, tropical forests and vegetation cover, water sources, etc. (OECD, 1990). With respect to environmental problems, the analysis may focus on, for instance (Myers, 1989): the extent to which, scale at which and type of environmental degradation that is taking place; extent to which and over what time horizon environmental thresholds or critical levels are being approached; occurrence of absolute and relative natural resource scarcity; uncertainties and possibilities of surprises with regard to future developments. If possible, actual resource use and sustainable resource use levels are compared and expressed in cardinal or ordinal sustainability indicators. Environmental policies and their impact can be presented. Again, the distributional dimension is important: where do problems occur, who are affected and at what pace?
- From a sustainability point of view the links between socio-economic and environmental systems are extremely important. These links can be approached from various angles:
    o the dependency of production and consumption (differentiated by social groups) on the environment: what is the use of renewable and non-renewable resources, waste disposal levels, etc. To what extent is substitution within economic and ecological production functions feasible? What is the share of natural

resources, directly and indirectly, in imports and exports? How do socio-economic policies affect natural resource use?

o what are the economic explanations (poverty, distribution of resources, population growth, economic policies, etc.) for environmental problems? To what extent do market prices reflect ecological costs?

o what are the economic consequences (for specific groups and levels) of environmental problems: how do population groups respond to environmental decay, what is the impact on possibilities for income-generating activities? How is long-term welfare affected?

Analysis of these links may result in an outline of level-specific sustainable development patterns and their short-term and long-term economic costs and benefits. This should inter alia provide insight into the question of how and to what extent sustainability prospects at various levels are interrelated. Special attention should be given to critical success factors (see Nijkamp et al,, 1990). They simultaneously provide insight into the environment as a potential means for development and the environment as a set of constraints on human activities. Critical success factors determine the boundaries of feasible projects in the project area. Projects that influence environmental parameters which are already close to critical levels are less attractive than projects which operate in less sensitive areas.

In the context of a flexible approach to sustainability, which emphasizes location-specific conditions, information about the project setting may be used to change elements of the policy framework, particularly the choice of policies regarding the sustainability constraint. If, for instance, in a particular area, substitution possibilities within production functions are considered feasible, a weaker sustainability condition may be formulated than when such opportunities are ruled out a priori.

4. SUSTAINABILITY AND IMPACT ASSESSMENT

4.1. Format of the Impact Matrix

Impact assessment starts from an impact matrix with several possible dimensions: alternatives, criteria, intratemporal distribution among social classes or spatial levels, time and uncertainty. In traditional impact assessment studies, this matrix usually has two (alternatives, criteria) or three (time added) dimensions. In table 2 the basic format of a traditional impact matrix is shown in two dimensions (criteria and time), assuming that such a matrix would be prepared for each alternative.

Table 2. Traditional impact matrix (to be prepared for all alternatives).

criteria	time (years)
	0 1 2 3 4 5 30
efficiency	
	effects
intratemporal equity	

As was argued above, efficiency and, to a somewhat lesser extent intratemporal equity, have been key appraisal criteria. In general, effects were assessed from a project and/or the national point of view, ignoring cross-border effects. The time horizon has usually been confined to a period of ten to thirty years. In principle, environmental effects were part of efficiency analysis. The attention paid to environmental effects has usually been confined to local impacts directly affecting production or productivity. In practice, at least within the framework of cost-benefit studies, environmental effects tended not to be included at all because of measurement or valuation problems, or because they were considered not specific to the project.

Quiggin and Anderson (1990) describe the traditional treatment of risk and uncertainty. They argue that analysts generally have been satisfied with best-estimate or even best-case (most favourable) outcomes. Projections appear to be "surprise-free", assuming that nothing unexpected will happen. A more data demanding approach, viz. probability analysis, has less often been applied (see for instance Reutlinger, 1970). Quiggin and Anderson found that expected values are generally calculated on the basis of unskewed, especially normal distributions. In the final appraisal stage, estimates tend to be subjected to partial sensitivity analysis to show the dependency of the outcomes of the appraisal on assumptions.

Because sustainability-oriented project appraisal starts from a different policy framework, the format of the impact matrix needs to be adjusted in several respects. The basic structure is shown in table 3.

Table 3. Sustainability-oriented impact matrix (to be prepared for all alternatives).

criteria	time	
	present generation	future generations
efficiency		
equity	effects/risks	
sustainability (resource use/ availability)		

The following adjustments have been made:
- An obvious change is that three instead of two criteria should be included, sustainability being added. Scores on the sustainability criterion involve a comparison between actual and normative resource use at specified levels. From these scores, changes in environmental stocks (i.e. resource availability) can be calculated.
- The time dimension changes considerably. A distinction should be made between short-term and long-term effects, and more in particular between effects on the present and on future generations. The period encompassing the present generation may still be accounted for in terms of years and cover scores on all criteria. With respect to impacts on future generations, two approaches may be applied. In the least data-demanding approach, only scores on the sustainability criterion are shown, i.e. actual availability of natural resources compared to threshold levels, possibly expressed in cardinal or ordinal indicators. Hence, this approach does not comprise efficiency and equity impacts on future generations. It can be justified if the interests of future

generations are considered to be fully accounted for if normative resource levels are respected. If not, a second, much more data-intensive, approach needs to be followed, involving estimates of scores on all criteria.

- The traditional treatment of risk and uncertainty does not meet requirements in sustainability-oriented impact assessment. Apparently, environmental effects often involve surprises, especially unpleasant ones. Moreover, probability distributions may often be skewed to the left. Uncertainty is significant, and probabilities may not be known at all or only be available in the form of beliefs people may have on ranges or intervals of probabilities for an event (Quiggin and Anderson, 1990). From the start, impact assessment should therefore be in terms of effect-risk combinations. This particularly refers to the sustainability criterion, which would have two attributes, viz. relative resource use, and risk and uncertainty involved. Particular attention should be given to "worst-case" outcomes, their probability and their consequences. The possibility of unfavourable surprises should also be acknowledged. Instead of presenting only one "best-case" impact matrix, several matrices may be added showing outcomes under extreme scenarios. One way of incorporating risk and uncertainty associated with environmental effects was proposed by Markandya and Pearce (1987). Through "certainty equivalence procedures" decision-makers express how much net benefit they would be willing to sacrifice in order to avoid the risk associated with expected values.

- The sustainability criterion requires a more differentiated analysis as far as spatial levels are concerned. Much more than in the past, contributions to supra-national environmental and welfare changes need to be taken into account.

4.2. Measurement Scales for Efficiency, Equity and Sustainability

Now that the basic issues regarding the format of the impact matrix have been identified, the dimension of the effects themselves needs attention. What are the measurement scales for efficiency, equity and sustainability? What are critical levels, dividing the set of alternatives into a group with a positive score on a criterion and a group with a negative score?

Efficiency has mostly been assessed on the basis of basically similar CBA indicators that share the monetary dimension. In the previous section reference was made to one of them, viz. the net present value (NPV). Under certain assumptions the NPV can be used to rank alternatives, a higher NPV being more attractive than a low NPV. The critical value on the NPV scale is 0. Similar critical values would be the value of the rate of discount if the internal rate of return (IRR) is applied, and 1 on a benefit-cost ratio (BCR) scale. Assessing the score on the efficiency criterion in terms of NPV, IRR or BCR indicators, however, assumes that all efficiency attributes, including the use of environmental resources, can be assigned a monetary value (valuation techniques are described in Dixon et al., 1988; Hufschmidt et al., 1983). When the scores on one or more attributes cannot be measured on a monetary scale, by definition no comprehensive efficiency score in monetary terms can be assessed (see section 2.1.). One possibility is to standardize scores on all attributes and consequently weigh the standardized scores, resulting in a dimensionless efficiency score.

Intratemporal equity scales are not unequivocal. With respect to distribution of income, inequality measures used at the national level may be taken as a starting point (see Gilles et al., 1987). The national distribution of income is often shown in the Lorenz curve. It shows the percentage of total income accounted for by any cumulative percentage of recipients. From such a Lorenz curve, the Gini concentration ratio can be calculated on a scale between 0 and 1.

In the case of perfect equality the Gini ratio is 0, whereas it is 1 in the case of perfect inequality. An alternative is the Kuznets ratio, which sums up the differences between income shares and population shares of all the cells into which a population might be divided. An even more crude approach is to focus only on the income share of the group that is considered to comprise the poor (bottom 20 or 40%).

In the context of project appraisal, the question is which part of net benefits accrues to specific groups. Distribution patterns could be shown in simplified types of Lorenz curves. The degree of inequality in principle might be assessed through Gini or Kuznets coefficients, but these approaches are likely to be too cumbersome in general. A relatively straightforward approach might be to measure the part of the net efficiency gains that accrue to specified target groups, for instance: landless labourers, people with a below average income level, etc. [5]. Alternatives might be ranked according to the score on this equity indicator. Critical values may be established for equity. For instance, a project may be considered attractive from an equity point of view if at least 40% of efficiency gains accrue to the target groups. Actual scores on this measure might be represented on a scale with 40 as a target level. Alternatively, the outcomes might be transformed to a dimensionless scale with -1 (0% of benefits accruing to target groups) and +1 (all benefits accruing to target groups) as extremes, and 0 (threshold part of benefits accrue to target groups) as centre value.

Possible sustainability scales have already been referred to. In its simplest form, it is measured on a binary scale: a project either does or does not satisfy resource use constraints. It was proposed to use cardinal or ordinal sustainability indicators if sufficient data are available. In any case, the critical value on the sustainability scale would be the threshold level discussed above. If measured on cardinal or ordinal scales, scores would show the relative distance between actual and normative resource use levels at a scale with 0 as critical level. The sustainability criterion may furthermore include a long-term risk attribute, whereby risk and uncertainty are expressed on a dimensionless scale.

Principal measurement scales and critical levels (NPV: 0; share to poor: x%; resource use: sustainable level) are summarized in table 4. For all criteria, scores might also be expressed in terms of relative distances between actual and threshold levels on a scale with 0 as centre value. With respect to presentation, scores of all alternatives might be gathered in a three-dimensional Euclidian space, with the three criteria at the axes. Differences between actual and threshold levels might also be gathered in a Möbius triangle (see Nijkamp et al., 1990) or in circles (with threshold levels at the edge of the circle and actual scores within or outside the circle), (cf. the AMOEBA model of ten Brink, 1991).

The conclusion from this section is that principal effects in sustainability-oriented project appraisal are likely to be in different dimensions. Often, a mixture of monetary, quantitative and qualitative data may result [6].

4.3. Problems in Measuring Scores on Environment and Sustainability

As scores on the environment criterion provide the basis for an assessment of the score on the sustainability criterion, we will devote some attention to the measurement of environmental

---

[5] The income distribution measures mentioned thus far are inequality measures in terms of *relative* incomes. Poverty measures relate to attributes of specified target groups in some *absolute* sense (income, nutrition) and are often considered a better criterion for measuring equity at low levels of income.

[6] For an example of impact matrices with mixed scores see Nijkamp and van Pelt (1989). Using a simple qualitative system model for Bhubaneswar (India) and its surrounding region, various scenarios for urban development in the region have been analyzed. Impacts were assessed on very different criteria, such as poverty, employment, social climate and migration. Time was accounted for by introducing ten stages, without a direct link to particular years.

effects 7).

Table 4. Principal measurement scales and critical levels.

criterion	scale	appreciation (critical level)		
efficiency	NPV	negative	(0)	positive
equity	% to target groups	negative	(40%)	positive
sustain-ability	resource use	positive	(sustain-able use)	negative

Measuring environmental effects focuses on the environment system and is known as environmental impact assessment (EIA) (for a discussion centred on developing countries see Biswas and Geping, 1987). An EIA focuses on the difference between expected environmental changes: a) in the absence of the project under consideration (see project setting profile) and b), if the project is implemented, at the project and higher levels. To measure b), an analysis is required of how the project influences the environment both during the construction phase and during normal operation. Here Hufschmidt et al. (1983) distinguish between:
- projects involving management of natural systems to produce certain outputs,
- projects that affect natural systems off-site,
- projects that eliminate a natural system and replace it with an alternative
  human-built system that might have important off-site effects, and
- projects that modify or replace the on-site ecosystem with a more or less
  artificial system on the site through alteration of the existing natural system.

Regarding the type of ecological effects, a distinction should be made between several groups of receptors (resident and migratory fish, resident and migratory animal species, natural vegetation, materials in structures, materials in vehicles, agricultural and forestry activities, industrial and commercial activities, other activities, residences, humans; see Hufschmidt et al., 1983). Effects should be assessed both on-site and off-site, including the transboundary effect. Chains of effects should be assessed, and possible time-lags identified. Besides assessing short-term costs (or benefits) of environmental effects, particular attention is required for possible irreversibilities (Toman and Crosson, 1991). Due attention should be given to risk and uncertainty. From the start, the EIA should have a distributional focus: who benefits from environmental improvement and who faces the social costs of environmental degradation?

Systematic research on a wide scale on environmental effects of human activities has only recently started in developing countries. The data base is still rudimentary. Many difficulties in measuring environmental effects tend to arise. Hufschmidt et al. (1983) give the following ones: 1) discharges of material and energy residuals into air, water and land are of many different types; 2) a wide range exists for both the rate of change in environmental quality and for the geographical area of influence of residual discharges on environmental quality; 3) there is a wide range in the time rates of effects on receptors from changes in environmental quality; 4) a large element of randomness exists in the levels of environmental quality over time because of

---

7) Parts of this section were published in van Pelt, Kuyvenhoven and Nijkamp, 1990.

differences in the time pattern of discharges and of the assimilative capacity of the environment; 5) residuals discharged from human activities are not the only factors affecting the quality of the environment.

But there are several more specific problems. Sustainability has a long-term focus and forecasting environmental effects over periods of several decades usually involves considerable degrees of uncertainty. Assessing environmental decay which is strongly localized may often be easier than environmental problems that tend to spread in space. Contributions to global environmental problems such as the greenhouse effect are a clear example. Pearce et al. (1990) mention effects on life support systems, such as contributions to geochemical cycles. Nijkamp et al. (1990) emphasize the importance of synergetic environmental impacts. This refers to the process that numerous, individually small impacts on the environment, together can have significant environmental effects. Nentjes (1989) stresses the importance of stock-type environmental problems. This refers to the cumulative effect of annual emissions. In such cases, overall problems increase even if annual emissions are decreasing. The most well-known example is acid rain. Such problems increase until the last year that the annual discharge of effluents is positive.

A particular problem with respect to impact assessment at the project levels refers to the question of which environmental effects should be attributed to the project [8]. In general the focus will be on "forward links", i.e. effects which result from the decision to start an activity. Environmental effects associated with the production of goods and services that are consumed by the projects will usually not be considered, if only because of the severe measurement problems involved.

With respect to the assessment of scores on the sustainability criterion, a distinction has been made between estimating only long-term ecological effects and estimating all welfare effects for future generations. In the former case, i.e. the assessment of the difference between actual and threshold levels of long-term resource use, the EIA might provide all necessary information. Threshold levels being determined in the policy framework (see section 2), they can be compared with actual resource use as established in the EIA. Actual resource use should particularly be estimated for the level at which sustainability is defined. Hence, if sustainability is defined at the project level (see Klaassen and Botterweg, 1976), the resource use of individual projects needs to be known. If it is defined at higher levels (regional, national, global), one should estimate how a project would contribute to resource use at these levels. The higher the level concerned, the more problematic measurement of the environmental effect and hence sustainability becomes for an individual project. What would be the score of a project on sustainability in view of its contribution to national acidification, let alone depletion of the ozone layer? Assessing the score on sustainability might be based on the following guidelines:

- if the supra-project level of sustainable resource use is already exceeded without the project, a project using more resources should be considered non-sustainable. The degree of non-sustainability increases with the relative size of the resource use by the project.
- if the supra-project level of sustainable resource use is not exceeded without the project but will be exceeded with the project, the project is not-sustainable if resource use cannot be reduced elsewhere (opportunity costs).
- if the supra-project level of sustainable resource use is not exceeded both with and without the project, the project may be considered sustainable. The degree of sustainability decreases with the relative size of the project's use of natural resources.

In general, assessing scores on the sustainability criterion is more difficult under "weak sustainability" than under "strong sustainability" strategies. Whatever the sustainability

---

[8] For a discussion on this issue from the perspective of a country, see Opschoor and Reijnders, 1989.

condition, a "no trade-off" position implies much weaker methodological and data demands than a "trade-off allowed" policy (see section 2.1.).

If not only ecological but also efficiency and equity impacts on future generations should be assessed, data problems become extremely large. Whatever the ultimate time horizon, ecological-economic models would need to be applied. Although much progress has been made in this field in recent decades, such models are still in their infancy (Braat and van Lierop, 1987; van den Bergh, 1990). The format of a comprehensive sustainability study is illustrated below through the example of the construction of large dams.

The first level at which sustainability should be analyzed is the project level. This involves an investigation of how environmental parameters may affect expected direct (i.e. within the project area) net benefits of the project in the short and long run. Sustainability analysis may then focus on the possibility of sedimentation. Although not a problem in the short run, this could in the long run lead to much higher than expected operation and maintenance costs and fewer benefits in the form of electricity and irrigation water. This would negatively affect welfare in the project area.

The second step analyses national sustainability effects. The issue here is whether the project would have sustainability-relevant effects at the regional and national level. Hence, to what extent does the project have a negative impact on the social welfare development potential, either through the possibility of producing goods and services or through environmental amenities, in the country? This step in particular requires paying special attention to distribution effects and poverty-environmental decay relations. The construction of the dam could, for instance, have a strong negative effect on water availability for downstream farmers. This would negatively influence income generation possibilities for those farmers. If these farmers were already near a subsistence level, they might have to turn to environmentally unsound agricultural practices. If such a process occurs, the sustainability of previously environmentally sound economic activities is threatened, although the project itself need not be rejected on sustainability grounds. Another possibility would follow a different chain. The dam could greatly improve the competitiveness of upstream farmers who benefit from the dam. Because of considerable cost reductions and roads constructed within the framework of the project, they can now sell at much lower prices than the traditional downstream farmers. Again, their income may drop below minimum levels, which may lead to unsustainable agricultural practices.

Finally, cross-border sustainability effects may be evaluated. Sustainability-relevant effects across national borders would need to be traced in a similar way. An example would be the flooding of rivers and consequent negative income and behaviour effects in neighbouring countries, due to erosion in the project country as a result of deforestation.

5. SUSTAINABILITY AND EVALUATION

5.1. Constraints and Adjustment of Project Design

In this section we explore possible consequences of the transformation of sustainability into a veto criterion. This implies that no project is accepted that involves resource use in excess of levels commensurate with normative sustainable levels. A strategy which turns sustainability into a goal, i.e. allows trade-offs between sustainability and other criteria, is considered in section 5.2.

Assuming that sustainability is a veto criterion, EIA outcomes provide the means to divide the initial set of alternatives into two groups: non-sustainable projects and sustainable projects. A project in the first group might be the construction of a dam aimed at electricity generation and irrigation improvement, with unacceptable long-term consequences for ground and surface water availability, water quality and sedimentation. Such a project may immediately be rejected, but a more constructive approach would involve an analysis of what will be called constraint-satisfying activities. The aim of such activities, which should actually be implemented, is to ensure that the adjusted project proposal would comply with the sustainability constraint. Constraint-satisfying activities might be classified as follows:

- changing the design of the project itself. The timing, site or technology might
  be adjusted. Measures may be included to prevent or mitigate negative
  environmental effects (defensive expenditures). A dam could be made lower,
  special filters could be installed or reforestation activities could be conducted
  to avoid sedimentation.
- additional activities may be embarked upon to compensate for negative environmental
  effects of the original project. Such activities have been called "compensating projects"
  (Pearce et al., 1990) or "shadow projects" (Klaassen and Botterweg, 1976). In both
  cases as much environment (in physical terms) should be "created" as will be lost
  due to the original project. Hence, environmental damage is allowed provided similar
  quantities of environmental resources of similar quality can be created by men. It
  should be noted that Klaassen and Botterweg propose to apply the sustainability
  constraint at the project level, whereas Pearce et al. favour the "programme" approach.
- other activities should reduce their use of resources (Winpenny, 1990). In this
  way the negative sustainability impact of the proposed project is compensated for
  externally.

If constraint-satisfying activities are included in the project proposal, the impact of the project should be reassessed. This includes environmental effects, which now should comply with sustainability conditions. It should be assessed whether the constraint-satisfying activities themselves need any resource input which might affect sustainability. In addition, costs need to be reestimated. Adjusting project design may often involve larger outlays, whereas additional activities by definition raise costs. Whether it is feasible to assign costs of additional activities to a project depends to a great extent on the level at which sustainability is defined. Following Klaassen and Botterweg's approach, the sustainability condition should be satisfied at the project level. Consequently, the costs of shadow projects can be attributed to and directly affect the economic feasibility of resource-using projects. Such a link cannot be established unequivocally if the sustainability constraint is defined at the programme level, as proposed by Pearce, Barbier and Markandya, or even higher levels. There is no straightforward way to assign individual projects the full costs of environmental resource use. As a consequence, the appraisal mechanism does not provide an incentive to prevent or mitigate environmental damage. Moreover, Pezzey (1989) poses the question of who will pay for the economically unattractive constraint-satisfying activity.

If constraint-satisfying projects are implemented on the supra-project level, a tentative solution to attributing costs to individual projects might involve the following steps:

- an ex ante estimate of total environmental damage in a year in a specific area,
- ex ante determination of corresponding shadow projects and their aggregate costs,
- determination of shadow project costs per unit environmental damage,
- assign unit shadow project costs to projects in proportion to environmental damage.

It should be acknowledged that the notion of compensating projects focused at sustainability constraints implies a fairly optimistic view on men's capability to "build" natural capital. In other words, possibilities for substitution in the environment production function are stressed. Not all ecologists would agree with such a view. Irreversible environmental and synergic effects by definition cannot be compensated.

Besides the impact on environment (sustainability) and costs, distributional aspects should be reconsidered after including constraint-satisfying activities. If compensating projects are not implemented at the same site as the resource-using project, a transfer in space of environmental capital takes place. Similarly, the social groups benefiting from a shadow project need not be the same as those that take the burden of the resource-using project. A recent proposal of Dutch suppliers of electricity may serve as an example. To compensate for emissions of greenhouse gases by a new Dutch power station, a contribution to reforestation in Brazil was offered. In theory, global environmental stabilization might be achieved in this way (although many ecologists will think otherwise), but this might be of little comfort to people living close to the power station. One might justify this particular transfer on the basis of intragenerational redistribution goals. But it is easy to think of examples of constraint-satisfying projects where the environmental burden will fall on poor groups or nations.

Constraint-satisfying activities may also raise questions of intergenerational equity. Preservation or rebuilding of natural capital can have significant short-term opportunity costs, whereas the benefits (actual compensation) often occur after several years. The present generation might thus be affected negatively in two ways: they experience the environmental burden (which will in effect be compensated only after some time), whereas they also face the bulk of the constraint-satisfying projects costs.

In the examples given above, it was assumed that the "strong" approach to sustainability applies. Constraint-satisfying activities then by definition involve saving or building natural resources. If the "weak sustainability" approach, which aims at maintaining the total stock of resources (whether man-made or environmental) is applied, constraint-satisfying projects could be of two types. The first possibility would be to implement an environmentally constraint-satisfying project, as above. The other possibility is to conceive an economic constraint-satisfying activity, which would involve investments in factors of production other than the environment.

The discussion above centred on activities aimed at satisfying ecological sustainability constraints. Efficiency and equity, however, may also be interpreted as veto criteria. If a proposed project fails to satisfy an efficiency precondition (for instance, NPV > 0), constraint-satisfying activities might be designed in similar ways. Activities might also be embarked upon to ensure that a sufficient part of the net project benefits accrues to target groups. The relocation of tribes who are negatively affected by large dams is an example.

5.2. Integrated Comparison of Alternatives

In the final appraisal stage, the overall performance of all project alternatives that have passed veto criteria requirements is assessed on the basis of their scores on efficiency, equity and sustainability and the criteria weights. Basically, two approaches may be followed. The former takes one criterion as a point of reference and subsequently adjusts the score on this criterion for scores on the other criteria. This single-indicator approach is represented by the CBA technique. Social CBA involves adjusting net cost and benefit flows in terms of economic efficiency for income distribution objectives. Through what may be termed a "social-sustainability" CBA, one might attempt to integrate sustainability objectives into a social CBA. The "rod of money" would continue to be the "numeraire", and the result would still be in terms of indicators such as an IRR or NPV.

The alternative approach does not aim at such a transformation whereby one criterion (and hence one measurement scale) is taken as bench mark. The starting point involved separate scores on key criteria. Through arithmetical operations, the combinations of weights and criteria scores are used to arrive at a ranking of project alternatives. This is the multi-criteria approach (MCA; see Voogd, 1983; Nijkamp, Rietveld and Voogd, 1990; Petry, 1990; van Pelt, Kuyvenhoven and Nijkamp, 1990).

From the above it follows that the difference between the two approaches does not necessarily refer to the number of objectives. In theory, CBA might cover not only efficiency but also equity and sustainability, and in that sense be considered a "multi-criteria" approach. The major difference between CBA and MCA concerns the integration of criteria, and in particular the use of multiple denominators (and hence data requirements).

The findings of this paper may be considered an appropriate means to explore the extent to which CBA and MCA can tackle the specific analytical problems associated with sustainability-oriented project appraisal. Such an analysis, however, is beyond the scope of this paper.

When sustainability is incorporated in project appraisal in ways outlined above, the outcome in terms of projects accepted and rejected will change. Differences will be most significant in the case of strong-sustainability concepts based on risk-aversive strategies. In principle, a shift can then be expected towards projects in the field of renewable energy, energy conservation and efficiency, recycling of non-renewable resources, development of substitutes for non-renewables resources, etc. The probability that projects like large dams and deforestation for agricultural development are accepted will be considerably smaller.

6. CONCLUSIONS

Incorporating sustainability in project appraisal for developing countries raises a number of issues which have been outlined in this paper. Fields in which the most far-reaching changes are required include the following:
- Sustainability parameters (what are sustainable levels of resource use? what are acceptable ecological risks? at which spatial levels?) should be based on explicit views of policy-makers regarding the responsibility of the present to future generations and the possibility of substituting man-made for natural capital.
- Adding sustainability to efficiency and equity adds to weighting problems in project appraisal. In addition to conflicts between efficiency and equity, conflicts between sustainability and efficiency and between sustainability and (intra)temporal equity need to be tackled.
- Impact assessment will need a long-term time horizon covering more than one generation.
- Instead of emphasizing "best-case" impacts, combinations of effects and associated risk, especially with respect to ecological variables, need to be presented.
- In view of measurement and valuation problems, accounting for several types of environmental effects in the determination of the efficiency criterion may be impossible. In such cases only partial efficiency outcomes can be arrived at.

On the basis of further research on these topics, appraisal methods may be selected that offer the best opportunities to tackle specific issues in sustainability-oriented project appraisal. The basic choice is between integrating sustainability concerns and other priorities in a single indicator (like the NPV or IRR in CBA), or treating various criteria separately and the application of policy weights (MCA).

With respect to appraisal processes, we expect that much more than in traditional analysis an interactive approach is required. Combinations of trade-off regimes, sustainability concepts (for instance strong and weak sustainability), risk attitudes (for instance various degrees of risk-aversion; no-regret strategies), and outcomes in terms of short-term and long-term economic and ecological effects and uncertainty may first be presented to decision-makers. In subsequent steps project redesign and constraint-satisfying activities may be considered and their impact assessed. On the basis of the response of decision-makers, further adjustments may be made. In the course of this process, the number of alternatives is likely to reduce.

ACKNOWLEDGEMENTS

This paper was prepared with financial support from the Foundation for the Promotion of Research in Economic Sciences (ECOZOEK), which forms part of the Netherlands Organization for Scientific Research (NWO).

REFERENCES

Bergh, J.C.J.M. van den (1990), "Aggregate dynamic economic-ecological models for sustainable development", Paper presented at the international congress "The ecological economics of sustainability", World Bank, Washington DC, 21-23 May 1990.

Biswas, A.K. and Q. Geping (eds.) (1987), Environment impact assessment for developing countries, Tycooly Publishing, London.

Bojö , J., K.G. Mäler and L. Unemo (1990), Environment and development: an economic approach, Kluwer, Dordrecht.

Braat, L.C. and W.F.J. van Lierop (eds) (1987), Economic-ecological modelling, North Holland, Amsterdam.

Brink, B.J.E. ten (1991), A quantitative method for description and assessment of ecosystems; the AMOEBA approach, Marine Pollution Bulletin.

Daly, H.E. (1990), "Towards environmental macroeconomics", paper presented at the international congress "The ecological economics of sustainability", World Bank, Washington DC, 21-23 May 1990.

Dixon, J.A., R.A. Carpenter, L.A. Fallon, P.B. Sherman and S. Manopimoke (1988), Economic analysis of the environmental impacts of development projects, Earthscan Publications, London.

Foy, G. and H. Daly (1989), "Allocation, distribution and scale as determinants of environmental degradation: case studies of Haiti, El Salvador and Costa Rica", Environment working paper No.19, Environment Department, World Bank, Washington.

Gillis, M., D.H. Perkins, M. Roemer and D.R. Snodgrass (1987), Economics of development, Second edition, Norton, New York.

Hedman, S. (1990), "Reversibility as a weighting factor in integrated least cost planning methodologies", reversed version of paper presented at the international congress "The ecological economics of sustainability", World Bank, Washington DC, University of Maryland, College Park.

Hufschmidt, M.M., D.E. James, A.D. Meister, B.T. Bower and J.A. Dixon (1983), Environment, natural systems and development: an economic valuation guide, Johns Hopkins University Press, Baltimore.

Hueting, R. (1980), New scarcity and economic growth, North Holland, Amsterdam.

Hueting, R. (1990), "Correcting of national income for environmental losses: a practical solution for a theoretical dillema", paper presented at the international congress "The ecological economics of sustainability", World Bank, Washington DC, 21-23 May 1990.

Klaassen, L.H. and T.H. Botterweg (1976), "Project evaluation and intangible effects: a shadow project approach", in P. Nijkamp (ed), Environmental economics, vol 1, theories, Martinus Nijhoff, The Hague.

Little, I.M.D. and J.A. Mirrlees (1974), Project appraisal and planning for developing countries, Heinemann Educational Books, London.

Mäler, K.G. (1985), "Welfare economics and the environment", in A.V. Kneese and J.L. Sweeney (eds), Handbook of natural resource and energy economics, vol 1, North Holland, Amsterdam.

Markandya, A. and D.W. Pearce (1987), Environmental considerations and the choice of the

308

discount rate in developing countries, report prepared for the Overseas Development Administration, London.

Ministry of Foreign Affairs, Directorate-General for International Co-operation (1990), Een wereld van verschil. Nieuwe kaders voor ontwikkelingssamenwerking in de jaren negentig, Tweede Kamer der Staten Generaal, 1990-1991, 21813, 1-2, SDU, The Hague.

Myers, N. (1989), "Natural resource systems and human exploitation systems: physiobiotic and ecological linkages", Environment working paper No. 12, Environment Department, World Bank, Washington.

Nentjes, A. (1989), "An environmental policy for sustainable growth", paper presented at the 45th congress of the International Institute of Public Finance, Buenos Aires.

Nijkamp, P. (1979), Theory and application of environmental economics, North Holland, Amsterdam.

Nijkamp, P. in collaboration with J. van den Bergh and F. Soeteman (1990), "Regional sustainable development and natural resource use", paper presented at the World Bank annual conference on Development economics, April 26 and 27, 1990, Washington D.C.

Nijkamp, P. and M.J.F. van Pelt (1989), "Spatial impact analysis in developing countries", International Regional Science Review, 12(2), 211-228.

Nijkamp, P., P. Rietveld and H. Voogd (1990), Multicriteria evaluation in physical planning, North Holland, Amsterdam.

OECD (1990), Development co-operation 1990 report, Paris.

Opschoor, J.B. and L. Reijnders (1989), "Duurzaamheidsindicatoren voor Nederland", draft paper discussed at IVM/RIVM workshop, October 30, 1989, Utrecht.

Pearce, D.W., E.B. Barbier and A. Markandya (1990), Sustainable development: economics and environment in the Third World, Edward Elgar, Aldershot.

Pelt, M.J.F. van (1991), "Measuring sustainability", draft paper, Wageningen Agricultural University, Wageningen.

Pelt, M.J.F. van, A. Kuyvenhoven and P. Nijkamp (1990), "Project appraisal and sustainability: methodological challenges", Project Appraisal, 5(3), 139-158. Shorter version of paper presented at the international congress "The ecological economics of sustainability", World Bank, Washington DC, 21-23 May 1990.

Pétry, F. (1990), "Who is afraid of choices? A proposal for multi-criteria analysis as a tool for decision-making support in development planning", Journal of International Development, 2 (2), 209-231.

Pezzey, J. (1989), "Economic analysis of sustainable growth and sustainable development", Environment working paper No. 15, Environment Department, World Bank, Washington.

Quiggin, J. and J. Anderson (1990), "Risk and project appraisal", paper presented at the World Bank annual conference on Development economics, April 26 and 27, 1990, Washington D.C.

Reijnders, L. (1990), "Normen voor milieuvervuiling met het oog op duurzaamheid", Milieu, 1990-5, 138-140.

Reutlinger, S. (1970), "Techniques for project appraisal under uncertainty", World Bank staff occasional paper No. 10, Johns Hopkins University Press, Baltimore.

Seneca, J.J. and M.K. Taussig (1984), Environmental economics, third edition, Prentice Hall, London

Squire L. and H.G. van der Tak (1975), Economic analysis of projects, Johns Hopkins University Press, Baltimore.

Squire, L. (1989), "Project evaluation in theory and practice", in H. Chenery and T.N. Srinivasan (eds), Handbook of development economics, vol II, Elsevier Science Publishers, Dordrecht.

Toman, M.A. and P. Crosson (1991), "Economics and sustainability: balancing tradeoffs and imperatives", working paper ENR-91-05, Resources for the Future, Washington D.C.

UNIDO (1972), Guidelines for project evaluation, United Nations, New York.

Voogd, H. (1983), Multicriteria evaluation for urban and regional planning, Pion, London.

Winpenny, J.T. (1990), "The relevance of global climatic effects to project appraisal", Project Appraisal, 5 (4), 213-219.

World Commission on Environment and Development (1987), Our common future, Oxford University Press, Oxford.

Zionts, S. (1989), "Multiple criteria mathematical programming: an updated overview and several approaches", in B. Karpak and S. Zionts (eds), Multiple criteria decision making and risk analysis using computers, Springer-Verlag, Berlin.

310

Diagram 1. Overview of main stages in project appraisal.

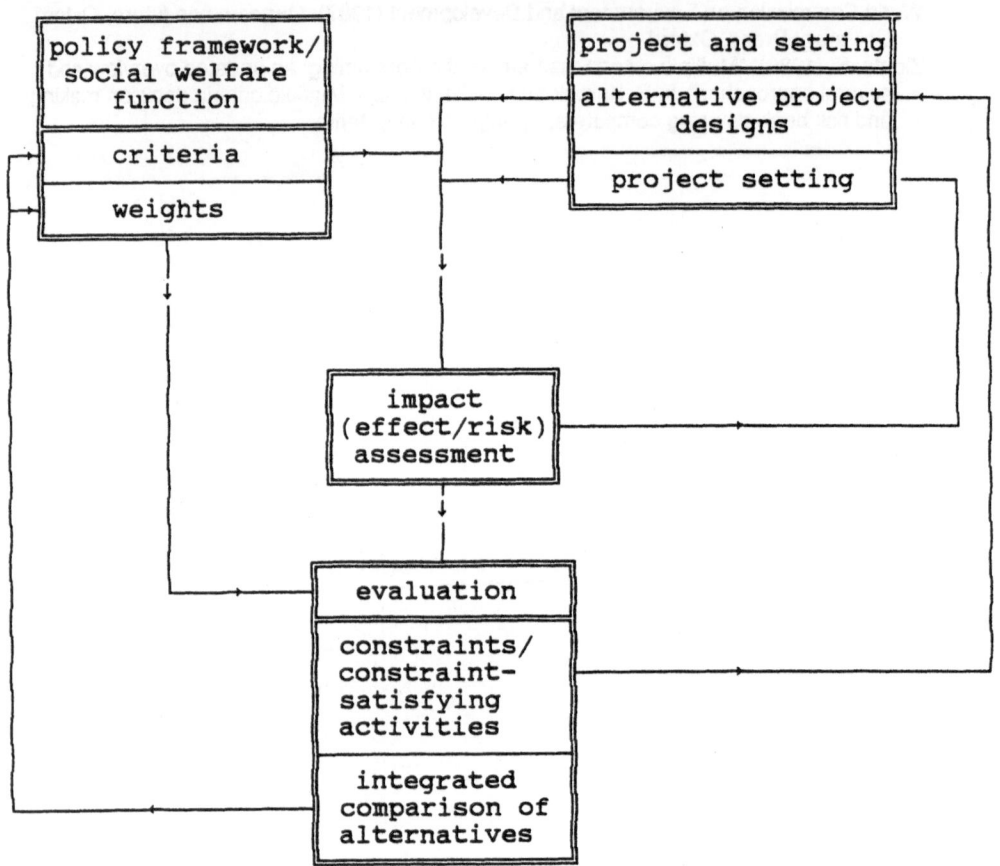

311

Diagram 2. Key criteria appraisal.

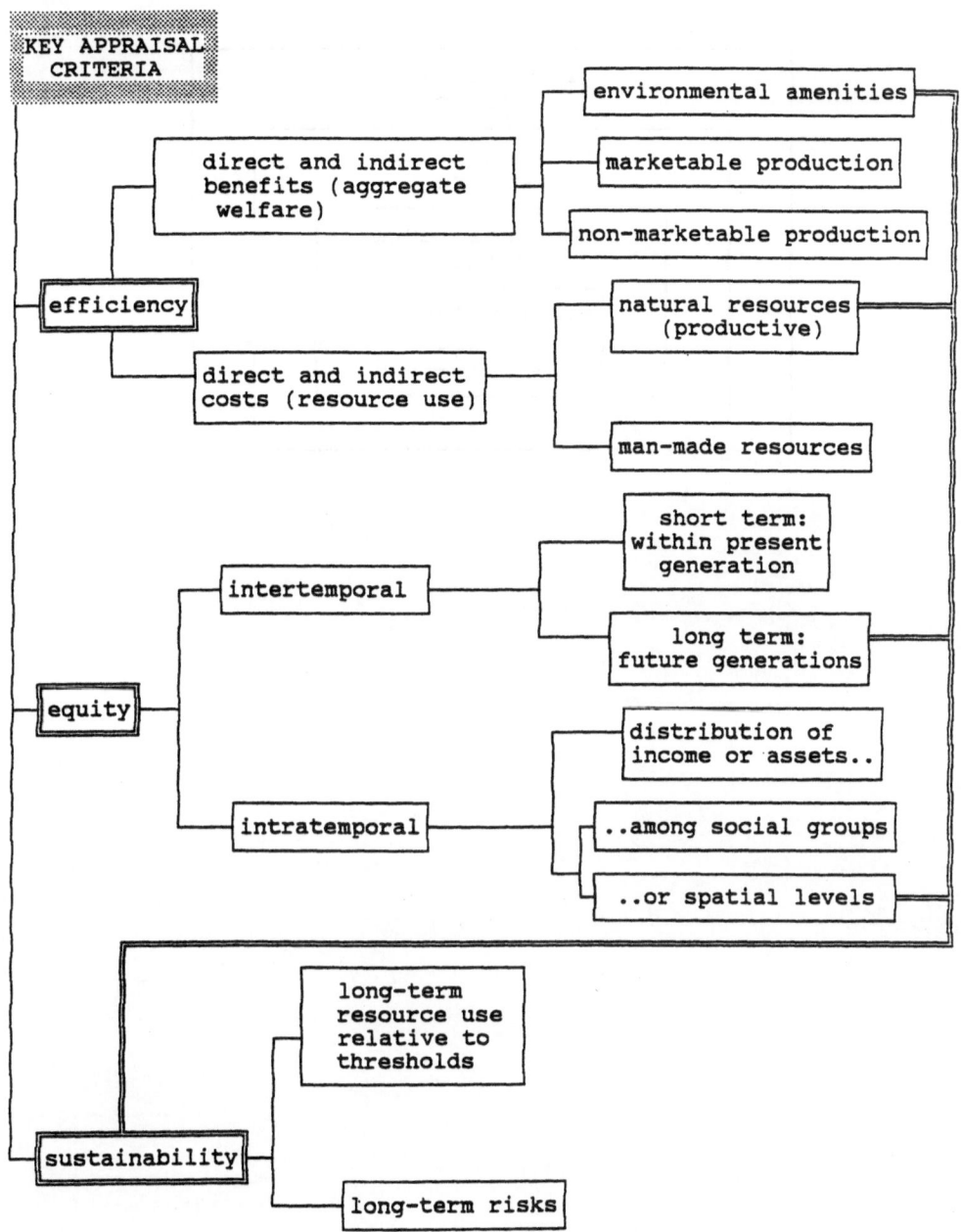

312

Diagram 3. Interlevel welfare linkages

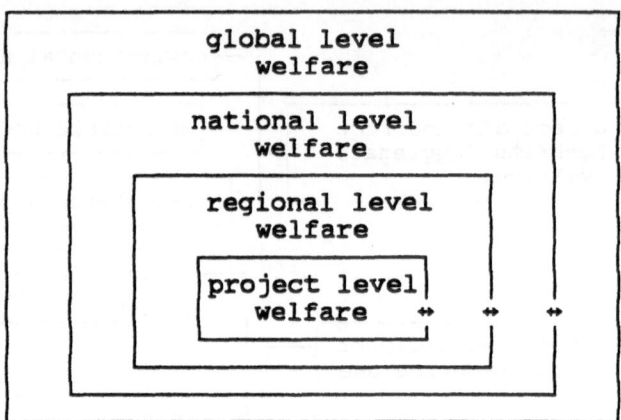

SOCIOLOGICAL ASPECTS
OF ENVIRONMENTAL IMPACT ASSESSMENT

L. Pellizzoni
Institute of International Sociology
via Mazzini 13
I - 34170 Gorizia
Italy

ABSTRACT. The paper discusses some sociological aspects connected with Environmental Impact Assessment (EIA), with particular reference to the procedure established by the EC directive 85/337. The attention focuses first on the field of Social Impact Assessment (SIA), on its relations with EIA and on the sociological implications of different EIA models. The associated problems concern the different functions of participation, the identification of the components of the interested public, their involvement in the procedure and the employment of objective and subjective indicators in the measurement of social impacts. A typology of the social actors facing the procedure and one of the media utilized in the communication flows between public and EIA authorities are proposed. Finally, some examples drawn from EIA legislation in Europe and America are briefly examined from a sociological point of view and some hints are presented for an implementation of European EIA.

1. INTRODUCTION

In our times there has been a radical change in the relationship between society and the environment. In the past the natural environment was considered to influence cultures, behaviour, institutions: the social systems answered environmental conditioning by adjusting themselves to it, i.e. by trying to obtain the resources necessary for their survival from sources so large that, compared with the needs, they appeared inexhaustible, and at the same time trying to control in some way the natural agents in order to limit disastrous consequences.

Today the relations of force between nature and society are reversed. Now it is not the environment that influences society but the latter that, thanks to science and technology, has subdued nature. We have passed from adjustment to dominance, a destructive dominance accompanied by the increasing scarcity of available resources compared with the needs.

The success of ecology as a science and the diffusion of ecological themes in politics and general culture is therefore explained by rapid technological and industrial growth and by the impact of human activities on nature. But it is also explained by the rise of new needs in advanced societies (thanks to the satisfaction of the primary needs), which may be summed up in the demand for a quality of life that is not simply provided by a high level of consumption but is largely linked to the quality of the environment. The rise of ecological movements, the institution of Ministries of the Environment and the increasing general attention to environmental problems are due to a number of political, economic, ethical, social, philosophical factors (are we due for a revival of romantic naturalistic philosophies?), but all this can be seen as a corrective reaction of complex societies to a growth that in its size and characteristics seems to

A. G. Colombo (ed.), Environmental Impact Assessment, 313–334.
© *1992 ECSC, EEC, EAEC, Brussels and Luxembourg.*

overstep the bounds of compatibility with nature.

2. ENVIRONMENTAL IMPACT ASSESSMENT
AND SOCIAL IMPACT ASSESSMENT

In this sense EIA is not only a technical or administrative procedure: it is above all a social and cultural fact, one of the greatest efforts made by society to exploit nature more rationally and to control its own modes of growth, without being seduced by an improbable return to the past. The idea implied in EIA is to rationalize the decisions that have important environmental consequences as much as possible, but also to offer the general public a possibility to control decisions or, better, to participate in evaluation and decision processes. EIA on the whole assumes a sociological relevance for at least the following reasons:

   a) the point of reference for assessment is man. Protection of the environment is
      essential for his survival and for a satisfying quality of life. But the assessment
      commonly concerns not only the natural environment but also material goods
      having economic, historical, aesthetic or social values, or aspects of natural
      environment with a marked social and cultural character, such as landscape;
   b) impacts on physical environment can be linked to "social" impacts, i.e., to effects
      related to man as a social being as well as a living being: effects on occupation or
      services, for instance, or effects on the identity or cohesion of a community or on
      people's perception of the acceptability of a risk. It should be remarked that some
      of the aspects mentioned cannot be investigated by reference only to material
      - or "objective" - data: it is necessary to analyze values, attitudes, opinions, interests,
      i.e., sociopsychological and cultural - or "subjective" - factors;
   c) EIA is a procedure that, under various titles, involves public and private organizations,
      experts and ordinary citizens: these are all social actors who act according to
      prescribed roles (sometimes violating them) and introducing ideas, skills, values,
      personal or group interests into the process. EIA is in itself a social institution, since
      it is a complex structure of roles and procedures organized around a particular
      activity.

Therefore, to consider EIA from a sociological point of view means examining two distinct prospects which are strictly linked at the same time:

   a) the prospect of the social impacts related to impacts on the physical environment;
   b) the prospect of the assessment procedure, as a social process as well as a
      political, administrative and technical process.

Some legislation on EIA soon understood the importance of the social aspects linked to environmental assessment: first of all US legislation with the National Environmental Policy Act (NEPA) and the Regulations of the CEQ (US Council on Environmental Quality, 1978). Social Impact Assessment (SIA) was therefore added to EIA.

SIA is one of the various assessments developed in the wake of EIA, such as Technology Assessment (TA) or Risk Assessment (RA), whose boundaries are not always easy to trace.

According to Freudenburg (1986), SIA "refers to assessing ... a broad range of impacts ... that are likely to be experimented by an equally broad range of social groups as a result of some course of action", while Finsterbusch (1985), notices that "Social impact assessment is a current buzz word, and is used to describe almost any type of research that looks at social factors", adding that the term in a technical sense is to be understood as a preventive evaluation of impacts undergone by individuals, groups, organizations, communities and institutions as a consequence of the realization of policies, programmes or specific projects (this is the range of action of EIA in the USA).

SIA's history is, as already said, closely interlaced with EIA's history, since both are connected with the NEPA and it is within the context of EIA that most SIAs have been conducted in the USA (Freudenburg, 1986). However, if we look at the definitions of EIA's subject existing at a legislative as well as a scientific level, we see that these definitions only sometimes also refer to social or socioeconomic impacts and that therefore the United States' model, by which everybody has been somehow inspired, has not always been received in full.

Although the NEPA, the institutive act of EIA in the USA, required an integrated use of natural and social sciences in preparing EISs (Environmental Impact Statements), that actually seldom happened in the first years of application of the regulations; later the intervention of CEQ, the agency responsible for overseeing the implementation of NEPA, served to state more precisely the relation existing between environmental and social impacts: "Social and economic effects by themselves do not require preparation of an EIS", but if social and biophysical impacts appear interrelated, EIS must face "all of these impacts on human environment" (US Council on Environmental Quality, 1978). However, the use of social sciences in EIA has not assumed a leading position. In EISs "socioeconomic" impacts have often been considered, but mostly with reference to strictly economic and demographic variables, not to the sociocultural ones, even if today the prevailing opinion is that SIA must have an interdisciplinary character, involving the whole of the social sciences (Freudenburg and Keating, 1982; Freudenburg, 1986; Carley and Bustelo, 1984; Branch et al., 1984). In this sense, a special importance has been given to psycho-social assessment (Lounsbury et al., 1983), that focuses on the individual subjective level of social impacts, i.e., on attitudes, beliefs, values, interests, intentions and consequent behaviours. On the other hand, there are more restrictive definitions of social impacts referring only to socioeconomic effects. For instance, Canter et al. (1985) include among social impacts only those related to housing, transportation, urban land use and land ownership.

SIA has sometimes been concerned with territorial or energetic planning, or new technologies, or developing projects in the third world, but till now it has mainly considered resource developments or large scale construction projects. For example, in Europe there have been some SIA investigations in the UK, in relation to the development of the North Sea resources, while in the USA and in Canada SIAs have had for their subject such problems as urban development, power generation, transmission line corridors, coal mining, new communities. SIA is therefore "a mode of analysis that could be used for any potential government or non-government action" (Finsterbusch, 1985) and that can be involved in any decision process even outside the sphere of application of EIA, for instance in the field of TA or RA. The link with EIA remains preferential, but it has been noticed that, even where EIA is conceived as an assessment of physical and social impacts, EIA and SIA are often seen as different and separate activities (Carley and Bustelo, 1984). This probably reflects not only the separation between social and natural sciences, but above all the difficulties which are met in integrating data regarding social impacts with those regarding biophysical impacts.

3. SIA: OBJECT AND METHODS

According to Finsterbusch (1985), SIA must examine or foresee such aspects as the dimensions of a project, the socioeconomic conditions of the community affected, economic and demographic changes as well as changes in demand and supply of houses and private or public goods and services, in the structure of the community and in the quality of life. A list of similar objects is proposed by Carley and Bustelo (1984); according to them, SIA must at least deal with: demographic impacts (e.g. labour force, population shift, displacement and relocation problems, changes in population make-up), socioeconomic impacts (e.g. income, employment rates, taxation), institutional impacts (e.g. demands on local financial and administrative services: housing, schools, health and welfare, recreational facilities), psychological and community impacts (e.g. changes in social integration, community and friendship networks, sense of place, community cohesion). According to Andrews (1987), three main types of social information showed be used in EIA: information about socioeconomic consequences of impacts on biophysical environment, about objective impacts on population and about subjective perceptions and meanings of impacts.

In conclusion, we can say that all possible effects deriving from the realization of a project or a plan or the application of a law that have to do with man as a social being fall in the field of application of SIA. I.e., we can think that effects concerning man as a biological being (essentially regarding his health) are included in EIA as an assessment of biophysical impacts, while SIA includes effects which are not strictly biological but linked to biophysical impacts - for instance those related to the aesthetic value of a landscape, to the historical and social value of a settlement, to the recreative value of a part of a territory - and effects not linked to biophysical impacts but equally depending on the realization of the project (or plan or policy) - effects on income, employment, services and   modifications of the social order of impacted communities.

It is a very broad field. In fact, SIA deals with one of the principal subjects of sociology: social change. Therefore, what characterizes SIA is not so much the originality or particularity of its object of study as the fact that, unlike most research on social change which analyzes slow and widespread processes, it deals with a very precise, specific, planned, localized change, concentrated in time, managed or at least controlled by public authority, and it deals with it not only for description purposes but for anticipatory ones. The nature of SIA as "anticipatory research" (Wolf, 1983) has important methodological implications because the techniques used must be able not only to describe the existing situation, but also to foresee its developments.

In this broad field (which corresponds to a variety of concrete experiences) some see a factor of uncertainty, a negative factor. For some people SIA represents "a mysterious object ... whose vagueness appears clear to experts for whom 'social' may mean everything or nothing" (Amendola, 1989), and has actually had a residual characterization, whose unique exclusive sectors have regarded the analysis of changes in social structure, above all of the breaking of social equilibrium (with consequent deviant behaviours) and impacts on particular groups (drop-outs, young and elderly people). But, above all, the concrete experience of SIA does not form a model that can be exported everywhere: in the USA much research is devoted to boom towns, i.e. to a phenomenon which is unknown or far less important in most European countries.

There is no space here for a detailed discussion about the methods used in SIA, for which a wide literature is available (see the note at the end of the References). However, some hints will be given here. A good level of agreement has been reached above all about the basic elements of methods for SIAs inserted in EIAs of specific projects. The methodological planning of SIA obviously refers to the EIA's. Wolf (1983) suggests a ten stage mode: scoping, problem identification, formulation of alternatives, profiling, projection, assessment, evaluation, mitigation, monitoring and management; among these the fundamental ones are (Finsterbusch, 1985): profiling, projection and assessment. Also Dietz (1987) points out three main stages: identification, analysis, evaluation. For Carley and Bustelo (1984), SIA and EIA foresee a similar process that includes a stage in which the present situation is described, a stage in which "means of describing change related to the project" are developed and a stage in which "changes in the base situation with and without the given project" are forecast. For Canter et al. (1984), the fundamental stages involve "procuring pertinent information, describing the current situation, predicting future conditions" with and without the project "and assessing the significance of the impacts which are predicted".

The vastness of the concept of social impact implies that methods from very different fields (sociology, psychology, social psychology, economy, anthropology, decision theory and so on) meet in SIA. The use of one or the other depends on the concrete situation to be dealt with and on the aims to be reached. Several general methods have also been developed sometimes exclusively centred on SIA, some times oriented to an integration of socioeconomic factors with environmental factors. Of these we can mention the Water Resource Assessment Methodology (WRAM) (see Carley, 1983a for a synthesis and a discussion), which organizes the variables into four groups: environmental quality, economic development, regional development and social well-being. These are subdivided into several categories and subcategories described by a wide number of variables (e.g., for education, health and safety, income, services, cultural opportunities: the WRAM even proposes an index of quality of life).

Another extensive approach is the Battelle Social Indicator Model (Olsen et al., 1981), which relates economic and demographic changes depending on the proposed policy, programme or project, through changes in community structure and in services, with changes in social well-being permitting to foresee the impacts. Another integrated methodology is the Group Ecology Method (GEM) (Flynn and Flynn, 1982; Flynn et al., 1983), which attributes a central position to functional groups present in a community, whose characteristics and interrelations are studied; the effects of the project are consequently analyzed as effects on them and therefore on the social structure of the community.

One of the most interesting approaches centred on SIA is the Social Organization Model (SOM) (Branch et al., 1984), which analyzes a community considering the relations between the inputs of the project, plan or policy and the resources of the local social system, its organization (structures and processes), and the subjective and objective indicators of quality of life.

Within or outside general models, the methods used in SIA are generally not innovatory; more often than not they are applications of well known techniques, from surveys with questionnaires to the use of social indicators, from in-depth interviews with leaders and experts to projective techniques (Delphi method, scenarios, etc.), to the innumerable methods used in psychology and social psychology for the measurement of attitudes, values, meanings, preferences, satisfaction levels and environmental and risk perception.

4. THE SOCIAL DISTRIBUTION OF IMPACTS

It is worth considering some problematic aspects concerning the theme of social impacts which assume clear relevance in the development of SIA inside EIA: the problems of the social distribution of impacts, of the dichotomy between subjective and objective impacts and of the nature of environmental assessment. The first two are often pointed out as the aspects that must be more developed in SIA, on a theoretical and empirical level, while the third intersects with political and normative aspects of EIA.

According to Wolf (1983), SIA must always try to answer the question: "Who benefits and who loses?" If impacts are a kind of social change, we must not only evaluate them in comparison with changes that would occur anyway without the projected intervention, but we must also analyze how impacts are distributed inside the impacted population. The analysis of the differentiation of impacts has not yet found adequate space in the literature (but see Elkind-Savatsky and Kaufman, 1986) thus remaining one of the themes requiring more investigation, even if they have been considered in some methods: the GEM, for instance, attempts "to identify groups that will experience differential effects (economic, demographic, housing, or fiscal) as a result of the project, or for which the evaluation of project-related effects is unique ... To the extent that the number of groups, group profiles, or group interaction patterns are different from the baseline projection, the proposed facility will be said to have caused a change in the social organization of the study area" (Flynn et al., 1983).

The analysis of the differential impacts must consider not only foreseeable effects in the different groups of a community in terms of employment, settlement, services, health and so on, but also their peculiar evaluation of impacts, based on different values, attitudes, opinions (about the proposed intervention but also about progress, technology, environment, community, etc.). The principal problem lies in the identification of indicators that can disaggregate impacts:

a problem common to the studies on the quality of life, a field from which it is certainly possible to obtain suggestions. Linked to the social differentiation of impacts, there is also the problem of differentiated participation to the EIA procedure by the various social groups. In fact, it is to be expected that, in the context of the forms of participation permitted by the law of a country, the demand for participation is not the same for every group, as it is influenced by the different interests and by the social position of each group: ecological sensitivity, which today seems to be spread beyond distinctions of rank or class, may be for instance influenced by the existence of more pressing problems (such as employment), while access to the forms of participation permitted by the law may be conditioned by a greater or lesser habit of participation acquired in the course of social life. The groups which scarcely utilize the means of participation offered by democracy should be somehow stimulated to become active subjects of EIA. The selection of adequate means of communication between authorities and public is in this sense very important: for instance, the compulsory or predominant use of the written form established by many laws in practice often excludes most of the population from the "dialogue" (see paragraph 9.3.).

5. OBJECTIVE AND SUBJECTIVE IMPACTS: AN ALTERNATIVE THAT CAN BE OVERCOME

The problem of subjectivity and objectivity of impacts is not new in the field of social sciences, since it concerns the alternative between an observer involved in the situation and an external one. Its centrality for SIA has been confirmed by many (e.g. Finsterbusch, 1985; Freudenburg, 1986; Andrews, 1987; Dietz, 1987; Beato, 1989), in answer to the insufficient attention paid by SIA and EIA to subjective impacts. As in the studies on the quality of life, the use of objective indicators (education and income levels, services, housing, pollution and criminal levels, etc.) is now generally considered insufficient if not accompanied by subjective indicators (satisfaction levels, expectations, perception of urban characteristics, etc.), so today there is a widespread opinion according to which in SIA it is necessary to make use of both objective and subjective indicators at the same time (Carley, 1983b; Land, 1983; Andrews, 1987). In fact, on the one hand it is a question of improving participation processes necessary for the formation of consent or at least for the democratic management of the procedure; on the other hand, as I have already said, there are impacts which cannot easily be evaluated in an objective way: this is the case, for instance, for landscape impacts whose evaluation leaves a wide margin of subjectivity, in spite of the efforts made to standardize judgements (for instance through statistical techniques for the analysis of visual preferences). Moreover, subjective impacts can constitute the basis for attribution of importance to objective factors (Andrews, 1987) and are indispensable in forecasting the effective reactions of a population to the realization of a project or plan or programme.

An assessment of subjective impacts does not imply a loss of scientific rigour for EIA, as shown by a brief reflection about the distinction between objective and subjective impacts, which is really rather ambiguous. For objective impacts we can in fact mean:
   a) impacts assessed in an objective way, thanks to the precision of analytical
      instruments and to knowledge about the consequences of an event for man:
      this is the meaning given to the objective indicators of quality of life;
   b) impacts assessed by an external observer who, as such, can objectively evaluate
      them. This second perspective, which is concerned with the social and situational
      detachment of the observer, is supported, for instance, by Dietz (1987) and Beato
      (1989).
Subjective impacts may likewise mean:
   a) impacts assessed without the necessary criteria or analytical instruments;

b) impacts assessed by a subject somehow involved in the problem and therefore probably lacking the required detachment and objectivity of judgement.

The first perspective considers the way of assessment, the second the subject that assesses. The latter appears debatable: no impact can ever be completely objective in the sense of a complete objectivity or absence of observer value judgements, nor can it be completely subjective in the sense of a complete arbitrariness of his judgements of reality. This perspective certainly makes sense in relative terms, but it obliges us to define the exact position of the actors in order to establish their degree of objectivity, under the not always true assumption that a minor degree of involvement means greater objectivity. On the contrary, the distinction between objective and subjective impacts acquires a greater shareable meaning in the first perspective, if objective impacts mean scientifically assessed effects; but then all impacts usually considered subjective fall within the category of objective impacts, as long as we can respect the scientific criteria of analysis (if, for instance, we can survey attitudes and opinions in a scientifically reliable way).

Therefore, we can call objective impacts those evaluated by a "scientific" observer, whose criteria and means of survey however are not neutral but form a filter through which reality is observed and its developments foreseen. In the case of subjective impacts, there is an observer who reads reality in a "natural" way, i.e., filtering it according to his beliefs, knowledge and value orientations almost never expressed previously: it can be the task of a second "scientific" observer to describe and analyze such a reading and the reasons for it. In fact, individuals or groups have a certain perception of reality (we may call it a real scenery) (Gasparini, 1987, 1988) that can prove to be more or less satisfactory in relation to their expectations based on needs, values, interests (we may call it a latent scenery) (Gasparini, 1987, 1988); then information and data available about the project permit them to foresee a future situation, when the project is realized, which will be judged negatively if the distance of this projected scenery (Pellizzoni, 1991a) from the latent scenery exceeds the distance which separates it from the real scenery. The task of the "scientific" observer is to define the three sceneries, specifying the distance which separates them and thus assessing the subjective impacts. Moreover, as satisfaction level and quality of life are strictly linked together, we can notice that assessing subjective social impacts means measuring a variation of well-being of the people involved.

In short, it is confirmed that the subjective and the objective ways of assessing impacts are not reciprocally interchangeable inside EIA. The practice of neglecting the effects of a project of social and psychological (and therefore political) relevance leads, as a consequence (Beato, 1989), to ignoring the factors which influence or determine peoples' opposition and rejection or consent and acceptance. We must not forget that environmental assessments (EIA, SIA, RA, TA) have a twofold nature: of technical and of political procedures. Moreover, the question of subjectivity does not concern only the assessment of social impacts but the environmental assessment as a whole, because it is linked first to the act of assessing and only secondly to the object of assessment. The selection of important variables or the setting of thresholds beyond which an impact is to be considered significant, are decisions which appear influenced by individual values and opinions in proportion to the lack of scientific agreement and certainty or possibility of obtaining reliable measurements and foresights. In this perspective, however, subjective indicators often appear weaker than the objective ones: the various questions linked to them - validity, reliability, scalability of measurement, predictability of the relation between attitudes and behaviour - perhaps explain why there has not yet been a sufficient consent on modalities of employment of psycho-social variables in EIA/SIA (Albrecht and Thompson, 1988). Anyway, we cannot do without subjective indicators at least in the case of factors whose character is markedly cultural. The already mentioned case of the landscape is obvious: we can correctly evaluate the aesthetic, historical and cultural value of a landscape only if we refer to the culture, evaluation criteria, fruition habits, meanings (aesthetic meanings are often mixed with affective or functional ones) of the local population or of visitors and tourists. Current methods, instead, tend to link abstractly the aesthetic quality (thus the "sensitivity") of a landscape to the values of the sublime, the rare, the ancient, the pure, while the analysis of the visual preferences generally utilize experts, whose opinions are very often different from those of the ordinary people (Canter, 1969; Kaplan, 1973; Groat, 1982).

6. ROLE OF SUBJECTIVE IMPACTS
AND NATURE OF THE ENVIRONMENTAL ASSESSMENT

Subjective indicators can form part of EIA in two ways: as data on which analysis and assessment by experts and decision makers are applied in the same way as on the objective indicators of biophysical and social impacts, or as acts of participation processes. I.e., impacts can be assessed through the assessment of opinions, attitudes and behaviour of people surveyed by the researchers, or these opinions, observations and evaluations can be directly acquired through participation. As data, subjective impacts are filtered by experts, who see them as elements to interpret and to link to other data. Participation acts acquire importance according to the political, economic and social weight of the subjects who perform them and according to the legal value ascribed to them by rules, but they can equally be utilized by experts as social impact indicators. In conclusion, the prospects of "scientific" and "natural" observers give rise to three different viewpoints: one offered by objective indicators of impacts; one offered by subjective indicators: in this case the assessment performed by the subjects becomes an object of inquiry and one offered by participation acts (advice, observations, protests, pressures) with which the interested subjects try to influence the decision making process. Each of these positions can be treated in various ways by EIA legislations.

These remarks about the role of subjective social impacts transfer attention to the problem of the nature of the environmental assessment. There are two different orientations, based in their turn on different experiences of EIA/SIA. Assessment is sometimes seen as a type of scientific activity performed by experts to improve a decision which will be taken by politicians and sometimes as a political activity in itself, having the task of modifying the normal decisional processes through the active participation of the population and groups with conflicting interests. So they talk of a "traditional" and an "emerging" pattern (Corigliano, 1989) of SIA, of "research" or "participatory" (Carley and Bustelo, 1984), "technical" or "political" (Boothroyd, 1982), "US" or "Canadian" (Freudenburg, 1986) approaches: the first approach is centred on the technical aspects of the procedure, which is developed within the normal decisional processes; the other is particularly interested in giving space to the direct participation of people in the evaluation and decision processes.

The debate arises from the "increasingly obvious paramount role of value judgements and value conflict in development decisions" (Carley and Bustelo, 1984). Along this line it has been observed (Torgerson, 1980, 1981) that the approach to SIA/EIA is often of a technocratic type, with the expert analyzing data and presenting objective results. On the other hand, if we acknowledge the value orientation of analysts and the often intuitive nature of judgements (a problem with every kind of assessment, even if perhaps more obvious in SIA) it becomes obvious that the participative processes must be exploited, as they are viewpoints which have the same dignity (even if not at a scientific level) as those of the researchers.

The general lines of EIA/SIA are then "decision-oriented" or "opinion-oriented" (Amendola, 1988, 1989): in the first case the objective is to help decision, to improve the quality of decision processes through the acquisition and elaboration of information and data; in the second case the objective is to verify and look for consent or to reduce and control disagreement, through the activation of suitable communication channels which allow a dialogue between people and institutions. Inside this second viewpoint some people distinguish between a version "after

Luhmann" and a version "after Habermas" (Moro, 1989), i.e., between a management of participation aiming at social control, at a reduction of possibilities of conflict, and a management that tries to elicit and compare in dialogue the different positions of the social actors involved.

In brief, we can say that in the prospect of a decision-oriented EIA the main function of SIA is technical, a data acquisition function, while in the prospect of an opinion-oriented EIA the function of SIA seems to be mainly political, a function of participation management, of consultation and manipulation of public opinion. The different collocation of subjective impacts (and therefore the different function of public consultation) in the two cases is quite clear: in the first there is a tendency to value subjective impacts as data to be added to objective impacts (biophysical or socioeconomic); in the second there is a tendency to consider them directly as participative acts. In any cases on consideration of socio-psychological factors one can fully realizes the participatory character of the assessment, not only, as has already been said, pointing out non-material aspects essential to the definition of impacts in an anthropocentric prospect, but also offering possibilities of participation that can serve to achieve good agreement on environmental choices, exceeding beliefs, prejudices, distrust between people and experts and politicians (Perussia, 1989), which are communication "noises" often provoking a strong opposition to these choices.

7. SOCIAL ACTORS: THE PUBLIC OF EIA

These last remarks transfer our talk to the theme of participation: this aspect constitutes, as has already been said, one of the reasons for the existence of EIA (according to some, the principal).

In general, the more the access to procedure is limited the more the number of participating subjects is reduced to those with a recognized institutional position, with a stable organization, with a socioeconomic or political weight (and it is not certain that the interests and viewpoints of these subjects deserve greater consideration): other interests and viewpoints come into play only if some of these subjects becomes their spokesman. On the other hand, thanks to the constraints imposed, it is possible to limit and control conflict among the parties and therefore quickly reach a balance and the subsequent decision. On the contrary, participation is top level if the procedure is developed as a dialogue and is open to requests coming from the most widely differing subjects and of the most various kinds and contents, but with the counterbalance of aggravating conflicts which do not always allow a conclusion of the assessment to be reached in a reasonable time, without remembering that in collective choices even if the time granted for discussion is unlimited, a unanimous and rational agreement does not necessarily follow (Elster, 1983).

The problem to be solved is in general that of obtaining the greatest number of participants without preventing EIA from reaching concrete results in a reasonable time. But there are at least three aspects linked to the notion of "interested public" which must be taken into account. The first regards the quantitative and qualitative variability of the public. The second regards the functions of participation in EIA procedure. The third (connected with the previous two) concerns the communication flows between authorities and public.

To outline the characteristics of the interested public is obviously a necessary condition for an effective management of participation. The public can be analyzed according to the position occupied by the actors facing the procedure. Thus we distinguish:

a) Institutional actors. They are public or private subjects that must be present: the authority in charge of the management of EIA, other authorities invited to give their opinions, the proponent of the project (or plan or policy);

b) Lawful actors. They are subjects who use the forms of intervention offered by legislation. They can be the whole population if forms of popular consultation are foreseen;

c) Indirect actors. They are subjects which intervene through the influence exercised on institutional and lawful actors: lobby groups, but sometimes even the currents of public opinion formed by the wave of diffused interest in environmental problems and through the spreading of news and debate conducted by mass-media;

d) Improper actors. They are the part of the interested public not included in the preceding categories. In the procedure they constitute the "social environment" on which SIA (as a process producing data) is exercised.

The actors forming the public of EIA can then be defined on the basis of (Pellizzoni, 1991b):

a) Their nature. They can be individuals or collectivities, public or private subjects, already existing or born in consequence of the presentation of the project;

b) Their interests. They can be grouped into political and economical interests and social and cultural interests;

c) Principal object of interest. It may be the problem related to the project or the project itself, or both;

d) Type of involvement. It may be direct if the effects of the project are immediately felt, indirect if such effects are felt in an intermediate way, as a consequence of the principal effects.

For instance, a committee of citizens formed spontaneously to oppose a dump project can assume the role of a lawful actor (if it presents written notes or observations) or of an indirect actor (if it acts as a pressure group on local politicians and managers with tactics varying from personal contacts to a protest march) and it is a collective, private, not pre-existing, directly involved subject, whose interest is social and concerns the project. In the same case, a national environmental association probably assumes the role of a lawful actor and is a collective, private, pre-existing, indirectly involved subject, whose type of interest is social and cultural and is concerned with the problem of dumps as well as the specific project under examination.

In this way, the difference between the two basic models of EIA/SIA is more clearly defined. The participatory or opinion-oriented approach tries to enlarge the space of lawful actors as much as possible, accepting many forms of participation, thus restricting the space of indirect actors. The research or decision-oriented approach does not permit an excessive enlargement of the debate and therefore it precisely limits the legitimation of actors through various constraints, with a possible consequent expansion of the indirect actors. Space can be given to improper actors through the use of objective and subjective indicators as sources of data for experts.

8. FUNCTION OF PARTICIPATION AND COMMUNICATION FLOWS BETWEEN PUBLIC AND EIA AUTHORITIES

With reference to the functions of participation within EIA, they can be summarised as follows (Pellizzoni, 1991a):

a) participation serves to raise the quality of EIA, making available more data and more opinions than those already at the disposal of authorities or supplied by the promoter of the project;

b) participation serves to assure transparency and credibility to the assessment and the subsequent decision, because it requires the clearness of the phases and contents of the assessment and of the criteria which guide the decision.

Therefore we can speak of:

a) participation as elaboration and evaluation when it increases knowledge and reflection on the impacts of the project, on possible alternatives and so on;

b) participation as information and control when its object is the exchange of
information (on the environment, population, procedure, impacts, etc.).

On the one hand, there is an increase of the absolute level of knowledge, on the other, a diffusion of knowledge among all the subjects involved. Of course, these functions are almost never realized independently. To give a piece of information to an actor (thus offering him an opportunity of control) often means that the actor himself will start an elaboration process which also assumes an informative value for the subject that had given the information. The simplest example is represented by the publication of information on a project submitted to EIA, against which a part of the public reacts presenting remarks that for the EIA authorities do not count only as elaborative contributions but also as information about the position taken by such subjects. Therefore, we can say that a participative act assumes a function which is mainly informative (or of control) or elaborative and evaluative depending on whether it has immediate consequences on the contents of the assessment or not, even if in the first case the consequences can be determined later (as in the example mentioned above).

Every participative act - this is the third aspect to be considered - is realized through communication flows between public and authorities in charge of EIA. These flows can make use of a variety of means, more or less suitable to the full realization of the aims of participation. A short discussion on this subject is therefore appropriate (see Pellizzoni, 1991b for more details).

A first category is that of the institutional means: it concerns all the means available under the regulations, some of them having a precise legal value (e.g. official publications, public hearings, compulsory documents, publicity forms required by law, observations submitted in conformity with the regulations, etc.). It is worth noticing that a participative, or opinion-oriented, legislation should try to broaden the category of institutional means. Then, there are the mass media (apart from those included in the institutional means): newspapers, radio, TV and advertising; the means at the level of interpersonal communication, managed by EIA authorities (public meetings, workshops, congresses, guided visits) or by the public (petitions, complaints forwarded through non-institutional channels, protests, sit-ins, permanent meetings, special committees); the scientific means: every scientific initiative developed outside regulations, by EIA authorities or by other public or private subjects.

Each of these means has its own advantages and disadvantages. The institutional ones allow regular and predictable communication flows, but they are also rigid in form and content. Mass media have the advantage of a broad diffusion of information, which becomes a disadvantage when information is too general (often because of the difficulty of explaining a subject) or when a subject has been discussed too much, causing habituation in the audience or indifference in the same mass media. The advantages of using interpersonal communication (public hearings are also institutional means) are obvious: active participation of the public, direct relations with authorities and experts with immediate feedback, possibility of "aiming" the contents at the characteristics of a specific public. But it is often difficult to involve a large part of the public in these initiatives, above all if it is a large community and if the discussion is very specialized: that favours only the positions of the most interested and competent; it is also difficult to choose the best form (meeting open to everybody or reserved to agents of the economic world and to environmental associations? Just one person speaking or many?). As for the means used by the public, the most traumatic forms of participation should be excluded or limited if there are other channels of expression of one's own opinions. Leaving room for mixed forms of participation may avoid these problems and may give information not obtained through the institutional means, but it may be difficult to arrange and place these inputs inside the procedure and to value their reliability. Also scientific means present the problem of inserting their results in the assessment procedure, besides their low diffusion among the public if not in simplified or potentially distorted forms. The problem of mediation between technical, scientific languages and common language, between rigour and comprehensibility is, incidentally, very serious, even if it is often solved by laws simply referring to a "non-technical summary" of a project.

The characteristics of each of these means should lead to one being preferred over the others at least according to the following factors: a) size and composition of the public directly or indirectly involved; b) intensity and nature of interests already shown by the subjects dealt with; c) nature of contents to communicate. We can say, for instance, that the increase of: size and heterogeneity of public, physical and psychological distance of most people from the problem, presence of indirectly involved subjects, little active involvement already shown, non-technical contents or "translated" into non-technical terms, should increase the presence of the mass media inside the participation activities planned by EIA authorities. The opposite characteristics of public and situation should lead to increasing the space for interpersonal means.

9. EIA LEGISLATIVE PATTERNS: A READING FROM A SOCIOLOGICAL VIEWPOINT

The previous discussion gives us the possibility of examining from a sociological viewpoint some of the existing legislative models of EIA, and first of all the EC directive's model. The talk will be grouped by problem areas as follows: a) sphere of application of EIA and its sociological implications; b) consideration of social or socioeconomic impacts; c) actors of the procedure, function and space given to the participation processes.

9.1. Sphere of Application of EIA

As regards the sphere of application of EIA, we can say there are two main models: the "American" one, which submits project, plans and legislative acts to EIA and the "European" one, accepted by the EC directive 85/337 and inspired by the French solution, which submits to EIA only projects (even if the EC directive does not prevent each member country from extending the application of EIA to plans and legislative acts). The reason of this choice is often identified in the remarkable differences of the planning systems existing in the EC countries, which would cause many difficulties in applying a uniform regulation (Cutrera, 1987). However, the opportunity for the EC to extend EIA to plans, programmes and laws has often been pointed out, because the choice and location of public and private activities which may have environmental consequences depend on them. It may be added that the directive allows the possibility of simplified forms of EIA, following the French model of the "notice d'impact" and other tendencies towards procedural simplification and concentration on projects of great environmental importance.

Another important difference between American and European EIA is due to the fact that, in the former, planning or project alternatives (including the "do nothing" one) and mitigation measures are assessed, and a monitoring phase is contemplated. The EC directive requires only a concise description of the principal alternatives and measures of mitigation foreseen, while there is no provision for a monitoring system. On these points the member countries have adopted various solutions. For instance, Italy (DPCM 377/1988) and Spain (RD 1131/1988) foresee plans of environmental monitoring, but neither Italy (DPCM 559/1988) nor Spain (RD 1131/1988), nor France (Decree 77-1141/1977) require an evaluation of project alternatives (they require only a specification of the possible choices and the reasons for the solution adopted). Possible compensating measures are foreseen even if not strictly related to the project: but that introduces the idea (Hebrard, 1981) of bargaining in terms of compensation offered to the population so that the project is accepted.

These restrictions of application and content of EIA cause limits to participation and SIA, such as the following:

    a) the activation of the public involved takes place only at the presentation of the project (and is sometimes limited to the population living in the area directly concerned with its realization: see paragraph 9.3) and regards only the project (not the planning and programming and not different project alternatives, including the "do nothing" one);

    b) it is difficult to forecast the objective social impacts reliably, both because they depend on the evolution of the whole interested area and not only on the project and because it is not possible to compare different alternatives;

    c) it is difficult to describe the real, latent and projected sceneries very well; in other words it is difficult to point out, explain and forecast the changes in values, beliefs, attitudes, behaviour of the population, because of the time required by the necessary research, often incompatible with the time permitted by an EIA of a single project and because the limitation of the involved population and the lack of discussion of alternatives handicap the comprehension of the sociocultural dynamics underlying certain social effects.

9.2. Assessment of Social and Socioeconomic Impacts

There are two legislative orientations about the object of EIA: according to the first, EIA must limit itself to the physical environment while social or socioeconomic effects must be assessed separately; according to the second, the assessment must be as comprehensive as possible. As regards the economic effects, for instance, some authors remark (Muraro, 1987) that a unified assessment allows errors of omission or of double calculation to be avoided and to reduce assessment costs, preventing the political overestimate of environmental aspects. On the contrary, applying the cost-benefits assessment to physical aspects would make EIA less rigorous and would extend its costs and time; an all-embracing EIA would appear as a necessary and sufficient document for decision making, practically depriving the existing decisional processes of power. As regards non-economic effects, the considerations can be the same: an integrated assessment allows a reduction of costs and errors and a linking together of physical and social impacts, but there are problems of time and of integration of very different data in a single table.

As has already been said, in the USA, NEPA and CEQ Regulations established the obligation to develop a SIA when social implications of environmental impacts appear relevant; in Canada too the "public review" of a proposal must include the study of its environmental and social effects. The EC directive is more ambiguous. Its purpose consists in realizing one of the EC objective in the field of environment and quality of life: the effects of a project must be assessed to protect health and to contribute through a better environment to the quality of life. The factors by which the effects of a project must be assessed are: man and those factors belonging to natural environment (fauna, flora, ground etc.), then the elements belonging to the "artificial" environment (material goods and cultural patrimony) and the landscape, which is an environmental aspect of clearly sociocultural relevance. No economic aspects are mentioned. The ambiguity consists in the fact that the environmental impact point of view is not very clear. In my opinion, the directive may be read in two different ways:

    a) a reading according to which man is one of the subject factors of EIA as a biological being and all the effects on natural and artificial environments which must be assessed in any case regard the physical environment. This reading is also proposed in the sociological field (e. g. Amendola, 1989) and is supported by the wording of the list of factors: man appears under an entry which contains fauna and flora; material and cultural goods are listed without any comment, thus meaning that they must be considered as such and not in their social use;

b) a reading according to which man is seen above all as a social being and there are several aspects which EIA can deal with only from a sociocultural viewpoint. This reading is supported by the presence of factors such as the landscape and the cultural heritage about which aesthetic judgement is fundamental, and by the references to the quality of life as the main mark of EIA: that means the social and economic implications of environmental effects, together with those on health, are the main object of EIA.

If we examine the regulations of the EC countries we often find, with reference to the object of EIA, a formulation similar to the directive: thus for instance in France (Decree 77-1141/1977), in Italy (DPCM 559/1988) and in Spain (RD 1131/1988). Even in this case there are some ambiguities. For instance, the Italian legislation speaks about "morphological and cultural aspects of the landscape, identity of the human communities involved and relative cultural goods" and of "visual or cultural-semiotic study of the relation between subject and the environment, as well as of the roots of transformation and creation of landscape by man" (DPCM 559/1988): it seems therefore on the one hand to outline an objective analysis of impacts on landscape and historical and cultural patrimony, on the other the analysis seems to be concerned with the value and meaning they have for individuals and populations, with subjective impacts. Moreover, in the recent act of the Italian Region Friuli-Venezia Giulia (LR 43/1990) it is said that environmental impacts must also be assessed in relation to the "socioeconomic environment" and that the impact study must describe the impacted environmental components with particular reference, among others, to "socioeconomic factors".

In conclusion, we must say that at this moment the position of SIA in the European EIA is not clear: i.e., it is not clear whether it must be considered an integral part of EIA, or whether it must be excluded (and then carried out apart, without a legislative obligation), or whether it must be limited only to some aspects (landscape and cultural goods). An obstacle to the application of the American solution is the fact that:

a) in the US model the necessity of a SIA appears in the preliminary assessment, not contemplated by the EC model;

b) in the EC model the impact study is carried out by the project's proposer, who, especially if a private individual, will not be much inclined to undertake the responsibility of a SIA and then will probably tend to minimize the social implications of the environmental effects.

9.3. Public Participation

As regards public participation, although it is not possible here to go into the details of each country's solution, which depends on very different legal and administrative systems, some examples will show that the two basic types of EIA procedure - the technical, decision-oriented procedure and the participatory, opinion-oriented one - are both outlined by legislation.

In the USA, participation may be exercised in every phase of the procedure, from the preliminary assessment of the need for an EIS to the discussion of the final text of the EIS. The opening of the scoping process to the public with the aim of fixing the essential points of the EIS (public and agencies to involve, environmental aspects and alternatives to examine, time and size limits for the drafting of the EIS), and the extension to EIA of the public hearing system, which is based on the information contained in the draft statement prepared by the Federal Agency in charge of the EIA are particularly important. It must be said that, over the last few years, there is a spreading philosophy of "negotiation" aiming at preliminarily defining the principal interested parties and their willingness to renounce prejudicial positions, and to make the rules of the dialogue between them less formal. Through negotiation techniques they want to make a preliminary comparison between interests and viewpoints in order to reduce conflicts during the assessment procedure, for instance by reaching a formalized and recognized agreement on the fundamental points of the assessment.

In Canada, participation exercises an influence even on the decision to submit a project to EIA, a decision taken also on the basis of reported public concerns. The definition of guidelines to draw up the EIS makes use of the public's suggestions obtained through meetings arranged by the commission responsible. Then, the discussion on the EIS passes through a "public review" managed by the commission, which develops public hearings conducted according to rules less rigid than those of the judicial proceedings. It has been observed (Scovazzi, 1989) that in the Canadian system participation is not included in a formally neutral procedure that offers citizens only a vague possibility of intervention, but rather it appears as a concrete objective that authorities must try to realize.

The EC directive was compelled to take into consideration the very different situations of the member countries as regards the legislative and administrative structures in which EIA was to be inserted. Its indications are therefore of a general character.

As to the actors that take part in the procedure, we find (institutional actors) the private or public project proposer, the authority that can give the permission (it may be the same as the authority in charge of the EIA), other authorities to consult according to the national regulations as well as the possible consultation of other member states of the EC, and (lawful actors) the interested public: each country defines the forms of consultation and information, identifying it on the basis of the characteristics of projects or of their places, specifying where the information can be taken, the way people can be informed, how the consultation must take place, and fixing the times of the procedure in order to guarantee that the decision will be taken within a reasonable time. The information given by the promoter of the project or collected in other ways, including that supplied by the population, must be taken into consideration by the authorization procedure, but the directive does not specify how, while the national legislations are often vague about it: for instance, Italian regulations say only that each or each group of observations and requests must be "considered", but in that way they run the risk of being avoided or of having little importance in the assessment.

In any case, a triple right is assigned to the public: the right of being informed, of being consulted and of being taken into consideration, therefore the purpose of making the assessment and decision processes transparent and democratic is clear, giving space to the information and consultation of people, i.e., evaluating the request for knowledge, control and participation which is one of the reasons of success of the ecological movements.

As to the duty of information, the directive outlines some ways such as posters or newspapers and assigns the promoter the task of preparing a non-technical summary on: characteristics of the project, measures to contrast its negative effects and data related to the forecast of principal effects. It is clear that the majority of the interested public will form an opinion just on the basis of this document: only particularly prepared or interested subjects (technicians, environmental organizations, economic groups) will refer to the technical documents. In this way the directive shows that it does not want to limit the debate to experts but wants to give everyone the chance to understand the problem and to form his own opinion.

As regards the ways of consulting the public, the directive points out two possibilities, written consultation and public inquiry, based both on the American model and on those already existing in some European countries (France, UK). In some countries, as in Spain, the regulations clearly prefer the solution of written remarks (RD 1131/1988); elsewhere the situation is more complex.

Let us examine two cases: France and Italy. In France the regulations of the "enquete publique" (L 83-630/1983 and following decrees) are crossed with those on EIA (Decree 77-1141/1977), that foresee various forms of assessment ("étude d'impact", "mini-étude d'impact", "notice d'impact"): the possibilities of participation pass, when a public inquiry is required, through access to the documents and through the presentation of remarks and counterproposals that are inserted in the file of the inquiry. The "commissaire enqueteur", according to the importance or nature of the project or other circumstances, can call a public meeting ("réunion publique"). The advertising of these activities takes place through posters and local and national newspapers. When the file is closed the "commissaire" draws up a report on the inquiry, on the remarks and counterproposals presented and on the replies of the promoter of the project, while in another document he draws reasoned but not binding conclusions. According to some critics (Scovazzi, 1989), the French system of public inquiry is a rigorous procedure in which people can obtain a lot of information and are sure that their observations are transmitted to the appropriate authority, even if it contemplates very formal and detailed procedures that do not offer the certainty of real and direct involvement of less prepared people or of those not belonging to some organization.

Italy is still waiting for a law that will rearrange the national regulations on EIA (a bill is awaiting parliamentary debate). At this moment there are two forms of participation. The first (L 349/1986), is that of the written remarks and opinions presented to some Ministries or to the interested Region (the presentation of the project is advertised in local and national newspapers); the second (DPCM 559/1988), limited to projects of electric power plants of ENEL (the national electric power utility), is that of the public inquiry, to which the public can participate with written notes. Public hearings can be organized with institutions and private people who have presented some notes (which are admitted only if technical and scientific and regarding the environmental consequences of the proposed power plant) but, according to some critics, there are no real possibilities for direct expression of the interests of the local communities (Luciani, 1990), which in practice are represented by the government of the interested Region. At the end, the president of the inquiry transmits the notes and ENEL replies to the Ministry of Environment with a brief report. Also in the act of the Friuli-Venezia Giulia Region (LR 43/1990) the consultation of public takes place through written remarks, but hearings can be held in cases that will be specified by the forthcoming rules for the enforcement of the law.

From what has been said we see that:

   a) Italian and French legislations discipline participation in a rather rigid way: this could be an obstacle for the majority, above all when the contents must have a technical-scientific character (it is not by chance that in both countries a privileged role is recognized by law to the environmental associations, which seem to become the guardians of the collective interests). Then, the present Italian public inquiry seems to offer fewer possibilities for participation than the French one: this allows the public to speak directly with the "commissaire" (one can ask to be heard, to ask questions, to explain one's proposals orally), while the Italian does not;
   b) The public inquiry in both countries utilizes hearings only in some cases (in Italy for only one type of project): the principal means of participation is therefore the written document, which requires a work of preparation and drafting beyond the reach of most people;
   c) The possibility of dealing with socioeconomic impacts, as we have already seen (paragraph 9.2) is doubtful. As regards the economic aspects, in Italy they seem to be excluded only in the public inquiry, while the remarks presented to Ministries and Regions have as their object "the project" submitted to EIA (therefore, so it seems, even DPCM 559/1988 requires consideration of aspects such as the economic evaluation of the project). In France, the wording of the L 76-629/1976 (that speaks of "harmonious balance between urban and rural populations") seems to involve, at least in some cases, the inclusion of socioeconomic aspects in the impact studies, which can then be participated in.

A further aspect allows us to distinguish the various national solutions to the problem of participation: the procedural phase where public involvement occurs. For instance in the USA, as said above, the public already intervenes at the levels of the scoping and of the draft statement. In Europe, the public generally intervenes on an already prepared impact study.

One more problem is that of the identification of the "interested public". The adopted solutions vary from a maximum of "openness", when the access to participation is permitted to any citizen (e.g. Italy, L 349/1986) to restriction of access to the residents in the affected area (e.g. Italy, LR 43/1990). The first case accepts the principle of involving all possible interests, the second accepts the idea of giving voice only to the most immediately involved citizens. In this case it should be necessary to fix the criteria for the identification of the area involved, but this is not always a simple task: for instance, for an electric power plant such an area includes a wide "risk basin" and not simply the town area where it will be built.

In conclusion, we can say that of the two functions of participation within EIA - to make available contributions that will increase its qualitative level and to offer information and control of the assessment - the second is now more protected than the first, which appears "checked" by various constraints introduced into the legislation.

10. TOWARDS AN IMPLEMENTATION OF THE EC REGULATIONS ON EIA

We can sum up the existing models of EIA in two opposite ideal types.

1. "Restricted" type: only for projects and without discussion of project alternatives, it deals only with the physical environment and rigidly limits participation access to a small number of actors who must utilize written notes.

2. "Enlarged" type: for projects, plans, programmes and laws, with discussion of alternatives including "do nothing", it deals with all impacts (physical, social, economic) enlarging the number of lawful actors, permitting a certain formal elasticity of participation processes and giving space to direct relations among public, authorities and promoters.

The present EC model tends, as we have seen, towards the "restricted" type. The directive in fact gives the minimal requirements of the procedure: limits often respected in a rigorous enough way even if they could be passed by the national regulations. Also for this reason an implementation of the EC regulations seems desirable, and that - from a sociological point of view - chiefly in three directions: a) field of application; b) participation processes; c) object of EIA:

a) The extension of the EIA procedure to laws and planning and to a discussion of project alternatives seems suitable for many reasons, the principal of which is the chance for people to examine programmes of territorial development before examining single projects, thus intervening first on choices of policy and then on technical aspects. An assessment on a planning level also allows, as has already been said, large scale surveys, which form a necessary basis in assessing the social impacts of a project;

b) Larger participation accesss desirable: it is necessary that everybody should be permitted to participate and not only an "interested" population often difficult to pick out, even if it is clear that a privileged position, with a larger quantity of means (interpersonal above all) should be attributed to people residing in the area more directly affected by the project. One may then have a better specification of the use of the participation means. A considerable literature is available on the advantages and disadvantages of each of them (see, for example, Burdge, 1983; Daneke et al., 1983; Zube, 1984; Pellizzoni, 1991b), without attaching importance only to the written form and giving more space to meetings and hearings which represent a way of involving the (usually large) part of the population that otherwise is obliged to grant a kind of "proxy" (as happens when citizens sign agreement to an already prepared opinion form to those (institutions, environmentalists, trade-unions, trusts) that are able to draft the remarks. The recent "negotiation" approach seems to move in a similar direction. It is then suitable to fix the contents of the non-technical summary, the public's main references, and to be more precise about the way in which the EIA authorities must consider requests and opinions of people. Public intervention before the drawing up of the impact study is certainly desirable. The reference model is here the US scoping process, but in Europe there is the problem of the conciliation between this preliminary assessment and the drafting of the study by a subject different from the authority in charge of the EIA;

c) An EIA limited to the physical aspects conceals the social and cultural nature of the assessment, does not completely answer the objectives of the directive and does not permit an adequate analysis of some environmental factors indicated by it. SIA must then rightfully form part of the European EIA: social impacts must be included in impact studies and be the subject of participative contributions. Of the questions put by this integration, the two most important are perhaps:Which are the indicators to consider? How can SIA be inserted in the EIA procedure? Here I can only present some hints. Having established that social impacts must be assessed through a combination of objective and subjective indicators, there are in my opinion two considerations that must guide the selection of the indicators. The first is that it is not possible to establish a definitive list of indicators: every situation has its own peculiarities that must be valued by the assessment team. The second is that in any case there is a variable that may constitute the mainstay of SIA: the quality of life (QOL). It is not a new idea: several methods, as we have seen, take this variable into consideration (e.g. Olsen et al., 1981; Guseman and Dietrich, 1978; Branch et al., 1984), but mainly utilizing objective indicators, just as several researchers (e.g. Olsen et al., 1985; Finsterbusch, 1985; Freeman et al., 1982; Carley, 1983b) underline its importance. But some authors point out (in my opinion rightly) that the centre of the whole SIA is this question: "Will there be a measurable difference in the quality of life in the community as a result of what the proposed project is doing or might do in the future?" (Burdge, 1983); in this case the social aspects that must be examined are those "identified by a relevant body of social science research as having specifiable implications for a social group's quality of life" (Freudenburg, 1986). Therefore, there is the research on the QOL, on the environmental quality (in particular on the perceived environmental quality) and on community needs (Needs Assessment) that offer the European EIA important impulses for the development of SIA.

Placing the QOL in the centre of SIA it is possible to find solutions to the problem of its link with the assessment of the physical environment. Here I shall only try to express a few ideas.

1. It is suitable to provide the execution of systematic profiles of the QOL of the communities set in an information system, periodically kept up to date also with the data of the SIAs already performed, as a baseline for the impact studies. That would prevent the promoter of a single project from making very big investigations: therefore he should assess the social impacts on the basis of the existing data and need make only restricted research based on limited population samples and on specific objects related to the project. A mapping of the territory based on the objective and perceived QOL can allow us to distinguish the different "social sensitivity" of the communities, meaning by this expression the degree of social risk of certain lines of development or of specific projects, because of the presence of particular situations (strong social disruption, strong perceived environmental deterioration, etc.), that make a community highly exposed to a lowering of its QOL/PQOL. The principle is that no plan or project should lead to a lowering of the QOL/PQOL of the population, but on the contrary to boosting it: every impact study should show the foreseen trend of the QOL/PQOL pointing out the measures to avoid its worsening or the project alternatives that could permit an improvement, and their costs.

2. With the QOL in the centre of SIA, objective and subjective indicators assume a complementary role. Subjective impacts can be assessed to coincide with the measurement of a variation in the expected QOL. Its determining elements are many: alienation, personal satisfaction, place identity, personal control, perceived loss of freedom, psycho-social transition and so on; they in their turn depend on knowledge, interests, beliefs, attitudes, values, intentions of behaviour (only some of which are directly linked to the project: the others are of a more general character, such as the attitudes towards progress, technology and industrial development). Subjective indicators can be used to pick out the most important objective indicators and control variables of them (and viceversa). For instance, it may be foreseen that a project does not cause a lowering of QOL and environmental quality (no negative objective impacts), but such a lowering is strongly felt by the population (subjective impacts); on the contrary, in a highly industrialized area people may not be very sensitive to the impact of a new plant (no negative subjective impacts), but the objective indicators can show an already compromised situation that would receive a heavy blow from this new plant. The agreement or disagreement between the two types of indicators also gives indications for the planning of communication flows with the population: in the case mentioned first the decision makers should give up the project or produce an informative programme trying to modify beliefs and attitudes.

3. The methods proposed to make the selection of the most important variables to assess subjective impacts less arbitrary are, in practice, the same as those used in the assessment of physical impacts: from checklists to relevance trees (Finsterbusch, 1980) or to the Delphi method (Singg and Webb, 1979). However, the margin of subjectivity remains wide. Subjective indicators are then obtained through surveys using questionnaires and interviews or through the participation inputs of people: the first solution is time consuming especially if the survey is exploratory; the second does not ensure that the intervening subjects correctly represent the opinions, attitudes and beliefs of the various social groups. There are several techniques to involve population groups in the definition of problems and objectives directly, from interactive methods such as the jury panel (Heberlein, 1976; Butler and Howell, 1980) or the nominal group process (Delbecq et al., 1975; Butler and Howell, 1980), to non-interactive methods such as the Delphi method. The problem is that experts' opinions cannot represent the subjective impacts of a population and that a sample of citizens is not competent enough to discuss the technical aspects of a project. Then, a possible solution may be to employ groups of citizens working specifically on the implications of the project on the future QOL of the community. An important aid in managing these working sessions can come from the use of expert systems (for a tentative proposal see Marzano and Pellizzoni, 1991).

The hints just given show, I hope, some directions towards which the EC legislation can move in order to integrate SIA and EIA. But my whole contribution shows, I think, the opportunity and the difficulty (all the more so if you must, as in Europe, harmonize quite different legislative systems and administrative traditions), but also the feasibility of translating into working terms the integrated and participated assessment of environmental and social impacts which constitutes the essential objective of EIA. This objective is indicated in the "birth certificate" of NEPA, but it is also highlighted, as I have tried to show, in the present EC directive on EIA.

REFERENCES

Albrecht, S.L. and Thompson, J.G. (1988), "The place of attitudes and perceptions in social impact assessment", Society and Natural Resources, 1.

Amendola, G. (1988), "Valutazione di impatto ambientale nella produzione energetica: studio di casi italiani. VIA/SIA in Italia: stato dell'arte e specificità ", paper presented at the International Congress IRIS - CNR on "Metodi di valutazione della pianificazione urbana e territoriale: teorie e casi di studio", Capri.

Amendola, G. (1989), "Prevedere per valutare. Gli spazi della sociologia nella valutazione di impatto ambientale", in Martinelli, F. (ed.), "I sociologi e l'ambiente", Bulzoni, Roma.,

Andrews, R. (1987), "Socio-economic aspects of environmental impact assessment: the United States experience", paper presented at the Congress on "Aspetti socio-economici dell'impatto ambientale", Bologna.

Beato, F. (1989), "La metodologia della valutazione di impatto sociale. Ricognizione critica sulla letteratura e problemi di ricerca sociologica", Sociologia e ricerca sociale, 29.

Boothroyd, P. (1982), "Overview of the issues raised at the international conference on social impact assessment", paper presented at the International Conference on Social Impact Assessment, Vancouver.

Branch, K., Hooper, D.A., Thompson, J. and Creighton, J. (1984), Guide to Social Assessment: A Framework for Assessing Social Change, Westview, Boulder.

Burdge, R.J. (1983), "Community needs assessment and techniques", in Finsterbusch et al., 1983 (see below).

Butler, L.M. and Howell, R.E. (1980), Coping with Growth: Community Needs Assessment Techniques, Western Rural Development Center, Corvallis.

Canter, D. (1969), "An intergroup comparison of connotative dimensions in architecture", Environment and Behavior, 1.

Canter, M.J., Atkinson, S.F. and Leistritz, F.L. (1985), Impact of Growth, Lewis, Chelsea.

Carley, M.J. (1983a), "A review of selected methods", in Finsterbusch et al., 1983 (see below).

Carley, M.J. (1983b), "Social indicators research", in Finsterbusch et al., 1983 (see below).

Carley, M.J. and Bustelo, E.S. (1984), Social Impact Assessment and Monitoring, Westview, Boulder.

Corigliano, E. (1989), "Analisi d'impatto ambientale: da tecnica analitica a stile di Planning. L'esperienza nord americana", in Martinelli, F. (ed.), "I sociologi e l'ambiente", Bulzoni, Roma.

Covello, V.T., Mumpower, J.M., Stallen, P.J.M. and Uppuluri, V.R.R. (1986) (eds.), Environmental Impact Assessment, Technology Assessment and Risk Analysis: Contribution from Psychological and Decision Sciences, Springer, Berlin.

Cutrera, A. (1987), "La direttiva 85/337CEE sulla valutazione di impatto ambientale", Rivista Giuridica dell'Ambiente, 3.

Daneke, G.A., Garcia, M.W. and Priscoli, J.D. (1983), Public Involvement and Social Impact Assessment, Westview, Boulder.

Deane, D.H. and Mumpower, J.L. (1981), "The Social Psychological Level of Analysis in Social Impact Assessment: Individual Well-Being, Psychosocial Climates, and the Environmental Assessment Scale", in Finsterbusch and Wolf, 1981 (see below).

Delbecq, A., Van De Ven, A. and Gustafson, D. (1975), Group Techniques for Program Planning: A Guide to Nominal Group and Delphi Processes, Scott Foresman, Glenview.

Dietz, T. (1987), "Theory and method in social impact assessment", Sociological Inquiry, 57, 3.

Elkind-Savatski, P.D. and Kaufman, J.D. (1986), Differential Social Impacts of Rural Resource Development, Westview, Boulder.

Elster, J. (1983), Sour Grapes. Studies in the Subversion of Rationality, Cambridge University Press, Cambridge.

Finsterbusch, K. (1980), Understanding Social Impacts: Assessing the Effects of Public Projects, Sage, Beverly Hills.

Finsterbusch, K. (1985), "State of the Art in Social Impact Assessment", Environment and Behavior, 17, 2.

Finsterbusch, K. and Wolf, C.P. (1981) (eds.), Methodology of Social Impact Assessment, Hutchinson Ross, Stroudsburg.

Finsterbusch, K., Llewellyn, L.G. and Wolf, C.P. (1983) (eds.), Social Impact Assessment

Methods, Sage, Beverly Hills.

Fishbein, M. and Ajzen, I. (1975), Belief, Attitude, Intention and Behavior: An Introduction to Theory and Research, Addison-Wesley, Reading.

Fitzsimmons, S.J., Stuart, L.I. and Wolf, C.P. (1977), Social Assessment Manual: A Guide to the Preparation of the Social Well-Being Account for Planning Water Resource Projects, Westview, Boulder.

Flynn, C.B. and Flynn, J.H. (1982), "The Group Ecology Method: A New Conceptual Design for Social Impact Assessment", IAIA Bulletin, 1, 4.

Flynn, C.B., Flynn, J.H., Chalmers, J.A., Pijawka, D. and Branch, K. (1983), "An Integrated Methodology for Large-Scale Development Projects", in Finsterbusch et al., 1983 (see above).

Freeman, D.M., Frey, R.S. and Quint, J.M. (1982), "Assessing resource management policies: a social well-being framework with a national level application", Environmental Impact Assessment Review, 3.

Freudenburg, W.R. (1986), "Social Impact Assessment", Annual Review of Sociology, 12.

Freudenburg, W.R. and Keating, K.M. (1982), "Increasing the impact of sociology on social impact assessments: toward ending the inattention", American Sociology, 17.

Gasparini, A. (1987), "Qualità della vita e simbolica degli spazi come base di costruzione di indicatori per la valutazione di impatto ambientale", in Schmidt di Friedberg, P. (ed.), " Gli indicatori ambientali", Angeli, Milano.

Gasparini, A. (1988), "Qualità della vita e informazione" in Gasparini, A., De Marco, A. and Costa, R. (eds.), " Il futuro della città", Angeli, Milano.

Gasparini, A. and Marzano, G. (1991) (eds.), Tecnologia e società nella valutazione di impatto ambientale, Angeli, Milano.

Greco, N. (1989), Processi decisionali e tutela preventiva dell'ambiente, Angeli, Milano.

Groat, L. (1982), "Meaning in post-modern architecture. An examination using the multiple sorting task", Journal of Environmental Psychology, 2.

Guseman, P.K. and Dietrich, K.T. (1978), Profile and Measurement of Social Well-Being Indicators for Use in the Evaluation of Water and Related Land Management Planning, Vicksburg, US Army Engineer Waterways Experiment Station.

Heberlein, T.A. (1976), "Some observations on alternative mechanisms for public involvement: the hearing, public opinion poll, the workshop and the quasi-experiment", Natural Resources Journal, 16, 1.

Hebrard, S. (1981), "Les études d'impact sur l'environnement devant le juge administratif", Revue Juridique de l'Environnement.

Kaplan, R. (1973), "Predictors of environmental preference: Designers and clients", in: Preiser, W.F. (ed.), Environmental Design Research IV. Proceedings of the Fourth Annual EDRA Conference (Vol. 1), Dowden, Hutchinson and Ross, Stroudsburg.

Land, K.C. (1983), "Social indicators", Annual Review of Sociology, 9.

Leistritz, F.L. and Ekstrom, B.L. (1986), Social Impact Assessment and Management: An Annotated Bibliography, Garland, New York.

Leistritz, F.L. and Murdokck, S. (1981), The Socioeconomic Impact of Resource Development: Methods for Assessment, Westview, Boulder.

Lounsbury, J.W., Sundstrom, E., Schuller, C.R., Mattingly, T.J. and De Vault, R.C. (1981), "Toward an Assessment of the Potential Social Impacts of a Nuclear Power Plant on a Community: Survey of Residents' Views", in Finsterbusch and Wolf, 1981 (see above).

Lounsbury, J.W., Van Liere, K.D. and Meissen, G.J. (1983), "Psychosocial Assessment", in Finsterbusch et al., 1983 (see above).

Luciani, M. (1990), "La localizzazione delle centrali elettriche. Problemi giuspubblicistici", Rivista Giuridica dell'Ambiente, 2.

Marzano, G. and Pellizzoni. L. (1991), "Measurement and exploitation of subjective social indicators in Environmental Impact Assessment", paper presented at the 11th European Congress on Operational Research, Aachen.

Moro, G. (1989), "La valutazione di impatto ambientale: i presupposti sociologici", PhD thesis, University of Catania.

Muraro, G. (1987), "Valutazione di impatto ambientale e analisi economica", Rivista Giuridica dell'Ambiente, 1.

Olsen, M.E., Melber, B.D. and Merwin, D.J. (1981), "A Methodology for conducting social impact assessments using quality of social life indicators", in Finsterbusch and Wolf, 1981 (see above).

Olsen, M.E., Canan, P. and Hennessy, M. (1985), "A value-based community assessment process: integrating quality of life and social impact studies", Sociological Methods Research, 13.

Pellizzoni, L. (1991a), "Partecipazione e valutazione di impatto ambientale", in Gasparini and Marzano, 1991 (see above).

Pellizzoni, L. (1991b), "Processi partecipativi e strumenti comunicativi", in: Gasparini and Marzano, 1991 (see above).

Perussia, F. (1989), Pensare verde. Psicologia e critica della ragione ecologica, Guerini e Associati, Milano.

Rossini, F.A. and Porter, A.L. (1983), Integrated Impact Assessment, Westview, Boulder.

Scovazzi, T. (1989), "La partecipazione del pubblico alle decisioni sui progetti che incidono sull'ambiente", Rivista Giuridica dell'Ambiente, 3.

Singg, R.N. and Webb, B.R. (1979), "Use of Delphi methodology to assess goals and social impacts of a watershed project", Water Resources Bulletin, 15.

Sundstrom, E.D., Costomiris, L.J., De Vault, R.C., Dowell, D.A., Lounsbury, J.W., Mattingly, T.J., Passino, E.M. and Peelle, E. (1981), Citizen Views About the Proposed Hartsville Nuclear Power Plant: A Survey of Residents' Perceptions in August 1975, Oak Ridge National Laboratory, Oak Ridge.

Tester, F. and Mykes, B. (1981) (eds.), Social Impact Assessment: Theory, Method and Practice, Detselig, Calgary.

Torgerson, D. (1980), Industrialisation and Assessment: Social Impact as a Social Phenomenon, York University Publications in Northern Studies, Toronto.

Torgerson, D. (1981), "SIA as a social phenomenon: the problems of contextuality", in Tester and Mykes, 1981 (see above).

US Council on Environmental Quality (1978), Regulations for Implementing the Procedural Provisions of the National Environmental Policy Act, US Council on Environmental Quality, Washington DC.

Wolf, C.P. (1983), "Social impact assessment: a methodological overview", in Finsterbusch et al., 1983 (see above).

Zube, E.H. (1984), Environmental Evaluation: Perception and Public Policy, Cambridge University Press, Cambridge.

Note. For a discussion about the methodologies used in SIA see, for example: Fitzsimmons et al., 1977; Finsterbusch, 1980; Finsterbusch and Wolf, 1981; Finsterbusch et al., 1983; Carley and Bustelo, 1984; Lestritz and Murdock, 1981; Soderstrom, 1981; Branch et al., 1984; Daneke et al, 1983; Rossini and Porter, 1983; Canter et al., 1985; Tester and Mykes, 1981; Covello et al., 1986. For bibliographic references see Leistritz and Ekstrom, 1986; Finsterbusch, 1985; Freudenburg, 1986; Beato, 1989 and the periodical section of Carley and Bustelo, 1984.